Lecture Notes in Biomathematics

Managing Editor: S. Levin

5

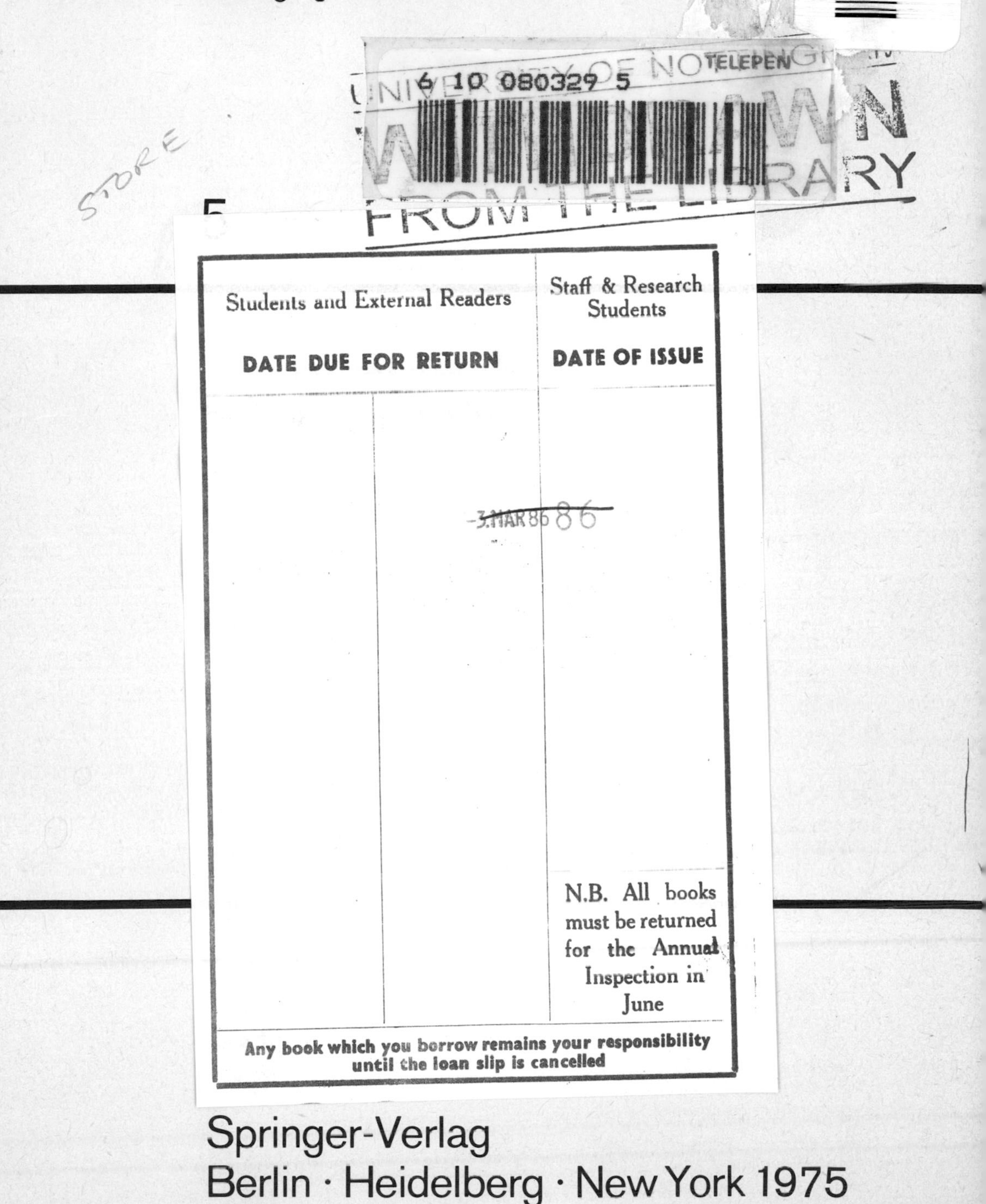

Springer-Verlag
Berlin · Heidelberg · New York 1975

Editors

Dr. A. Charnes
University of Texas at Austin
Center for Cybernetic Studies
Austin, Texas 78712/USA

Dr. W. R. Lynn
Hollister Hall
Cornell University
Ithaca, N.Y. 14850/USA

Library of Congress Cataloging in Publication Data

Mathematical analysis of decision problems in ecology.

(Lecture notes in biomathematics ; v. 5)
Bibliography: p.
Includes index.
1. Ecology--Mathematical models--Congresses.
I. Charnes, Abraham, 1917- II. Lynn, Walter R., 1928- III. North Atlantic Treaty Organization. IV. Series.
QH541.15.M3M36 301.31'01'84 75-19493

AMS Subject Classifications (1970): 62P10, 90-02, 90A15, 90C50, 90D45, 92A05, 92A10, 92A15, 92A25, 93A15, 93C20

ISBN 3-540-07188-1 Springer-Verlag Berlin · Heidelberg · New York
ISBN 0-387-07188-1 Springer-Verlag New York · Heidelberg · Berlin

Printed in Germany

Offsetdruck: Julius Beltz, Hemsbach/Bergstr.

PREFACE

The contents of this volume involve selection, emendation and up-dating of papers presented at the NATO Conference "Mathematical Analysis of Decision Problems in Ecology" in Istanbul, Turkey, July 9-13, 1973. It was sponsored by the System Sciences Division of NATO directed by Dr. B. Bayraktar with local arrangements administered by Dr. Ilhami Karayalcin, Professor of the Department of Industrial Engineering at the Technical University of Istanbul. It was organized by A. Charnes, University Professor across the University of Texas System, and Walter R.Lynn, Director of the School of Civil and Environmental Engineering at Cornell University.

The objective of the conference was to bring together a group of leading researchers from the major sciences involved in ecological problems and to present the current state of progress in research of a mathematical nature which might assist in the solution of these problems.

Although their presentations are not herein recorded, the keynote address of Dr. John A. Logan, President, Rose-Hulman Institute of Technology and the presentation of Dr. Michael J. L. Kirby, then Assistant Principal Secretary to the Premier of Nova Scotia, Canada, greatly influenced the discussion of papers and their subsequent emendations and revisions.

For convenience, this book is subdivided into four sections:

A. Problems of Land Use Management

B. Air, Noise and Water Pollution Control

C. Control of Infectious Disease

D. Social and Behavioral Analysis

In Section A, Charnes (mathematician), Haynes (geographer), Hazelton (economist), and Ryan (mathematical economist-engineer) present a model system developed for the Texas Coastal Land Use Project. The distinguished forester, Dr. F. H. Bormann of Yale University, presents an ecosystems approach derived from his vast experience in forest land studies and management. Dr. F. E. Wielgolaski of the Botanical Laboratory, University of Oslo, Norway, studies questions such as sizes of animal populations which plants on grazing lands can withstand without being destroyed. The section concludes with a paper by environmental engineer Dr. James Heaney, and student Hasan Sheikh, in an important application of new concepts of game theory to equitable regional environmental quality management.

In Section B, R. Abrams of Northwestern University details current progress (including his own) on optimization models for regional air pollution control. Dr. D. P. Loucks, Chairman of the Department of Environmental Engineering at Cornell University, presents a new approach to the analysis of environmental noise. Dr. L. Lijklema, Department of Chemical Engineering at Twente University of Technology, The Netherlands, demonstrates how physical chemistry can contribute to the modelling and analysis of biological and waste water treatment. The internationally distinguished sanitary engineer, H. B. Gotaas, of Northwestern University, and W. S. Galler, Department of Environmental Engineering, North Carolina

University exhibit the latest engineering approach to optimization of biological filter design.

In Section C, Dr. A. Arata and Dr. K. Dietz of the World Health Organization present some of this organization's current work, respectively, on vectors of human disease and on models for development of genetic control alternatives, e.g. as in the sterile male technique of insect control. Their mathematical development is of a classical nature which contrasts markedly with the highly sophisticated contemporary mathematical techniques of Costello and Taylor ("Mathematical Models of the Sterile Male Technique of Insect Control") and Hirsch and Nåsell ("The Transmission and Control of Schistosome Infections"). Most interestingly, the predictions of the stochastic process models of Costello-Taylor agree with those of K. Dietz to within 10 percent where comparable.

In Section D, the economist Martin Beckmann recasts the question of "Limits to Growth" from the deterministic feedback formulation of Meadows into a normative neo-classical form employing Cobb-Douglas functions for the production of "goods" and "bads". A less global approach is attempted by Osman Coskunoglu in "Dynamic Optimization of the Trade-off Between Production and Pollution". The volume concludes with the attempt by the Ecole Polytechnique group of Denis Bayart, Bertrand Collomb, and Jean-Pierre Ponssard to analyze and study behavioral (psychological) reactions to pollution by means of a multi-person "game".

The mathematical techniques apparently applicable and necessary for analysis of decision problems in ecology cover a tremendous range. The integro-differential equation systems of Hirsch and Nåsell, the reaction kinetics of Lijklema, the multi-goal, multi-level programming of Charnes, Haynes, Hazelton, and Ryan, the diffusion theory and geometric programming of Abrams and Beckmann, the stochastic processes of Costello and Taylor and the game theory approaches of Heaney and Sheikh and Bayart, Collomb, and Ponssard indicate the tremendous variety and flexibility of our mathematical resources for analysis of ecological decision problems. Conversely, they indicate the vital importance to the individual disciplines involved in ecology of a working acquaintanceship with these valuable tools.

A. Charnes and W. Lynn

CONTENTS

Editorial Assistant:
Fred Phillips

Typist: Holly McCall

A.

PROBLEMS OF LAND USE MANAGEMENT

AN HIERARCHICAL GOAL PROGRAMMING APPROACH TO ENVIRONMENTAL-LAND USE MANAGEMENT*

Abraham Charnes
Kingsley E. Haynes
Jared E. Hazleton
Michael J. Ryan

University of Texas, Austin

ABSTRACT

The purpose of the study is the development and application of a methodology for evaluating the economic and environmental effects of various policies for the management of the Texas Coastal Zone. The specific goals of the study are: (1) the establishment of operating criteria which are specific to the Texas Coastal Zone; (2) the development of a methodology or systematic approach for evaluating the economic and environmental effects of alternative policies for Texas Coastal Zone management, and (3) the testing of the systematic approach by the application of several hypothetical management policies to a specific region of the Texas Coastal Zone.

Part of the past year's efforts focused on the quantitative evaluation of the various land and water environmental units within the Texas Coastal Zone in terms of their geochemical, biological and physical properties which restrain or are affected by man's various activities. Particular emphasis was placed upon the establishment of water quality criteria for the maintenance of biological life in the estuaries.

The second objective during the past year was the development of the various components of a systematic approach to evaluate various possible Coastal Zone management policies in terms of their economic and environmental impact. As a coordination focal point, the Coastal Bend Planning Region including metropolitan Corpus Christi and surrounding bays was chosen. A regional input-output economic model and a demographic model were developed in conjunction with land use models for spatial allocation of projected industries and households. Water demands and residuals generation along with associated treatment and costs also were established for the region. In order to relate wastewater discharges to environmental impact, hydrodynamic and conservative and non-conservative water quality models were adapted and tested for this particular estuary.

---

*The authors wish to acknowledge the support of N.S.F. (RANN) Grant No. GI-34870X "Establishment of Operational Guidelines for Environment Management."

Information was compiled to evaluate the effects of projected activities and wastes upon the region's resource capability units and biological assemblages within which the activities occur.

A major objective of the study's second year is to integrate the various components of the methodology. Further testing of the methodology as regards its capability to provide economic and environmental impact significance will be undertaken using rather unrealistic hypotheses such as "no change in public policy" affecting the Coastal Zone and "no further development within 1,000 feet of waterfront." To demonstrate the applicability of the proposed methodology, two somewhat realistic although hypothetical coastal management schemes also have been chosen for evaluation.

INTRODUCTION: BACKGROUND

The Texas Coastal Zone comprises a two-county wide land area throughout the Coast of Texas together with the adjacent bay, shallow water and barrier island systems. The economic activity in this area has historically been based principally upon the extraction of oil and natural gas and associated refining and petrochemicals industries. The availability of these fuels and access to ports has also led to the development of steel and aluminum industries [1]. Major industrial complexes are located in the Beaumont-Port Arthur-Houston-Texas City-Galveston and Corpus Christi areas. Each of these cities is a port.

The major agricultural products of the coastal zone are cotton, sorghum, rice and cattle [2]. Much of this agricultural production requires irrigation with water being drawn both from surface and artesian sources. Networks of channels and reservoirs have been proposed by the Texas Water Development Board [3] to increase supplies of water for both industrial and agricultural use.

It is anticipated that land use management policies will be implemented by legislation based upon "environmental units."

"Environmental units" are being developed for both land and marine areas by the Bureau of Economic Geology at the University of Texas, and comprise broadly defined geological structures for which admissible types and intensities of use will be established based upon both surface and subsurface geological considerations. These units provide a direct means of incorporating zoning considerations into land use planning models and provide one major focus of this paper. A second focal point is that of ports. As noted above, port cities have been of major importance in the economic development of the coastal zone. Both as ports and as industrial and residential centers these cities affect bays, estuaries and littoral environmental units. The influence of ports on marine environments may manifest itself through depletion of fresh water inflows canalization, waste inflows and recreational demands. Construction on littoral areas may conflict with conservation policies derived from both political and geological considerations.

MULTI LEVEL MODELS

In order to evaluate the impact of land use management policies on the type, location and intensity of industrial and residential activity in the coastal zone, a series of coupled linear programming models are being developed. The overall model has a multi level structure with the variables at each level being coupled to those at the next by appropriate constraints.

This model draws upon the work of Charnes, Cooper, Niehaus *et al.*[4] on manpower planning problems as part of the research program of the U. S. Department of the Navy. These models were developed to aid in the management, projection and assignment of civilian and naval manpower. The central set of models are multi-level multi-period goal programs which incorporate input-output analysis in order to determine manpower requirements and Markov processes to model manpower transitions, e.g. geographical movements, skill changes and retirements of manpower, together with decision variables on new hires, involuntary retirements (RIF's), etc.

THE MODEL IN OUTLINE

In the present work we are considering a three level model. At the top level the allocation of population and industrial activity between areas within the coastal zone is determined. At the second level housing and industry are located spatially within each area and at the third level water, waste treatment and demographic models are included. These models are presently at an early stage of development and thus the following outline of their structure is directed primarily toward elucidating the nature of the coupling mechanisms between population, industrial output and land use within the multi-level structure.

THE COASTWIDE-COG LEVEL

At the top level we consider the allocation of industry (including agriculture) between coastal Council of Government (COG) areas. A COG is an agency set up under state legislation to coordinate plans and initiate policies for groups of counties. When COG's were set up it was attempted to make them conform as far as possible to local trading areas by basing the division upon Standard Metropolitan Statistical Areas. COG's thus provide a means of incorporating policy making bodies into the models in a way which is coherent with economic considerations.

For each COG an input-output table and final demand projections will be determined for each of a series of planning periods. We have for the rth COG in the tth planning period

$$x_i^r(t) - \sum_j L_{ij} x_j^r(t) + \sum_s e_i^{rs}(t) - \sum_s e_i^{sr}(t) + g_i^{r+}(t) - g_i^{r-}(t) = d_i^r(t) \quad (1)$$

where

$x_i^r(t)$ is the output of the ith economic sector in period t

L_{ij} is the input from sector j required to produce unit output sector i.

$e_i^{rs}(t)$, $e_i^{sr}(t)$ are dollar amounts of exports and imports respectively for COG r and area s which include coastwide, international and domestic trade.

$g_i^{r+}(t)$, $g_i^{r-}(t)$ are deviations from the final demand goals $d_i^r(t)$.

Restrictions on trade reflecting port capacities and balance of payments considerations may be represented schematically as:

$$\begin{aligned} E_{1S}^{rL}(t) &\leq \sum_{s\in S} e_i^{rs}(t) \leq E_{1S}^{rU}(t) \\ E_{2S}^{rL}(t) &\leq \sum_{s\in S} e_i^{sr}(t) \leq E_{2S}^{rU}(t) \end{aligned} \quad (2)$$

where $E_{1S}^{rU}(t)$, $E_{1S}^{rL}(t)$ are upper and lower bounds on exports for a subset S of sectors in COG r and E_{2S}^{rU}, E_{2S}^{rL} are upper and lower bounds on imports.[1]

---

[1] The formulation at this and at the COG level draws upon work in interregional linear programming. See Isard, et. al., [5].

COG LEVEL

At this level we consider the allocation of industrial and residential activity spatially within each COG. We have the coupling conditions:

$$x_i^r(t) - \sum_{k_r \epsilon \bar{K}^r} x_{ik_r}^r(t) = 0 \tag{3}$$

where $\bar{K}^r$ is the set of land tracts in the rth COG. For subsets K^r of land tracts[2] we have population projections $p_{hK^r}^r(t)$ for the hth type of household. We partition this population into labor force and non-labor force participating (e.g. retired people, second-home owners) components as follows:

$$p_{hK^r}^r(t) - \rho_h \lambda_{hK^r}^r(t) - \mu_{hK^r}^r(t) = 0. \tag{4}$$

where $\lambda_{hK^r}^r(t)$ is the amount of labor of type h in tract K^r in period t

ρ_h is the ratio of households to labor for type h.

$\mu_{hK^r}^r(t)$ is the number of non-labor force participating households in tracts K^r of type h in period t.

We have relationships between labor and output given by labor coefficients ρ_{ih}:

$$\sum_i \sum_{k_r \epsilon K^r} \rho_{ih} x_{ik_r}(t) - \lambda_{hK^r}^r(t) + \bar{g}_{hK^r}^{r+}(t) - \bar{g}_{hK^r}^{r-} = 0 \tag{5}$$

where $\bar{g}_{hK^r}^{r+}(t)$, $\bar{g}_{hK^r}^{r-}(t)$ are deviations from labor requirements of the type h in land tract K^r in period t.

Land constraints take the following form:

$$\sum_h b_{hk_r} p_{hk_r}^r(t) + \sum_i a_{ik_r} x_{ik_r}^r(t) \leq A_{k_r}^r \tag{6}$$

---

[2]These subsets may correspond to cities, counties, or other geographic or political units.

where b_h is the land required for a household of type h located in tract k_r and a_{ik_r} is the land requirement for a unit of output of sector i in tract k_r. These coefficients reflect the zoning considerations specified for the environmental unit k_r.

$$\text{Also} \sum_{k_r \in \bar{K}^r} p^r_{hk_r}(t) = p^r_{hK^r}(t) \tag{7}$$

That is, each subset of tracts K^r for which population projections are made must house that population.

Water constraints are derived in a similar way to those on land and take the form

$$\sum_h \sum_{k_r \in \bar{K}_r} e_{hk_r} p^r_{hk_r}(t) + \sum_i \sum_{k_r \in \bar{K}_r} f_{ik_r} x^r_{ik_r}(t) \leq q_{\bar{K}_r}(t) \tag{8}$$

where e_{hk_r}, f_{ik_r}, are unit water requirements for households and industry respectively and $q_{\bar{K}_r}(t)$ is the available water to a subset $\bar{K}_r$ of land tracts in period t. This will generally include water from both surface and artesian sources.

POPULATION, WATER, AND WASTE MODELS

The population within the coastal zone may be coupled both spatially and temporally by means of Markov transition matrices to reflect the migration and aging respectively of the population (See description of the Cohort Migration Projection Method). These couplings are made at a third level within the model structure to the variables $\mu^r_{hk_r}(t)$ and $\lambda^r_{hk_r}(t)$ for non-labor force participants and labor force participants respectively.

Water and waste models will be developed based upon the Ph.D work (under Charnes, Logan, and Gotaas) of Heaney [6], Lynn [7] and Deininger [8] which are variations and specializations of the original (unpublished) 1958 multi-page models of Charnes, Logan and

Pipes [9].

We may couple the variables $q^r_{\bar{K}_r}(t)$ to a water distribution system in the following way.

Let $q^r_{\bar{K}_r}(t) = W^r_{\bar{K}_r}(t) + Z^r_{\bar{K}_r}(t)$ and $W^r_{\bar{K}_r}(t)$ are ground water availabilities in the set of tracts $\bar{K}_r$ in period t.

For the surface water distribution system the tracts $\bar{K}_r$ may themselves correspond to environmental units as illustrated in Figure 1. The quantities $Z^r_{\bar{K}_r}(t)$ would then correspond to the net quantities of water withdrawn as shown in the figure. Seasonal demand and supply variation, storage and precipitation, and return flows are considered in detail in Heaney's work and need not be elaborated upon here.

Once the type, intensity and location of industrial and residential activities are known we turn to consideration of waste treatment models. Deininger's work provides an approach to the determination of the scale and location of treatment facilities in order to meet prescribed quality at minimum cost. Lynn considered the design and staged expansion of waste treatment plants over time in order to provide treatment for prescribed quantities of waste inputs at minimum cost including considerations of financing and service charges to the people served.

THE OBJECTIVE

We seek to minimize a weighted sum of deviations from regional growth goals $g_i^{r+}(t)$, $g_i^{r-}(t)$ subject to the constraints (1) - (8), (see Figure 2). We also have goals of minimizing unemployment $\bar{g}^{r+}_{hk_r}$ and job vacancies $\bar{g}^{r-}_{hk_r}$ on a local basis. The objective may then be

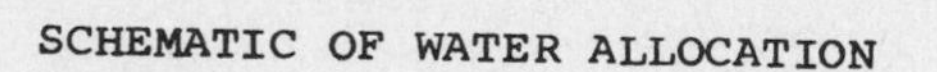

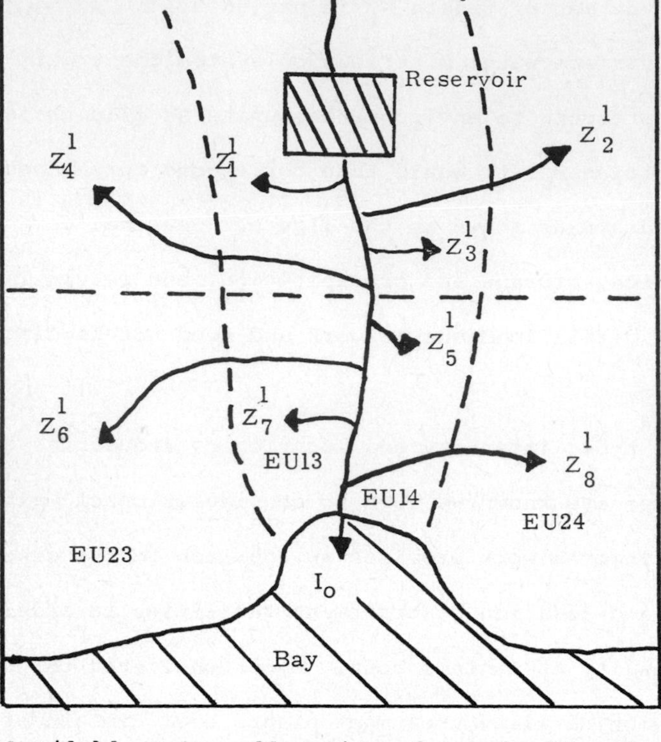

Available water allocations determined by the Texas Water Rights Board

Figure 1

EU12 represents an environmental unit of type 1 in location 2 within the COG.

Minimize $\sum c_i^{r+}(t)\, g_i^{r+}(t) + c_i^{r-}(t)\, g_i^{r-}(t) + \bar{c}_{hK^r}^{r+}\bar{g}_{hK^r}^{r+}(t) + \bar{c}_{hK^r}^{r-}\bar{g}_{hK^r}^{r-}(t)$

LEVEL 1

Input Output: $x_i^r(t) - \sum_j L_{ij} x_j^r(t) + \sum_s e_i^{rs}(t) - \sum_s e_i^{sr}(t) + g_i^{r+}(t) - g_i^r(t) = d_i^r(t)$ (1)

Trade Restrictions: $E_{1S}^{rL}(t) \leq \sum_{s \in S} e_i^{rs}(t) \leq E_{1S}^{rU}(t)$

$E_{2S}^{rL}(t) \leq \sum_{s \in S} e_i^{sr}(t) \leq E_{2S}^{rU}(t)$ (2)

LEVEL 2

Spatial Distribution of Industry: $x_i^r(t) - \sum_{k_r \in \bar{K}^r} x_{ik_r}^r(t) = 0$ (3)

Demographic Relationships: $p_{hK^r}^r(t) - p_h \lambda_{hK^r}^r(t) - \mu_{hK^r}^r(t) = 0$ (4)

Employment: $\sum_i \sum_{k_r \in \bar{K}^r} \beta_{ih} x_{ik_r}(t) - \lambda_{hK^r}^r(t) + \bar{g}_{hK^r}^{r+}(t) - \bar{g}_{hK^r}^{r-} = 0$ (5)

Spatial Distribution of Households: $\sum_{k_r \in \bar{K}_r} p_{hk_r}^r(t) - p_{hK^r}^r(t) = 0$ (7)

Land Restrictions: $\sum_i a_{ik_r} x_{ik_r}^r(t) + \sum_h b_{hk_r} p_{hk_r}^r(t) \leq A_{k_r}^r$ (6)

Water Restrictions: $\sum_i \sum_{k_r \in \bar{K}_r} f_{ik_r} x_{ik_r}^r(t) + \sum_h \sum_{k_r \in \bar{K}_r} e_{hk_r} p_{hk_r}^r(t) - q_{\bar{K}_r}(t) \leq 0$ (8)

LEVEL 3

WASTE TREATMENT MODELS	DEMOGRAPHIC MODELS	WATER SUPPLY MODELS

FIGURE 2

represented schematically as:

$$\text{Minimize } \sum c_i^{r+}(t)\, g_i^{r+}(t) + \sum c_i^{r-}(t)\, g_i^{r-}(t) + \sum \bar{c}_{hK^r}^{r+}(t)\, \bar{g}_{hK^r}^{r+}(t) + \sum \bar{c}_{hK^r}^{r-}(t)\, \bar{g}_{hK^r}^{r-}(t)$$

where the weights $c_i^{r+}(t)$, $c_i^{r-}(t)$, $\bar{c}_{hk_r}^{r+}(t)$, $\bar{c}_{hk_r}^{r-}(t)$ may reflect both political and economic factors.

SUMMARY AND EXTENSIONS

The model reported here is directed toward environmental policy appraisal. The multi-level structure which has been outlined was developed in response to the need to model the micro level implications of macro level zoning policy decisions. With this structure varying degrees of detail and of emphasis may be incorporated within the overall model. In particular we wish to focus upon subareas within the coastal zone due to their geological and ecological significance. For example, marsh and salt flat environmental units which are adjacent to bays and estuaries are major sources of nutrients for marine species.

Subsets of environmental units may become economically important due, for example to the growth of port cities. In view of this, more detailed research is being directed toward the development of a dynamic model for urban growth.

REFERENCES

1. Appendices to the Interim Report, Volumes I and II, The Coastal Resources Management Program of Texas Interagency Resources Council, Office of the Governor, (Austin, Texas, 1971).

2. Ibid.

3. The Texas Water Plan Summary, Texas Water Development Board, (Austin, Texas, 1968).

4. Charnes, A., Cooper, W. W., and Niehaus, R. J. "Studies in Man-Power Planning," Office of Civilian Manpower Management, Department of the Navy, Washington, D. C. (July 1972).

5. Isard, W., _et al_. _Methods of Regional Analysis_, Cambridge, Mass., M. I. T. Press (1960).

6. Heaney, J. P. _Mathematical Programming Model for Long Range River Basin Planning with Emphasis on the Colorado River Basin_, Unpublished Ph.D. Dissertation, Northwestern University (1968).

7. Lynn, W. R. _Process Design and Financial Planning of Sewage Treatment Water_, Unpublished Ph.D. Dissertation, Northwestern University (1963).

8. Deininger, R. A. _Water Quality Management: The Planning of Economically Optimal Pollution Control Systems_, Unpublished Ph.D. Dissertation, Northwestern University (1965).

9. Charnes, A., Logan, I. A. and Pipes, W. "Multi-Page Water and Waste Water Models," Systems Research Group, Northwestern University, unpublished mimeograph (1958).

AN ECOSYSTEM APPROACH TO PROBLEM SOLVING

F. Herbert Bormann
School of Forestry
Yale University

Forest ecosystems exert considerable control over chemical, hydrologic, and meteorologic relationships of regional landscapes. To quantify man's effect, we have for the past ten years intensively studied chemical, hydrologic and biologic aspects of both natural and man-manipulated ecosystems in Northeast U.S.A. The data show that forests not only regulate amounts and flow rates of streamwater, but also closely regulate streamwater chemistry and erosion. Clear-cutting severely upsets natural relationships mainly through its effects on hydrology and on microbial populations. Large quantities of nutrients are lost from the ecosystem, erosion is greatly accelerated, and streamwater is eutrophied. However, nature has designed responses which tend to lessen these adverse effects.

Narrow approaches to natural resource management such as those often employed in agriculture, forestry, or land development tend to ignore basic interrelationships demonstrated by our study, and thus lead to unexpected, undesirable ecological imbalances. Many of these imbalances could be anticipated and dealt with if land management techniques were based on a better understanding of ecosystem function.

Multiple-use management implies that in many circumstances, the total net benefit from forest lands can be maximized through some combination of two or more uses. Thus, multiple-use entails not one end product but several such as wood, water, wildlife, and recreation plus hard-to-measure values such as aesthetics and biological preserves.

Given the basic abiotic and biotic complexity of the land, the ecological processes of succession and retrogression, the multiplicity of managerial goals that multiple-use implies, and a desire for efficient long-term use of the land, it is obvious that some theoretical framework upon which we can assemble and interrelate these diverse components is a necessity.

The ecological system or ecosystem concept provides such a framework. It allows us to see more clearly how nature works and how managerial practices affect the working of nature. It spotlights basic ecological realities which in the long run determine whether or not economic policy is wise.

Eugene Odum has defined the ecosystem as the basic functional unit of nature which includes both organisms and their nonliving environment, each interacting with the other and influencing each other's properties, and both necessary for the maintenance and development of the ecosystem.

For the purpose of this discussion, we might visualize a forest ecosystem as a 1000 acre stand of northern hardwood forest. The lateral boundaries of this ecosystem might be defined as the biological edge of the stand, or the property lines of the stand. Vertical boundaries of the ecosystem could be the tops of the trees and the deepest depths of soil at which significant biological activity occurs. In other words, this hypothetical forest ecosystem would occupy a volume of space at the air-soil interface.

Our forest ecosystem would be composed of a series of living and nonliving components intimately linked together by an array of natural processes. These relationships are portrayed in a model that we have found to be conceptually useful in research on nutrient cycling (Fig. 1). Components or compartments of the

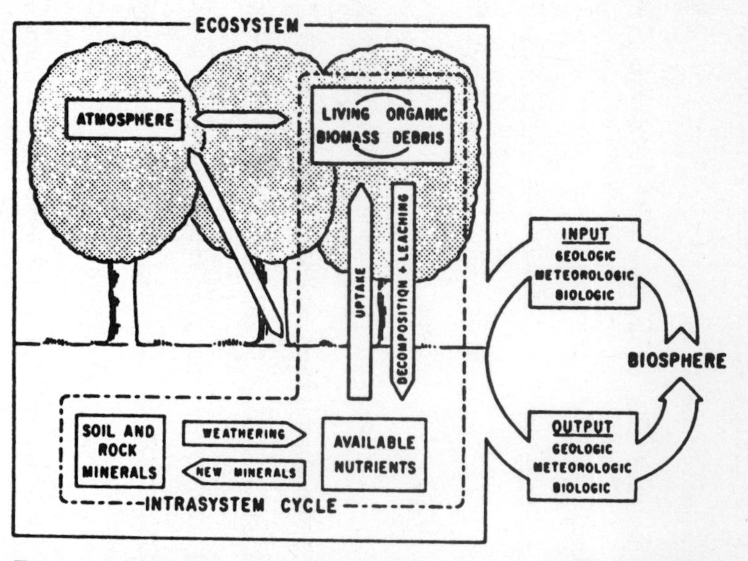

Figure 1.—Nutrient cycling model of the ecosystem showing compartments (boxes) where nutrients accumulate and pathways (arrows) or processes whereby nutrients are moved between compartments. Major processes whereby nutrients are moved into or out of the ecosystem are geological (moving water and gravity), meteorological (precipitation and wind) and biological (animal power including man).

ecosystem are: 1) the organic compartment composed of the biota and its organic debris (The biota, the trees and other plants and animals, probably exceeds 2500 species in our 1000 acre ecosystem.) 2) the available nutrient compartment composed of nutrients held on the exchange sites of the soil and in the soil solution, 3) the soil and rock mineral compartment containing nutrients in forms unavailable to the organisms, and 4) the atmospheric compartment made up of gases of the ecosystem both above and below ground. These compartments are intimately linked together by a variety of natural processes. For example, nutrients are removed from the pool of available nutrients by uptake by the higher plants, they circulate within the organic compartment in highly complex food webs, and they are returned to the available nutrient compartment by ultimate decomposition of organic matter or by leaching. Primary and secondary minerals are decomposed by weathering reactions and previously unavailable nutrients in the soil-rock compartment are transferred to the available nutrient compartment. Some chemicals temporarily solubilized are returned to the soil-rock compartment through the formation of new minerals such as clays or insoluble compounds. Atmospheric gases are continually being taken up or released in the processes of photosynthesis and respiration. Knowledge of these components and their linkages leads to an understanding of the interrelationships within an ecosystem. Such knowledge is basic to understanding the ramifications of any managerial practice applied at any point within the ecosystem.

Our forest ecosystem is not an independent entity but rather is connected to surrounding ecosystems, forests, fields, and streams and to the biosphere of the earth in general by a system of inputs and outputs. Inputs and outputs may be in the form of energy, water, gases, chemicals, or organic materials moved through the ecosystem boundary by meteorological processes such as precipitation or wind, by geological processes associated with moving water or gravity, and by biological processes where materials are moved by animal power. Human activities such as fertilization or harvesting, are part of the latter process.

The important point here is to understand that an ecosystem can be strongly influenced by activities (both natural and man-made) originating outside the ecosystem. For example, there are numerous cases where an input of air pollutants has had devastating effects on forest ecosystems. We have only to look at Ducktown, Tennessee, Sudbury, Ontario, or some of the forests around Los Angeles to realize the potential seriousness of this matter. Equally, the manager must realize the intimate way in which his ecosystem is linked to others and that he can produce secondary effects beyond his system that will not appear in his ledgers, but will appear in overall ledgers of society. And, of course, it is by the ledgers of society that we judge the viability of social and political systems. Examples of practices that may extend well beyond individual forest ecosystems are practices that accelerate erosion, the use of persistent pesticides that find their way into regional and worldwide food chains, controlled burning that might make a major contribution to regional air pollution, and the development and spread of forest fertilization practices that may greatly exacerbate the already dangerously polluted conditions of our surface and gound waters.

The last point should be one of particular concern to forest managers. During the last three decades, the use of nitrogen fertilizer has risen from 0.5 million tons in 1949 to a use of eight million tons in 1970. Coupled with this is widespread and steadily increasing nitrate pollution of our surface and ground waters, part of which is directly attributable to increasing leakage of nitrate from agricultural ecosystems. Nitrate pollution of U.S.A. water supplies is a major national problem and probably will get worse. Forest managers would be wise to take this into account before embarking on policies of widespread forest fertilization with nitrogen compounds.

Thus far, I have spoken generally about the broad principles of ecosystem structure and function. Now I would like to spend some time examining the details of structure and function and how these may be altered by managerial practice. For the past ten years my colleague Gene Likens and I have been studying the relationships shown in Figure 1 in both undisturbed and experimentally manipulated northern hardwood forest ecosystems in central New Hampshire. To do this, we use the small watershed technique where small forested watersheds are the fundamental units of study. Because these ecosystems are small watersheds, geological input, as I previously defined it, is eliminated from consideration.

The fact that our watersheds are underlain by impermeable glacial till and bedrock causes all liquid water leaving the ecosystems to leave via small streams where its volume, chemical content, and particulate matter content can be accurately measured. Using this technique, we have constructed budgets for individual ecosystems showing the input and output of water and chemicals and their net loss or gain from the system. Because our watersheds have a more or less homogeneous geology, it has been possible to use our budgetary data to provide accurate estimates of weathering of primary minerals within the ecosystem.

Our study is being carried out in cooperation with the Northeastern Forest Experiment Station at the Hubbard Brook Experimental Forest in central New Hampshire, U.S.A. (Fig. 2) This is the major installation of the U.S. Forest Service in New England for measuring aspects of the hydrologic cycle as they occur in small, forested watershed.

Six watersheds, tributary to Hubbard Brook and 12 to 43 hectares in size, are being studied. Each watershed is forested by a well-developed second growth stand of sugar maple, beech and yellow birch. The area was heavily cut over in 1919 but since that time there has been no disturbance by cutting or fire.

Precipitation entering these watersheds is measured by a network of gauging stations and drainage water is measured by weirs anchored on the bedrock across the stream draining each watershed (Fig. 2). Since there is no deep seepage, the loss of water by evapotranspiration is calculated by subtracting hydrologic output from hydrologic input.

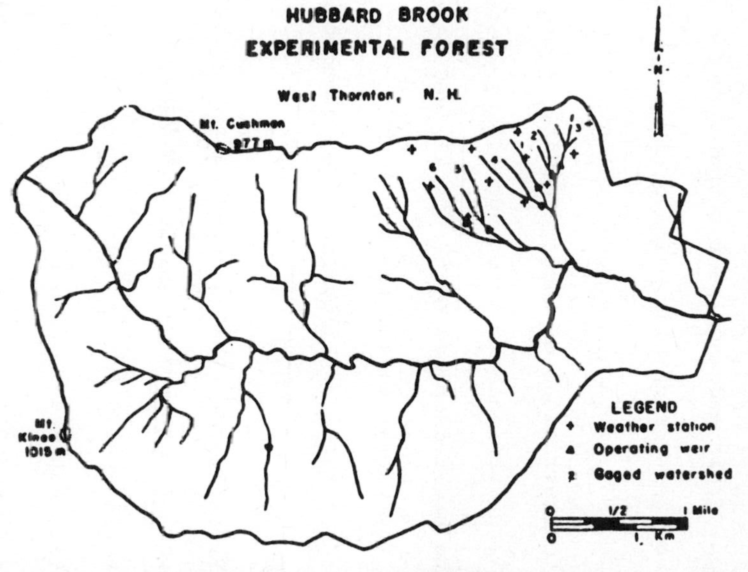

Figure 2.—The Hubbard Brook Experimental Forest. The major watershed research facility of the U. S. Forest Service in New England. The outline map shows the various drainage streams that are tributary to Hubbard Brook. The smaller watersheds under study are enclosed by dotted lines (Likens et al. 1967).

The average monthly water budget for the period 1955-69 indicates that precipitation is distributed rather evenly throughout the year whereas runoff is not. Most of the runoff (68%) occurs during the snowmelt period of March, April and May. In fact, 35% of the total runoff occurs in April. In marked contrast, only 0.7% of the yearly runoff occurs in August.

To measure chemical parameters of the ecosystem, weekly water samples of input (rain and snow) and output (stream water) are collected and analyzed for calcium, magnesium, potassium, sodium, aluminum, ammonia, nitrate, sulfate, chlorine, bicarbonate, and silicate. The concentrations of these elements in precipitation and in stream water are entered into a computing system where weekly concentrations are multiplied by the weekly volume of water entering or leaving the ecosystem and the input and output of chemicals is computed in terms of kilograms of an element per hectare of watershed.

Our data on input indicate that the chemistry of precipitation changes from year to year and may reflect the dominance of different kinds of air masses. For example, increases in the input of sodium and magnesium may result when our area is under the dominance of maritime air masses. Our data also indicate a steadily increasing trend of nitrogen input over the past six years. This may be a reflection of increasing air pollution.

Our data show that concentrations of metallic cations in drainage water from our undisturbed forests are remarkably constant even though the discharge of stream water from our watersheds ranges over four orders of magnitude, varying from summer trickles to spring torrents. This result was somewhat unexpected since during the spring melt period, when the bulk of liquid is discharged from our ecosystems, we thought there would be considerable dilution and that concentrations of elements in stream water would be very low. This was not the case.

The chemical stability of the stream water draining the mature forest ecosystem at Hubbard Brook is very likely due to the mature, highly permeable podzolic soils. Because of microtopography, loose soil structure, and absence of frozen ground during the winter virtually all of the drainage water must pass through the soil. The intimate contact afforded by this passage, plus the relatively large buffering capacity of the soil materials (relative to the quantities of chemicals lost to percolating water) and the relatively small range of temperatures within the soil mass ($10^{o}C$ at a depth of 91 cm) apparently succeed in buffering the chemistry of transient waters.

Because of the relative constancy of cation concentrations, total output is strongly dependent on volume of streamflow (Fig.3). Consequently, based on a knowledge of hydrologic output alone, it is now possible to predict chemical output, with a fair degree of accuracy, from the mature forested ecosystem. This relationship would seem to have considerable value for regional planners concerned with water quality. Knowing the input and output of chemicals, we have constructed nutrient budgets for 11 ions (Table 1).

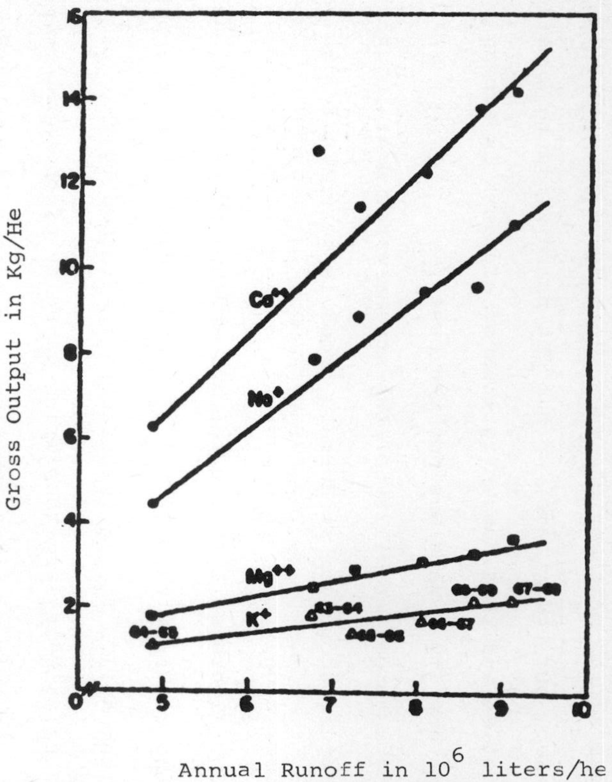

FIGURE 3. Relationship between the gross output of the major cations and annual water discharge during 1963-1969 for the Hubbard Brook watershed ecosystems. The year of occurrence is shown only for the potassium data, but applies to the other ions as well (modified from Likens *et al.*, 1971).

TABLE 1

Table 1. AVERAGE CHEMICAL INPUT AND OUTPUT FOR UNDISTURBED, FORESTED WATERSHED-ECOSYSTEMS W1-W6, HUBBARD BROOK EXPERIMENTAL FOREST, 1963-1969. (Year [e.g. '5 = 1964-1965] and watershed number [e.g., W2] indicated for minimum and maximum values; output for W2 after treatment not included)

Element	Input			Output			Mean values		
	Watershed years*	Range: Minimum	Range: Maximum	Watershed years*	Range: Minimum	Range: Maximum	Input	Output	Net loss or gain
		kg/ha-yr	*kg/ha-yr*		*kg/ha-yr*	*kg/ha-yr*	*kg/ha-yr*		
SiO_2-Si	0	....	—	10	9.7 ('5 W2)	22.3 ('8 W4)	†	16.4	—16.4
Ca^{++}	30	1.6 ('9 W2)	3.0 ('8 W6)	32	4.8 ('5 W6)	17.3 ('9 W4)	2.6	11.7	— 9.1
SO_4-S	15	9.7 ('5 W2)	16.0 ('8 W6)	10	9.8 ('5 W6)	19.6 ('8 W4)	12.7	16.2	— 3.5
Na^+	30	1.0 ('9 W2)	2.3 ('5 W6)	32	3.5 ('5 W1)	11.0 ('8 W4)	1.5	6.8	— 5.3
Mg^{++}	30	0.32 ('9 W2)	1.2 ('5 W6)	32	1.6 ('5 W3)	3.7 ('8 W1)	0.7	2.8	— 2.1
Al^{+++}	0	....	....	10	0.9 ('6 W4)	3.2 ('9 W6)	†	1.8	— 1.8
K^+	30	0.57 ('7 W3)	2.4 ('4 W4)	32	0.76 ('5 W3)	2.7 ('9 W1)	1.1	1.7	— 0.6
NO_3-N	15	1.5 ('5 W2)	5.2 ('8 W6)	10	1.1 ('5 W6)	2.9 ('8 W4)	3.7	2.0	+ 1.7
NH_4-N	15	1.6 ('5 W2)	2.6 ('8 W6)	10	0.08 ('9 W6)	0.7 ('6 W6)	2.1	0.3	+ 1.8
Cl^-	12	2.6 ('6 W2)	6.9 ('7 W6)	8	4.2 ('6 W4)	5.3 ('8 W4)	5.2	4.9	+ 0.3
HCO_3-C	0	—	—	4‡	2.4 ('6 W4)	3.3 ('7 W4)	†	2.9	— 2.9

* Number of watersheds times years of data.
† Not measured but very small.
‡ Watershed 4 only.

These inputs and outputs represent connections of the forest ecosystem with world-wide biogeochemical cycles. These data, from undisturbed forest ecosystems, also provide comparative data whereby we may judge the effect of managerial practices on biogeochemical cycles.

Net losses of calcium, sodium, and magnesium occurred in every year spanning a range of wet, dry, and average years. The potassium budget on the other hand was just about balanced each year. This suggests that potassium is a particularly sensitive element and that is is accumulating in the ecosystem relative to the other cations. Several factors may account for this. Part of the potassium may be retained in illitic clays developing within the ecosystem and/or potassium may be retained in proportionally greater amounts than other cations in the slowly increasing biomass of the system. Biomass of the forest is still increasing after the 1919 cutting.

Thus far, I have mentioned only chemical losses occurring as dissolved substances. Losses also occur when chemicals locked up in particulate matter, such as rock or soil particles and in organic matter like leaves and twigs, are washed out of the ecosystem by the stream. We have also measured these outputs and developed mathematical equations expressing particulate matter losses as a function of discharge rate of the stream. Time does not permit discussion of this portion of our study except for two points: 1) losses of dissolved substances account for the great bulk of chemical losses from our undisturbed forest ecosystems, and 2) while dissolved substance losses are largely independent of discharge rate, particulate matter losses are highly dependent on discharge rate. The latter point is of special interest since forest management practices can either increase or decrease discharge rates, and thus shift the balance between dissolved substance and particulate matter losses.

Using the small watershed technique, weathering, or the rate at which elements bound in primary minerals are made available for biological use, can be estimated. Based on net losses of elements from our ecosystems and a knowledge of the bulk chemistry of the underlying till, we estimate that the nutrients contained in 800 $\pm$ 70 kg/ha of bouldery till are made available each year by weathering.

We now have for the undisturbed northern hardwood ecosystems of Hubbard Brook estimates of chemical input in precipitation output in stream water, and the rates at which ions are generated by the weathering of minerals within the system. To complete the picture of nutrient cycling according to the model shown in Figure 1, it is necessary to obtain the nutrient content of the vegetation, litter, and the available nutrient in the soil, as well as estimates of annual rates of nutrient uptake by the vegetation and nutrient release from the biota and organic debris. The measurement of these parameters is under way. Measurements for the calcium cycle are presented in Figure 4. Some 2310 kg/ha are localized in the trees and litter, while 690 kg/ha represent exchangeable calcium in the soil. This gives a total of 3000 kg of Ca/ha in organic matter and as available nutrient. Our data indicate that the vegetation takes up a net total of about 49 kg/ha of calcium from the available nutrient or exchangeable pool. About the same amount is returned each year. Annual input is 2.6 kg/ha, output is 12 kg/ha, and 9 kg/ha are generated by weathering.

Data in Figure 4 suggest a remarkable ability of these undisturbed northern hardwood forests to hold and circulate nutrients for the net annual loss of 9 kg/ha represents only about 0.3 of the calcium circulating within the system. However, if percentage loss calculations were based on the actual amounts of calcium circulated annually (annual uptake and release) rather than total calcium, percentage losses would be much higher and perhaps less remarkable.

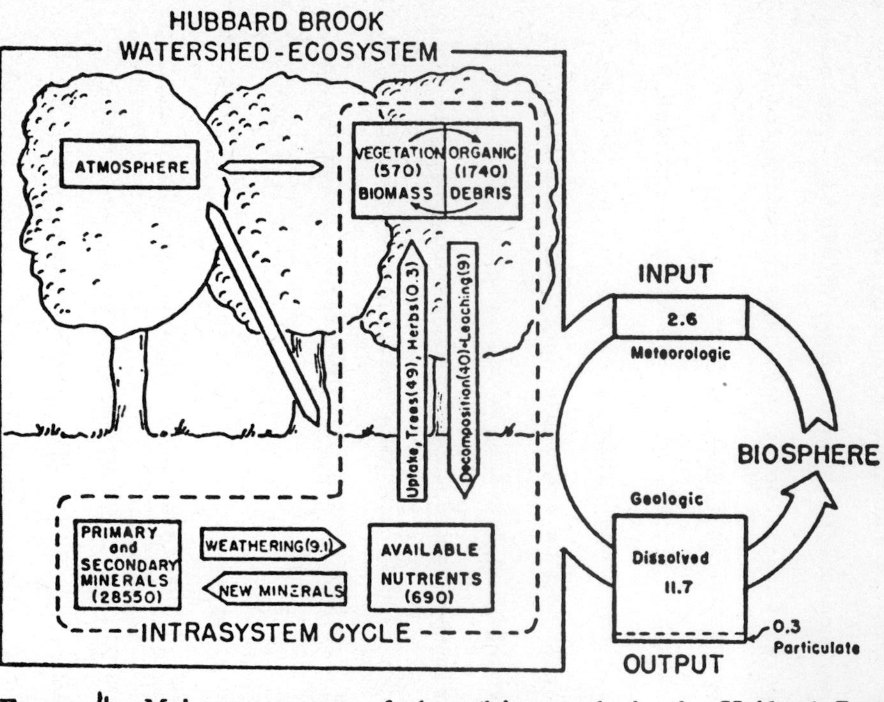

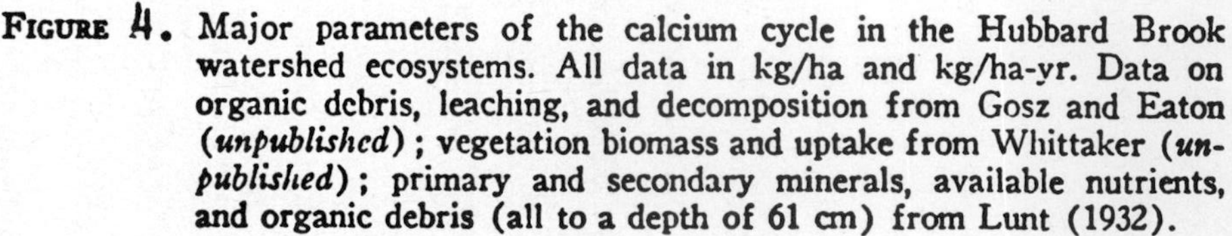
FIGURE 4. Major parameters of the calcium cycle in the Hubbard Brook watershed ecosystems. All data in kg/ha and kg/ha-yr. Data on organic debris, leaching, and decomposition from Gosz and Eaton (*unpublished*); vegetation biomass and uptake from Whittaker (*unpublished*); primary and secondary minerals, available nutrients, and organic debris (all to a depth of 61 cm) from Lunt (1932).

Knowledge of the nutrient relationships shown in Figure 4 has considerable value for long term forest management. Data on the nutrient condition of the mature forest and data on the quantity of nutrients removed from the ecosystem in forest products provide a means to evaluate the nutrient drain of harvesting on the nutrient capital of the ecosystem. Coupled with information on weathering, or the rate at which nutrients are made available from primary minerals, such data would provide a basis for the rational development of forest fertilization practices.

The small watershed approach provides a means by which we can conduct experiments at the ecosystem level. Using a watershed ecosystem calibrated in terms of hydrologic-nutrient cycling parameters it is possible to impose treatments and to determine treatment effects either by comparison with control watersheds or with predicted behavior had the watershed not been treated. This is, of course, the long-standing method employed by the U.S. Forest Service in its study of the hydrologic parameters of forests. This approach makes it possible to evaluate managerial practices in terms of the whole ecosystem rather than isolated parts and to test the effects of various land management practices (cutting, controlled burning, grazing, etc.) and to determine the effect of potential environmental pollutants (pesticides, herbicides, fertilizers, etc.) on the behavior of nutrients, water, and energy in the system.

In 1965, the forest of one watershed was clear-cut in an experiment designed (1) to determine the effect of clear-cutting on streamflow, (2) to examine some of the fundamental chemical relationships of the forest ecosystem, and (3) to evaluate the effects of forest manipulation on nutrient relationships and eutrophication of stream water.

The experiment was begun in the winter of 1965-66 when the forest of one watershed, 15.6 ha, was completely leveled by the U.S. Forest Service. All trees, saplings, and shrubs were cut, dropped in place, and limbed so that no slash was more than 1.5 m above ground. No products were removed from the forest and great care was exercised to prevent disturbance of the soil erosion. The following summer, June 23, 1966, regrowth of the vegetation was inhibited by an aerial application of 28 kg/ha of the herbicide, Bromacil.

One of the objectives of this severe treatment was to clock the ecosystem pathway, nutrient uptake by the higher plants (Fig. 1), while the ultimate decomposition pathway continued to function. Under these circumstances, we question whether or not the ecosystem had the capacity to hang on to the nutrients accumulating in the available nutrient compartment (Fig. 1).

Stream water samples were collected weekly and analyzed, as they had been for a two-year period preceding the treatment. Similar measurements on adjacent undisturbed watersheds provided comparative information.

The effect on water yield was astounding! Yields increased 31% over that of the undisturbed forest. Ten square miles of forest treated in this fashion would have yielded about two billion gallons of additional water per year. Based on New York City per capita consumption data this would be sufficient to supply the needs of 37,000 persons. This doesn't even consider recycling usage which would substantially increase the number of persons.

In the deforested ecosystem, there was a striking loss of nitrogen. This suggests that alteration of normal patterns of nitrogen flow played a major role in the loss of nutrients from the cutover ecosystem.

This loss is best understood by a consideration of nitrogen patterns in the undisturbed ecosystem (Fig. 5). Data from these ecosystems indicate a strong and reproducible seasonal cycle of concentration of nitrate-nitrogen in stream water. High concentrations are associated with the winter period from November through April, while low concentrations persist from April through November.

The decline of nitrate concentration during May and low concentration throughout the summer correlate with heavy nutrient demands by the vegetation and generally increased biologic activity associated with warming of the soil. The winter pattern of nitrate concentration may be explained in physical terms, since input of nitrate in precipitation from November through May largely accounts for nitrate lost in stream water during this period. In fact, yearly input of nitrate in precipitation exceeds losses in stream water for undisturbed ecosystems.

Figure 5.—Average monthly concentrations of selected cations and anions in stream water draining from forest ecosystems: undisturbed (solid lines) and deforested during the winter of 1965-66 (dashed lines) Bormann et al. 1968).

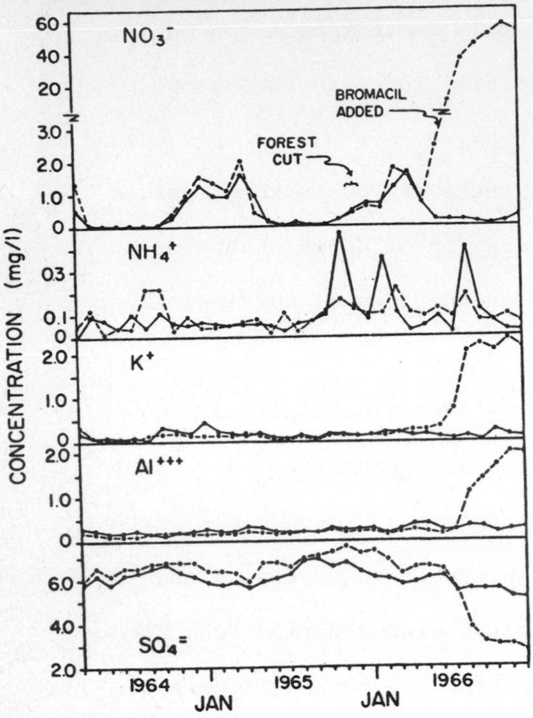

Nitrification is a microbial process where soil bacteria, Nitrosomonas and Nitrobacter oxidize ammonia-nitrogen to nitrate-nitrogen. Outside of nitrate input this process is the only source of the nitrate anion within the ecosystem. It is widely believed that growing, acidifying vegetation represses the process of nitrification and that the process of nitrification is of little importance in acid podzol soils similar to those found under mature forests at Hubbard Brook. The occurrence of nitrate in stream water draining from undisturbed forests is not conclusive evidence of the occurrence of the process of nitrification since it may be accounted for by input in precipitation.

Beginning June 7, 1966, 16 days before the herbicide application, nitrate concentrations in the deforested watershed showed a precipitous rise while the undisturbed ecosystem showed the normal spring decline (Fig. 5). High concentrations of nitrate have continued for three years in the stream water draining from the deforested system. The increase in nitrate concentrations is a clear indication of the occurrence of nitrification in the cutover ecosystems. There is no doubt that cutting drastically altered conditions controlling the nitrification process.

The action of the herbicide in the cutover watershed seems to be one of reinforcing the already well-established trend of nitrate loss induced by cutting alone. This is probably effected through the destruction of the remaining vegetation, herbaceous plants, and root sprouts by the herbicide. Even in the event of rapid transformations of all the nitrogen in the Bromacil this source could at best contribute 5% of the nitrogen lost as nitrate.

Average net losses of nitrate nitrogen were 120 kg/ha during the period 1966 through 1968. The annual nitrogen turnover in our undisturbed forests is approximately 60 kg/ha based on an equilibrium system in which annual leaf fall is about 3200 kg/ha and the annual loss of roots is about 800 kg/ha. Consequently, an amount of elemental nitrogen equivalent to double the amount normally taken up by the forest has been lost each year since cutting.

In the process of nitrification, hydrogen ions are produced and these replace metallic cations on the exchange surfaces. This is precisely what is seen in the cutover ecosystem. Calcium, magnesium, sodium, and potassium concentrations in the stream water increased almost simultaneously with the increase in nitrate. This was followed about one month later by a sharp rise in the concentrations of aluminum. Sulfate, on the other hand, showed a sharp drop in concentration coincident with nitrate rise (Fig. 5).

Net losses of potassium, calcium, aluminum, magnesium, and sodium were 21, 10, 9, 7, and 3 times greater, respectively, than those for an undisturbed watershed for the water years 1966-67 and 1967-68. This represents a major loss of nutrients from the ecosystem, and the data suggest that more attention should be focused on the effect of harvesting practices on the loss of solubilized nutrients.

Our results indicate that our ecosystem has a limited capacity to retain nutrients when the bulk of the vegetation is removed. The accelerated rate of nutrient loss is related both to the cessation of nutrient uptake by plants and to the larger quantities of drainage water passing through the ecosystem. Accelerated losses may be also related to increased decomposition rates resulting from changes in the physical environment, e.g. increased temperature, or an increase in organic matter available for decomposition.

However, the direct effect of the vegetation on the process of nitrification cannot be overlooked. In the cutover ecosystem, the increased loss of cations is correlated with the increased loss of nitrate anions. Consequently, if the intact vegetation inhibits the process of nitrification as workers in other regions have shown and removal of the vegetation promotes nitrification, the release of the process of nitrification from inhibition by the vegetation may account for major nutrient losses from the cut-over ecosystem.

Results of the particulate matter study also indicate a basic change in the pattern of losses. After two years of modest increase in erodibility, erodibility increased exponentially in the third year of treatment. Loss of inorganic material from the stream bed accelerated because of the greater erosive capacity of the now-augmented streamflow and because several biological barriers to stream bank erosion have been removed. The extensive net work of fine roots that tended to stabilize the bank is now dead as a result of cutting and herbicide treatment, and the dead leaves that tended to plaster over exposed banks are now gone.

Finally, the export of nitrate to the small stream draining watershed 2 has resulted in nitrate concentrations exceeding established pollution levels (10 parts per million) for more than three years. In general, the deforestation treatment has resulted in eutrophication of the stream ecosystem and in the occurrence of algal blooms. This latter finding indicates that under some circumstances forest management practices can make significant contributions to the eutrophication of our streams.

The set of interactions triggered by cutting is shown in Figure 6.

I have included this somewhat detailed description of the Hubbard Brook study not only to illustrate the structure and function of an ecosystem, but also to illustrate the utility of the ecosystem concept as an analytical tool. The ecosystem concept

FIGURE 6

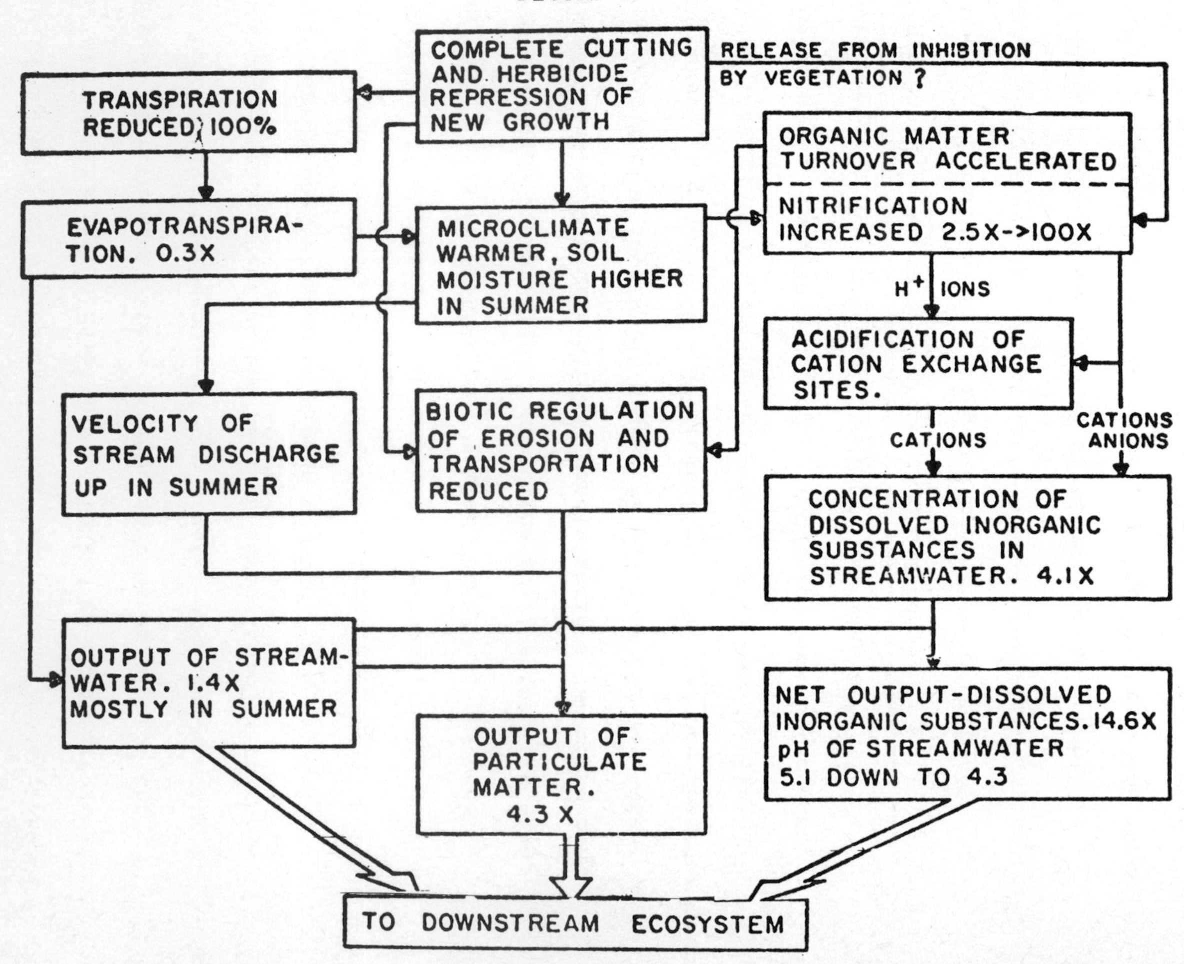

COMPLETE CUTTING AND HERBICIDE REPRESSION OF NEW GROWTH
RELEASE FROM INHIBITION BY VEGETATION ?
TRANSPIRATION REDUCED 100%
EVAPOTRANSPIRATION. 0.3X
MICROCLIMATE WARMER, SOIL MOISTURE HIGHER IN SUMMER
ORGANIC MATTER TURNOVER ACCELERATED
NITRIFICATION INCREASED 2.5X->100X
H+ IONS
ACIDIFICATION OF CATION EXCHANGE SITES.
VELOCITY OF STREAM DISCHARGE UP IN SUMMER
BIOTIC REGULATION OF EROSION AND TRANSPORTATION REDUCED
CATIONS
CATIONS ANIONS
CONCENTRATION OF DISSOLVED INORGANIC SUBSTANCES IN STREAMWATER. 4.1X
OUTPUT OF STREAMWATER. 1.4X MOSTLY IN SUMMER
OUTPUT OF PARTICULATE MATTER. 4.3 X
NET OUTPUT-DISSOLVED INORGANIC SUBSTANCES. 14.6X pH OF STREAMWATER 5.1 DOWN TO 4.3
TO DOWNSTREAM ECOSYSTEM

provides the necessary basis for multiple-use because it provides the best means for seeing the whole of nature. In practice, multiple-use is primarily an economic concept that involves valuation of various goods and services inherent in wild lands. Underlying good multiple-use practice is effective valuation, and underlying this is a thorough understanding of structure and function of the wild land ecosystem.

The ecosystem concept emphasizes the following points, all of which are fundamental to sensible land use planning and should be taken into account in the development of multiple-use management plans.

1. The forest ecosystem is a highly complex natural unit composed of all the organisms (plants, animals, microbes, and man) and their organic environment (air, water, soil, and rock).

2. All parts of a forest ecosystem are intimately linked together by natural processes that are part of the ecosystem, such as uptake of nutrients, fixation of energy, movement of nutrients and energy along food chains, release of nutrients by decomposition of organic matter, weathering of rock or soil particles to release nutrients and the formation of new minerals.

3. Individual forest ecosystems are linked to surrounding land and water ecosystems and to the biosphere in general by input and output connections with the worldwide circulation of air and water and with food chains. This means that in any multiple-use plan, the forest ecosystem should not be considered a strictly local basis. Potential regional or even world effects of the

management plan should be considered and evaluated in an overall social context.

4. Management practices applied to any single aspect of the system, such as cutting or fertilization, will be transmitted to other parts of the ecosystem by existing pathways and beyond the ecosystem by its output links with the surrounding biosphere. For example, deforestation of our watershed reduced the flow of ammonia-nitrogen to higher plants, increased the flow of ammonia-nitrogen to nitrifying organisms, decreased the output of water as a vapor, increased the output of water as a liquid, increased the rate of losses of nitrate-nitrogen and other ions from the ecosystem, increased the movement of hydrogen ions to cation exchange sites in the soil, increased movement of stored energy to respiration, increased fertility and eutrophication of stream water and so forth.

The complex linkages that exist in any ecosystem bring into question the reality of any single-use management plan. Although a manager may limit utilization to one good or service and may enter in his ledgers only money values related to that good or service, in actuality, by the very nature of the ecosystem, he is affecting all values of the system in some degree or other. Whether or not he chooses to valuate these changes is a management decision; nevertheless, changes occur and are ecologic realities. This applies even in cases where land is managed for the single-use of preservation. Often the operation of natural processes, like succession, acting through time will substantially change the nature of the ecosystem and its input-output relationships. In other words, single-use exists only as a method of economic bookkeeping, where the economics of other changes imposed by the management system are ignored.

5. Good management requires that managerial practices be imposed only after a careful analysis of all the ramifications of the proposed practice. A focus for this type of analysis can be achieved by utilizing the ecosystem concept.

Failures in environmental management often result because of a widespread lack of appreciation of the complexity of nature and the widespread acceptance of the naive assumption that somehow or other nature has the capacity to absorb all the types of manipulations we throw at her and still serve man in ways most desirable to him. The polluted condition of our air and water, and tragedy of Lake Erie, and the increasing burden of non-degradable pesticides on the world ecosystem are, in large part, the accumulated result of many small economic or political decisions made without due regard to all of the ecological factors involved.

A knowledge of ecology and the structure and function of ecosystems will help the forest manager to avoid these pitfalls.

REFERENCES

1. Bormann, F. H. and Likens, G. E. "Nutrient cycling," Science 155 (3761), 424-429 (1967).

2. Likens, G. E., Bormann, F. H., Johnson, N. M. and Pierce, R.S. "The calcium, magnesium, potassium and sodium budgets in a small forested ecosystem," Ecology 48(5), 722-785 (1967).

3. Juang, F. and Johnson, N. M. "Cycling of chlorine through a forested watershed in New England," J. Geophys. Res. 72(22), 5641-5647 (1967).

4. Bormann, F. H., Likens, G. E., Fisher, D. W. and Pierce, R.S. "Nutrient loss accelerated by clear-cutting of a forest ecosystem," Science 159, 882-884 (1968); also in Plant-Plant Conf. of the U. S. National Committee for the International Biological Program, Univ. of California, Santa Barbara, 109-112 (March 18-22, 1968).

5. Johnson, N. M., Likens, G. E., Bormann, F. H. and Pierce, R.S. "Rate of chemical weathering of silicate minerals in New Hampshire," Geochimica et Cosmochimica Acta 32, 531-545 (1968).

6. Fisher, D. W., Gambell, A. W., Likens, G. E. and Bormann, F.H. "Atmospheric contributions to water quality of streams in the Hubbard Brook Experimental Forest, New Hampshire," Water Resources Res. 4(5), 1115-1126 (1968).

7. Smith, W., Bormann, F. H. and Likens, G. E. "Response of chemautotrophic nitrifiers to forest cutting," Soil Science 106(6) 471-473 (1968).

8. Bormann, F. H. and Likens, G. E. "A watershed approach to problems of nutrient cycling in forest ecosystems," Proc. 6th World Forestry Congress, Vol. 2, Technical Comm. III, 2303-2306, Madrid, Spain (1969).

9. Bormann, F. H. "A holistic approach to nutrient cycling problems in plant communities," Symposium in Terrestrial Plant Ecology, Antigonish, Nova Scotia (1966), Nova Scotia Museum Publ., Halifax, 149-165 (1969).

10. McConnochie, K. and Likens, G. E. "Some trichoptera of the Hubbard Brook Experimental Forest in central New Hampshire," Canadian Field-Naturalist 83(2), 147-154 (1969).

11. Bormann, F. H. and Likens, G. E. "The Watershed-ecosystem concept and studies of nutrient cycles," in G. M. VanDyne, ed. The Ecosystem Concept in Natural Resource Management, Academic Press, Inc., New York, 49-76 (1969).

12. Eaton, J. S., Likens, G. E. and Bormann, F. H. "The use of membrane filters in gravimetric analyses of particulate matter in natural waters," Water Resources Res. 5(5), 1151-1156 (1969).

13. Likens, G. E., Bormann, F. H. and Johnson, N. M. "Nitrification: Importance to nutrient losses from a cut-over forested ecosystem," Science 163(3872), 1205-1206 (1969).

14. Bormann, F. H., Likens, G. E. and Eaton, J. S. "Biotic regulation of particulate and solution losses from a forested ecosystem," BioScience 19(7), 600-610 (1969).

15. Smith, W. "Techniques for collection of root exudates from mature trees," Plant and Soil 32(1), 238-241 (1970).

16. Johnson, N. M., Likens, G. E., Bormann, F. H., Fisher, D. W. and Pierce, R. S. "A working model for the variation in stream water chemistry at the Hubbard Brook Experimental Forest, New Hampshire," Water Resources Res. 5(6), 1353-1363 (1969).

17. Likens, G. E., Bormann, F. H., Johnson, N. M., Fisher, D. and Pierce, R. S. "Effects of forest cutting and herbicide treatment on nutrient budgets in the Hubbard Brook Watershed-ecosystem," Ecol. Monogr. 40(4), 23-47 (1970).

18. Bormann, F. H., Siccama, T. G., Likens, G. E. and Whittaker, R. H. "The Hubbard Brook ecosystem study: Composition and dynamics of the tree stratum," Ecol. Monogr. 40(4), 373-388 (1970).

19. Siccama, T. G., Bormann, F. H., Likens, G. E. "The Hubbard Brook ecosystem study: Productivity, nutrients and phytosociology of the herbaceous layer," Ecol. Monogr. 40(4),389-402 (1970).

20. Ledig, F. T. "Phenotypic variation in sugar maple in relation to altitude," in Proc. of 17th N. E. Forest Tree Improvement Conf., Univ. Park, Pa., 41-58 (1970).

21. Likens, G. E. and Bormann, F. H. "Chemical analysis of plant tissues from the Hubbard Brook ecosystem in central New Hampshire," Yale Univ. School of Forestry Bull., No. 79 (1970).

22. Likens, G. E. and Eaton, J. S. "A polyurethane stem flow collector for trees and shrubs," Ecology 51(5), 938-939 (1970).

23. Likens, G. E., Bormann, F. H., Pierce, R. S. and Fisher, D. W. "Nutrient-hydrologic cycle interaction in small forested watershed-ecosystems," in L. d'Andigni de Asis (ed.), Productivity of Forest Ecosystems, Proc. Brussels Symp., 1969, UNESCO Publ., 553-563 (1971).

24. Likens, G. E. and Gilbert, J. J. "Notes on quantitative sampling of natural populations of planktonic rotifers," Limnol. Oceanogr. 15(5), 816-820 (1970).

25. Pierce, R. S., Hornbeck, J. W., Likens, G. E. and Bormann, F.H. "Effects of elimination of vegetation on stream water quantity and quality," International Symposium on the Results of Research on Representative and Experimental Basins, New Zealand. Intern. Assoc. Sci. Hydrol., 311-328 (1970).

26. Bormann, F. H. and Likens, G. E. "The nutrient cycles of an ecosystem," Scientific Amer. 223(4), 92-101 (1970).

27. Botkin, D. B., Janek, J. F., and Wallis, J. R. "A simulator for northeastern forest growth: a contribution of the Hubbard Brook ecosystem study and IBM research," IBM note #14356 (1970).

28. Botkin, D. B., Janak, J. F. and Wallis, J. R. "Some ecological consequences of a computer model of forest growth," IBM Research Report #15799 (1971).

29. Botkin, D. B., Janak, J. F. and Wallis, J. R. "The rationale, limitations and assumptions of a northeast forest growth simulator," IBM Note #14604 (1970).

30. Hornbeck, J. W., Pierce, R. S. and Federer, C. A. "Streamflow changes after forest clearing in New England," Water Resources Res. 6(4), 1124-1132 (1970)

31. Bormann, F. H. and Likens, G. E. "The ecosystem concept and the rational management of natural resources," Yale Scientific Magazine, April, 2-8 (1971).

32. Likens, G. E. "Effects of deforestation on water quality," Proc. Amer. Soc. Civil Engineering Symposium on Interdisciplinary Aspects of Watershed Management, Bozeman, Montana, 133-140 (1972).

33. Siccama, T. G. "A computer technique for illustrating three variables in a pictogram," Ecology 53, 117-118 (1972).

34. Likens, G. E. and Bormann, F. H. "Nutrient cycling in ecosystems," in J. Wiens, ed. Ecosystems, Structure and Function, Oregon State Univ. Press, Corvallis, 25-67 (1972).

35. Johnson, N. M. "Mineral equilibria in ecosystem geochemistry," Ecology 52, 529-531 (1971).

36. Gosz, J. R., Likens, G. E., and Bormann, F. H. "Nutrient content of litterfall on the Hubbard Brook Experimental Forest," New Hampshire, Ecology, Vol. 53, 769-784 (1972).

37. Gosz, J. R., Likens, G. E. and Bormann, F. H. "Nutrient release from decomposing leaf and branch litter in the Hubbard Brook Forest, New Hampshire, in press.

38. Fryer, John H., Ledig, F. Thomas and Korbobo, Donald R. "Photosynthetic response of balsam fir seedlings from an altitudinal gradient," in Proceedings of the Nineteenth Northeastern Forest Tree Improvement Conference., Orono, Maine, 27-34 (1972).

39. Fryer, John H. and Ledig, F. Thomas "Microevolution of the photosynthetic temperature optimum in relation to the elevational complex gradient," Canadian Journal of Botany 50, 1231-1235 (1972).

40. Smith, W. H. "Influence of artificial defoliation on exudates of sugar maple," Soil Siol. BioChem. 4, 111-113 (1972).

41. Likens, G. E. and Bormann, F. H. "Biogeochemical cycles," The Science Teacher 39(4), 15-20 (1972).

42. Likens, G. E., Bornamm, F. H. and Johnson, N. M. "Acid Rain," Environment 14(2), 33-40 (1972).

43. Fisher, S. G. and Likens, G. E. "Stream ecosystem: Organic energy budget," BioScience 22(1), 33-35 (1972).

44. Likens, G. E. "Eutrophication and aquatic ecosystems," in G. Likens (ed). Nutrients and Eutrophication, Amer. Soc. Limnol. Oceanogr., Lawrence, Kansas, Special Symposia Vol. 1, 3-14 (1972).

45. Likens, G. E. and Bormann, F. H. "An Experimental approach to nutrient-hydrologic interactions in New England landscapes," INTECOL Symposium, "Interactions Between Land and Water," Leningrad (in press), (1972).

46. Marks, P. L. and Bormann, F. H. "Revegetation following forest cutting: Mechanisms for return to steady state nutrient cycling," Science 176, 914-915 (1972).

47. Pierce, R. S., Martin, C. W., Reeves, C. C., Likens, G. E., and Bormann, F. H. "Nutrient loss from clearcuttings in New Hampshire," American Water Resources Association. Proc. Nat'l. Symp. Watersheds in Transition, Urbana, Ill. (In press) (1972).

48. Botkin, D. B., Janak, J. F. and Wallis, J. R. "Rationale, limitations, and assumptions of a northeastern forest growth simulator," IBM Jour. of Res. and Development 16(2), 101-116 (1972).

49. Likens, G. E. "A checklist of organisms for the Hubbard Brook ecosystems," Ecology and Systematics, Cornell University, Mimeo (1971).

50. Johnson, N. M., Reynolds, R. C. and Likens, G. E. "Atmospheric sulfur: Its effect on the chemical weathering of New England," Science 177 (4048), 514-516 (1972).

51. Likens, G. E. "Mirror Lake: Its past, present and future?" Ecology and Systematics, Cornell Univ. Appalachia, (in press) (1972).

52. Burton, T. M. and Likens, G. E. "The effect of strip-cutting on stream temperatures in the Hubbard Brook Experimental Forest, New Hampshire. Submitted to BioScience (1973).

53. Fisher, S. G. and Likens, G. E. "Energy flow in Bear Brook, New Hampshire: An integrated approach to stream ecosystem metabolism". Submitted to BioScience (1973).

54. Bormann, F. H., Likens, G. E., Siccama, T. G., Pierce, R. S. and Eaton, J. S. "The effect of deforestation on ecosystem export and the steady-state condition at Hubbard Brook." Accepted by Ecol. Monogr. (1973).

55. Sturges, F. W., Holmes, R. T. and Likens, G. E. "The role of birds in nutrient cycling in a northern hardwood ecosystem." Submitted to Oikos (1973).

56. Hobbie, J. E. and Likens, G. E. "The output of phosphorus, dissolved organic carbon and fine particulate carbon from Hubbard Brook watersheds," Limnol. Oceanogr. (in press) (1973).

57. Eaton, J. S., Likens, G. E. and Bormann F. H. "Throughfall and stemflow chemistry in a northern hardwood forest," Submitted to J. Ecol. (1973).

58. Cataldo, D. A. and Duggin, J. A. "The rapid oxidation of atmospheric CO to CO_2 by soil microorganisms," Submitted to Science (1973).

A GRAZING LANDS SIMULATION MODEL

F. E. Wielgolaski
Botanical laboratory, University of Oslo, Norway

Grazing land is here taken to include all areas where grazing by vertebrates is important. This means that all vegetation types dominated by herbaceous plants have to be included. These range from tundra and temperate grassland to savanna and desert.

The model is developed to compare the influence of grazing on plants in different types of vegetation and to get better understanding of the carbon and nutrient flows through the system. The assimilation, translocation, death and decomposition of the plants are studied in the model as functions of environmental factors. Most of the relations are simple linear functions varying between zero and one; when the functions are combined, however, the new relationship is non-linear.

Since a wide applicability of the model is intended, the resolution is relatively low. Neither competition between plant species nor the preferences of the grazing animals for various plant species are taken into account, nor are the different regeneration rates after grazing considered. In spite of these simplifications, the model may help in predicting the size of the animal population which the plant communities can withstand under different environmental conditions without being destroyed. Optimisation techniques, however, are not used in the model.

INTRODUCTION

In most countries ecologists have just started employing mathematical dynamic models to help solve their problems on analysis of nature. Biologists often say that if the relationship between the various components studied is not obvious to visible inspection then the correlation between the components is too weak to be on any interest. This may be true in typical experimental biology but in field ecology the biological variation is usually too wide and the complexity of the environmental factors interacting with the biological parameters too high to find a clear causal relationship through visual inspection alone.

Static mathematical models such as regression models, sometimes stepwise multiple regressions, have been in use for some years. They may be of great value in ecology, especially, possibly to pinpoint the environmental factors of greatest importance to biological events. However, they do not express anything about the dynamics of the system. Simulation and optimization models have to be employed for the prediction of changes in an ecosystem due to changes in environmental factors. Resource management and control of system behaviour are two important objectives in the analyses of models of dynamic ecological systems (Patten [13]). By simulation models it is also possible to compare the behaviour of various ecosystems under different conditions. It is thus obvious that ecologists need to use this tool in their studies and that a close cooperation is necessary with cybernetic mathematicians.

There are several viewpoints on the development of simulation models in ecology. Models can be very simple and use only constant coefficients and linear relationships in transfers from one state variable (box, compartment) to another. The flow can be dependent or not on external factors. More or less of the physiology behind the processes between state variables may be included.

Typical for relatively simple compartment models often used in ecology (Gore & Olson [7], Dahl & Gore [3], Kelley et. al. [10]) is a set of equations with only intrinsic factors of the system included and with steady state situations. In its classical form this type of model is a purely descriptive or empirical one, based as it is on only measured compartments at a specific site and with no inclusion of the physiology underlying the changes. This type of model will usually be very site-specific, although use of various values of different parameters may give possibilities for comparison of various sites on a macroscale. It is often expected that this type of model is most useful for a total ecosystem study. However, Goodall [5,6], Van Dyne [14], Milner [12] and Timin et. al.[13a] have included also in models the influence on flows and compartments of extrinsic variables and interactions between various intrinsic factors.

This could be seen as a step towards a more mechanistic model for more or less total ecosystems. In a typical mechanistic model the flows should express fully the physiology behind the processes in the model. This often biochemical information is of course easiest to obtain on details of an ecosystem. Mechanistic models are therefore mostly developed for small parts of the ecosystem such as e.g. de Witt et. al.[4] for a photosynthetic system, Miller [11] for a plant-minerel uptake system or Bunnell [2] for a decomposition and nutrient flux system. However, a partly mechanistic approach has also been tried for a total ecosystem model. A model of this type is the ELM model under development by US/IBP Grassland Biome (Anway et. al.[1]). It is, in the described version, built up of five sections (abiotic, producer, mammalian consumer, decomposer and nutrient sections), most of them with some submodels.

Another partly mechanistic ecosystem model is under development in the US/IBP Desert Biome.* This model is built up of several

*Personal communication with Dr. Goodall and his coworkers

versions of plant, animal and climate-soil submodels within a main program. A simple version of one or more of the submodels may be combined with higher resolution versions of the other submodels. The same high flexibility is possible in the use of the model on single species within the system (plants or animals) or on small or large groups of species. These might again be subdivided into compartments dependent on the interest of resolution. A special feature of the model is the possibility of considering simultaneously within each of the compartments the amounts of nitrogen, ash elements (again subdivided into anions and cations if actual), protein carbon, reserve carbon (soluble carbohydrates and fats) and structural carbon (partly again subdivided into cellulosis and lignin together with other slow decomposing elements if actual). This subdivision into the building stones of plant and animal material, makes for more realistic handling of the physiological processes going on in flows between the state variables. The disadvantage of including such biochemical variables is that the functions describing their flows are not completely known.

More physiological studies have therefore to be carried out to achieve the full success of the method. In my opinion however, this type of model seems to be the most promising when enough knowledge has been obtained. Until then, lower resolution models may be of interest for total ecosystem studies, e.g. in grazing land systems.

The aim of this paper is therefore to introduce a relatively low resolution, but non-linear, general simulation model for grazing lands ecosystems. The model is designed to handle a wider range of situations and to be useful from relatively humid to dry ecosystems and from cold (polar and mountain tundra) to warm areas of the globe. Although animals are important factors of the ecosystems, they are only included in the present version of the model as grazers. Simulation models have earlier proven useful aids to synthesizing and projecting herbivore-forage relations (Walters & Bunnell [16]). The present model is based on the producer model developed at the International Grassland/Tundra Workshop of the International Biological Program in early fall 1972, and some of the general ideas in the model are also from this workshop in the USA. The model in the present version is run for Norwegian alpine tundra data, while the workshop producer version was run for temperate and tropical grasslands as well. The model utilizes the simulation compiler developed by the US/IBP Grassland Biome. A revised version (SIMCOMP IV) of the compiler described by Gustafson and Innis [8,9] has been used.

MODEL DESCRIPTION

The model is structured with a simple abiotic submodel, a relatively detailed producer carbon submodel, a decomposer submodel depending on the decomposition rate of different material and a nutrient submodel for plants. The model is constructed with "normal" climatic years and no grazing, but is sensitive to changes in the climate and to grazing.

The time step used in the model is one day and the simulation period is two to five years. A detailed functional description of the various submodels is given by Wielgolaski _et. al._ [17].

Abiotic submodel

Simple yearly sine surves were used to simulate daily radiation and air temperature which were used as driving variables in the model. The amplitudes of the curves were site-specific and a site-specific time lag was found necessary for temperature in relation to radiation, as in many places maximum and minimum temperatures occurred somewhat later than the radiation extremes.

The total soil depth of interest was divided into three layers. The relative depths of each of the layers were site-specific, dependent on the proportional root distribution in the soil. It was also necessary to introduce site-specific soil temperature sine curves for the three soil layers with specific amplitudes and time lags in relation to air temperature.

Site-water relations were modelled by simulated time trends in soil water content and not by considering precipitation patterns and soil water budgets. The water conditions at the sites were defined by the soil water contents in the three soil layers and by the site-specific maximum and minimum volumetric soil water content and the volumetric water content at wilting point and at field capacity. The water content of each layer was said to vary throughout the year by a piecewise linear function from zero to one defined by four time parameters for each of the three soil layers.

Plant-Carbon submodel

Only carbon assimilation and translocation in plants and death of plants were dealt with in this submodel. Four carbon compartments (live green, above ground reserves, below ground reserves, and live roots) were considered for each of four plant types (shrubs, perennial herbs, annual herbs, and cryptogams). The separation between above- and below ground reserves was described because of the structural and functional difference in e.g. seed reserves and bulb reserves. Often however, as e.g. in shrubs, there is no strict structural and functional difference between the reserve organs above- and below ground. It was maintained that roots are not reserve organs. In the field it may be difficult to distinguish between roots and below ground reserve organs. As seen in Fig. 1 all inputs into the producer carbon system were flowing through live green biomass, resulting in an increase in this state variable and/or reserves (above- and below-ground) and live roots, while no flow was modelled between above- and below ground reserve organs. By definition translocation from reserves to live green biomass and to roots was said to be possible. In the carbon submodel no flow was modelled from live roots to the other live compartments. The outputs from the system were the amount of dead material from all state variables and the respiration. The flows between plant compartments and the atmospheric source/sink and the location of the controls (modifiers) are seen in Fig. 1 as well as the numbering of the 16 producer state variables (X_{21} - X_{36}) for the four compartments of the four plant types.

Further subdivision of these compartments may be required to allow for the separate treatment of e.g. grasses, other monocotyledons and dicotyledons of both the annual and the perennial herb categories,

FIGURE 1

Elementary diagram of the producer subsystem.

X21, 25, 29, 33 are state variables for shrubs.
X22, 26, 30, 34 are state variables for perennial herbs.
X23, 27, 31, 35 are state variables for annual herbs.
X24, 28, 32, 36 are state variables for cryptogams.

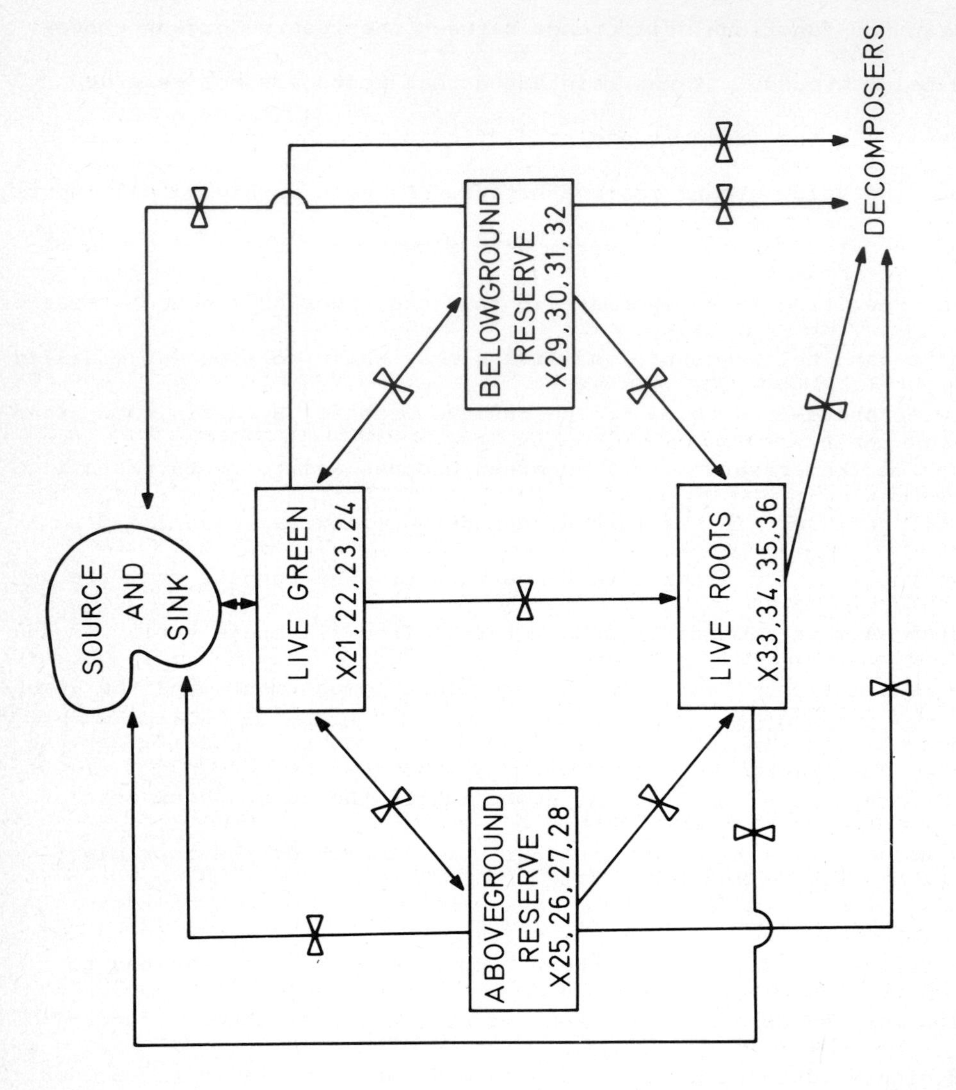

bryophytes and lichens for cryptogams and nitrogen and non-nitrogen fixers within most groups. In the nutrient submodel for nitrogen (see later section) a subdivision of each of the four plant types, shrubs, annual and perennial herbs and cryptogams is made whereby a proportion of the plant type is considered to have nitrogen fixing capacity.

The present model does not allow competitive interrelations to modify the carbon balance of plant types growing together. Each of the four plant types operate independently; thus the only way in which the response of a complex community can be approximated is by addition of the individual response of the component plant types.

1. Assimilation

When assimilation was higher than respiration, the net flow of the CO_2 into live green biomass from the atmospheric souce/sink was calculated. The net assimilation (NA) was said to be a function of radiation, air temperature, soil water and the amount of green biomass of the various plant types. The explicit form of this function may be written: $NA = PMGR \cdot PFT \cdot PFW \cdot PFS \cdot X_{21-24}$, where PMGR = maximum relative growth rate (g/g/day) when conditions for growth are optimal. PFT, PFW, PFR and PFS are site-specific factors varying between 0 and 1 expressing the modifying effects on the relative growth rate of temperature, soil water, radiation, and self shading, respectively, according to various defined functions. X_{21-24} are the green biomass of the plant types of interest.

Below the compensation point, respiration was modelled separately. A similar temperature-dependent respiration rate (R) was assumed for both shoots and roots. The response to temperature was said to be exponential, defined by two parameters, the respiration rate at $0^{o}C$ (PRR0) and at $20^{o}C$ (PR20). The temperature parameter (T) was air temperature for respiration of green material and soil temperature for roots. The respiration rate was said to be zero when $T < 0^{o}C$. The respiration was of course also dependent on the biomass (X) of shoots and roots respectively for the various plant types. The function may generally be written:

$$R = (PRR0 + (PR20 - PRR0)^{\frac{T}{20}}) \cdot X$$

2. Translocations

As seen in Fig. 1 there is a two-way translocation flux between green compartments and the reserve compartments aboveground (wood, shoot bases, seeds) and/or below ground (rhizomes, bulbs, woody roots). Only the net flow was modelled, and was said to be positive when going from green to reserves.

At the beginning of growth there is a negative "reactivation flow" (germination in the case of annuals) from all types of reserve compartments to green shoots. This flow (TSSA) was said to start when both soil water (PHW) and air temperature (PHT) were above certain levels and stopped when the green biomass (PHX21) exceeded a certain level. The "reactivation flow" may be expressed by the equation:

$$TSSA = -PACTA/B \cdot PHT \cdot PHW \cdot PHX21 \cdot (1-PKD) \cdot X_{25-32}$$

where PACTA/B = maximum rate (g/g/day) of reserve activation from above ground (A) and below ground (B), modified by the specific effects of soil water, air temperature, amount of green biomass and a factor PKD due to deteriorating growth conditions, all factors varying from 0 to 1. X_{25-32} are the biomass of the reserve compartments of the various plant types.

During the growing season a slow translocation flow (TSSB) was modelled from live green to reserves, taken to be a constant proportion of the green biomass (PKX21) when above a certain threshold, and also a constant proportion of the translocatable green biomass (PTN).

$TSSB = PTRMA/B \cdot PKX21 \cdot PTN \cdot X_{21-24}$, where PTRMA/B = maximum daily relative translocation rate from live green to above ground (A) and below ground (B) reserves and X_{21-24} = green biomass as earlier.

At the end of the growing season as temperature and/or soil water and/or radiation decrease and with shading by high green biomass, an accelerated "storage flow" (TSSC) was modelled from shoots to reserves. The flow was said to be proportional to the rate of a combined deterioration effect (PKD) on growth by the above mentioned factors as well as proportional to the translocatable green biomass (PTN). The maximum daily relative translocation rates (PSTMA/B) in this period were considerably higher than the rates during the middle of the growing season.

$TSSC = PSTMA/B \cdot PTN \cdot PKD \cdot X_{21-24}$.

The reserve to root translocations were said to occur only during the "reactivation" stage (defined earlier) at the beginning of the growing season, and were proportional to the "reactivation" flow to the shoots at the same time. There was said to be no back-flow from active roots at any time. Shoot to root translocations occurred during the growing season (i.e. when green biomass is above a certain minimum and growth conditions are not deteriorating) and the rate was a constant proportion of green biomass and translocatable material in the shoots.

3. Death

Death of green material (PDTH) was said to be a maximum rate (g/g/day) for death (PDLG) modified by site-specific air temperature (PMT), soil water (PMW) and a shading effect of the biomass itself (PMS). Both temperature and soil water have lower and upper limits for death which were taken into account in the model. The maximum death rate of green biomass due to shading, equals the maximum growth rate; and at the highest amount possible of standing green biomass, death rate equals growth rate.

$$PDTH = PDLG \cdot (1-(1-PMT)(1-PMW)(1-PMS)) \cdot X_{21-24}.$$

The death of the reserves (including the woody parts of shrubs) simply took place at a constant low rate in the model, only dependent on the living biomass of the compartments. Death of roots (PDTRO) was modelled as a constant daily rate (PCRDR) modified by site specific soil temperature (PMRT), the soil water (PMW) and a factor (PKRT) depending on the root reserve ratio, allowing roots to remain alive during the non-growing season (all varying between 0 and 1) and also dependent on the living root biomass of the various plant types (X_{33-36}).

$$PDTRO = PCRDR \cdot (1+PKRT(1-(1-PMRT)(1-PMW))) \cdot X_{33-36}.$$

Decomposition-carbon submodel

The submodel is supposed to take care of the further treatment by decomposers of dead material from the plant-carbon submodel. It has been based on dead herbaceous and woody material of above and below ground compartments respectively. Dead cryptogams were treated separately while the vascular plant types were combined. It was assumed that all above ground dead vascular plant material was transferred to litter via standing dead, although leaves, especially of shrubs, often go directly from live green to litter. Herbaceous and woody litter, dead cryptogams and below ground dead material were all subdivided into three decomposition types (see Fig. 2), viz. "easy", "medium", and "difficult" according to the ease of decomposition. "Easy" to decompose materials are, for example, soluble carbohydrates and proteins; materials "medium" to decompose are, for example, celluloses and hemicelluloses; while materials "difficult" to decompose are, for example, lignin, polyphenols, and silicon-rich materials. This subdivision is a very important detail of the model where chemical data of organic compounds are needed. By making this distinction however, it was possible to get realistically different decomposition rates for various types of material dependent upon the proportions of the above components. In particular, this method will define a higher decomposition rate for recently dead material relatively richer in easy-decomposed soluble carbohydrates compared with

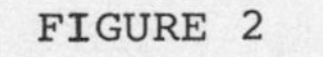

FIGURE 2

Box-and-arrow diagram for the decomposer submodel

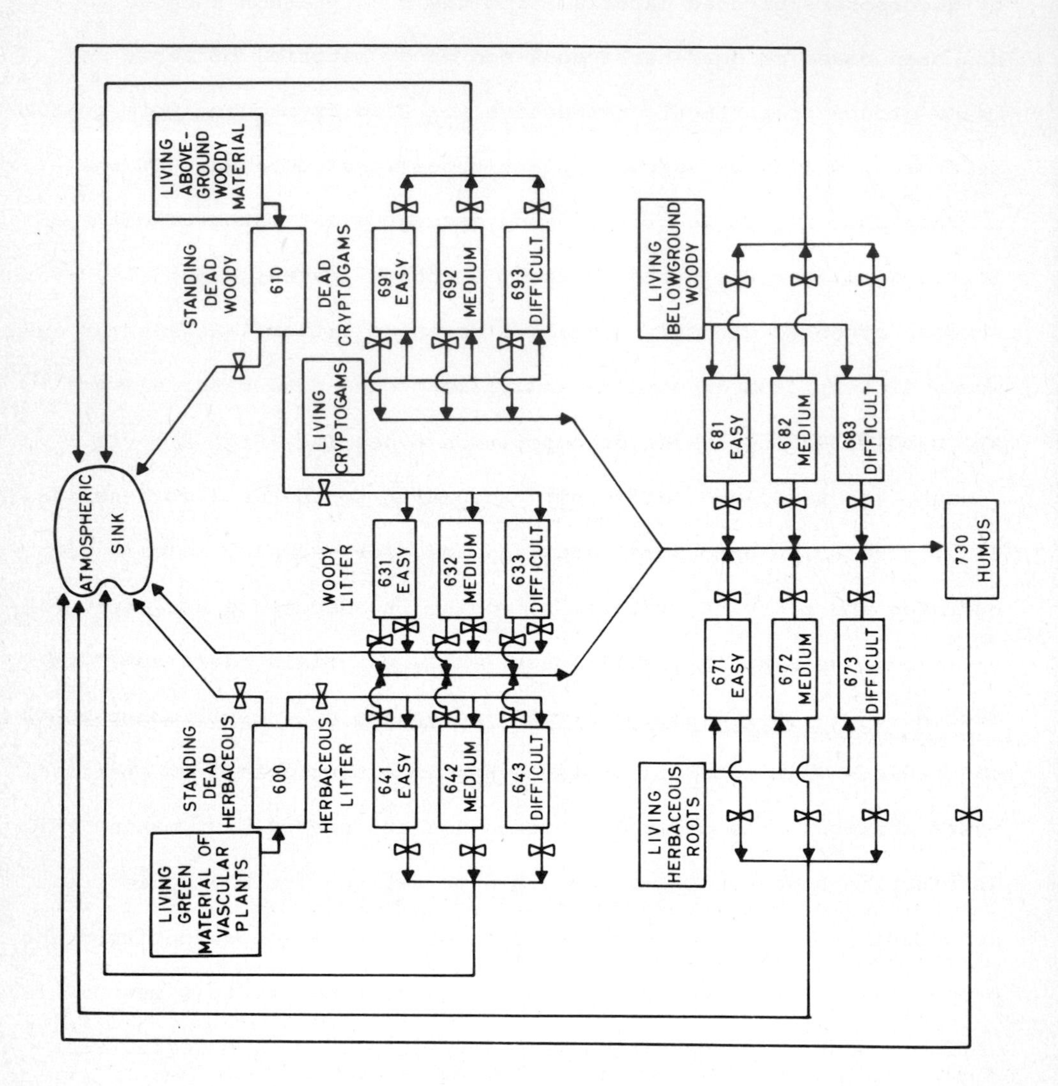

older material. By the decomposition, CO_2 was lost to the atmospheric sink and material transferred to humus. All organic material was said ultimately to be transferred to humus, to be decomposed by microorganisms as all other components in the submodel. When the dead plant material in the soil was no longer recognizable as to origin, it was referred to a s humus.

1. Decomposition of standing dead and transfer to litter.

Standing dead biomass of vascular plant, herbaceous and woody material was assumed to decompose at site-specific fixed relative daily rates (g/g/day) which were higher for herbaceous than for woody material. In a higher resolution model this flow may be driven by air temperature, precipitation and humidity. Since, however, death of material, according to the carbon submodel, varied with the climatic factors and thus time of year, the highest decomposition is to be found in the fall when the amount of standing dead is highest.

The transfer from standing dead vascular plants to litter was also assumed to take place at a site-specific fixed relative daily rate with different values for herbaceous and woody material. The material was separated into "easy", "medium", and "difficult" to compose litter according to an assumed proportion of these different types in the herbaceous and woody litter.

2. Decomposition of vascular plant litter and below ground dead and of dead cryptogams.

The decomposition (Flow) of the two vascular plant litter types (herbaceous and woody) and dead cryptogams was said to take place according to an assumed maximum daily relative rate (DMAXR) for the various proportions of "easy", "medium" and "difficult" to decompose material modified by site-specific effects, varying between 0 and 1 for temperature (DET) and water (DEW) in the upper soil layer (chosen to be only 1 cm) and by pH (DEPH) and the amount of nitrogen (DEN) in the dead compartments. The highest decomposition rate was assumed to be found at medium soil water and pH, and by increasing temperature and nitrogen content.

$$\text{Flow} = \text{DMAXR}_{1-3} \cdot \text{DET} \cdot \text{DEW} \cdot \text{DEPH} \cdot \text{DEN} \cdot X_{1-3}$$

where X_{1-3} = the dry weight biomass of the three types of material ("easy", "medium" and "difficult") within the actual compartments.

The functions for modifying the maximum relative decomposition rates of below ground herbaceous and woody material, respectively, were the same as described for above ground material, but site-specific average soil temperature and soil water for all three soil layers were used in the modifications in addition to pH and nitrogen content of the dead below ground compartments.

3. Transfer to humus and decomposition of humus.

From all compartments, dead material in an amount equal to that lost as CO_2 to the atmosphere by decomposition was transferred to humus. This is a simplification, but little is known of actual transfer rates from various plant compartments to humus. The state variable humus had no subdivisions. The decomposition of humus was made dependent on the same site-specific variables as were used for below ground dead plants to modify the maximum daily relative decomposition rate per gram humus.

Soil-plant nutrient submodel

The nutrient submodel, closely linked to the producer and decomposer submodels, is depicted in the box-and-arrow diagram in Fig. 3. The objective was to design a framework which could have general applicability to all plant nutrients and yet at the same time allow for characteristic differences that exist between the flows of various elements through the ecosystem. In the present study the nutrient submodel has been applied to both phosphorus (P) and nitrogen (N), and in this case an example of the difference between the relationships for those two nutrients is highlighted by the importance of the flows associated with biological fixation of atmospheric N_2. While the model included the possibility for fertilizer additions, these considerations are not allowed for. In the present version of the model there is no feedback from the nutrient to the carbon submodel.

Nutrients are taken up from soil solutions by the roots and translocated to above ground parts. For cryptogams it may be somewhat different, for although rhizoids of some mosses may be able to take up some nutrients, most of the nutrients in cryptogams will be taken up from water in lower parts of above ground organs (partly depending on precipitation, but this was ignored).

The nutrients were brought back to the soil via decomposition of the various main dead plant compartments used in the decomposer submodel. This neglects, as in the decomposer submodel, the importance of animal intake, especially of live above ground material. In principle however, this pathway could be treated similarly to live-dead

FIGURE 3

Box-and-arrow diagram for nutrients (phosphorus is used as an example).

	Shrubs	Perennials	Annuals	Cryptogams
Green biomass	103	104	105	106
Above ground reserves	107	108	109	110
Below ground reserves	111	112	---	---
Roots	115	116	117	(118)
Woody above ground standing dead	119	---	---	---
Herbaceous above ground standing dead	120	120	120	121
Woody litter	122	---	---	---
Herbaceous litter	123	123	123	121
Woody below ground dead	124	---	---	---
Herbaceous below ground dead	125	125	125	125
Humus (non-recognizable organic material)	126	126	126	126
Unavailable nutrients in soil		127		
Available nutrients in upper soil layer		100		
Available nutrients in middle soil layer		101		
Available nutrients in lower soil layer		102		
Nutrient source		198		
Nutrient sink (leaching)		199		

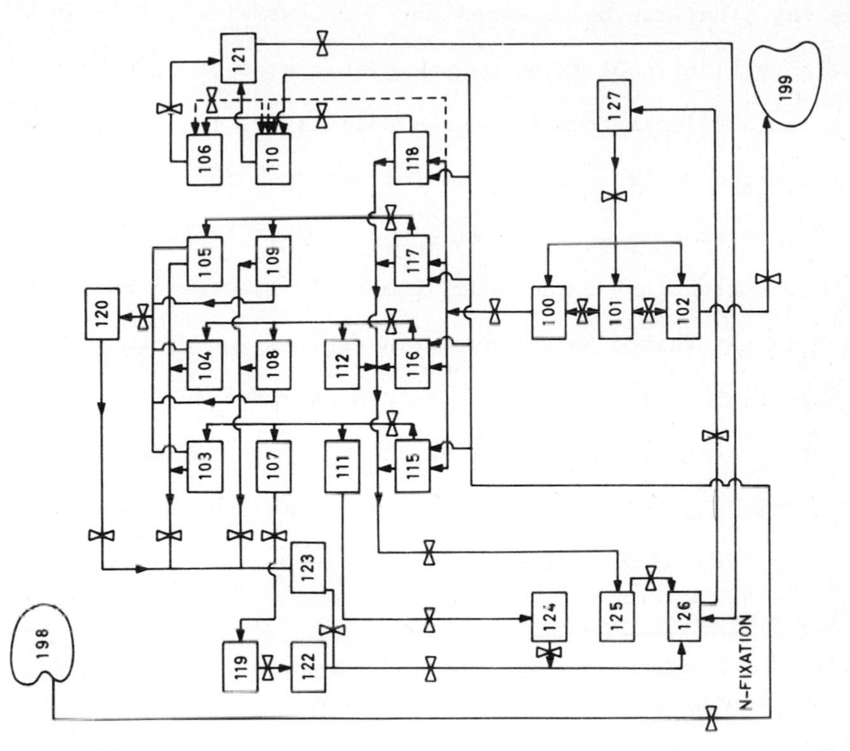

100
101
102
103
104
105
106
107
108
109
110
111
112
115
116
117
118
119
120
121
122
123
124
125
126
127
198
199
N-FIXATION

FIGURE 3

plant compartments giving transfer back to the system via urine and feces. Nutrients taken out of the system by grazing, harvesting, and leaching are then replaced by input from the source (X_{198}), including fertilizers.

The nitrogen fixed by root nodule organisms is taken to be easily available to the hosts and was fed directly into the roots. All nutrients may otherwise be released more or less slowly from an unavailable nutrient pool (X_{127}). The plants will take up part of the available nutrients from the three soil layers ($X_{100-102}$) during the growth period while some will be leached down the soil profile to the leachate sink (X_{199}), becoming unavailable to plant roots.

In the construction of the model the analogy between the flows of P and N is maintained in the numbering system for state variables and the nomenclature for flows and control parameters. The first number in the variable name indicates the mineral studies, e.g. X_{120} means P in standing herbaceous dead material, X_{220} means N in the same material, and so on. In the nitrogen submodel N replaces P in each variable name.

1. Release of nutrients from the unavailable pool (X_{127}).

The maximum release of nutrients (FPMX) from the unavailable pool in the soil was said to be dependent on the size of the pool (X_{127}) as well as on a site-specific character defining, for each of the nutrients, the maximum proportion that can flow into the available pool in each of three soil layers ($FP127_{1-3}$) in one day.

$$FPMX = X_{127} \cdot FP127_{1-3}$$

This maximum flow rate was modified by site specific effects, varying between 0 and 1 for average soil temperature (FPETP) and soil water (FPEWP) in the three layers as well as by the concentration in ppm of nutrients already in the available pool (FPECP).

$$\text{Flow} = \text{FPMX} \cdot \text{FPETP} \cdot \text{FPEWP} \cdot \text{FPECP}$$

The effects of temperature and soil water were expected to be defined by saturation functions and the effect of concentration of available nutrients in the soil by an inverse saturation function.

2. Leaching of nutrients in the soil.

The leaching of available nutrients from the three soil layers was assumed to be dependent on the amount (g/m^2) of available nutrients in each of the layers $X_{100-102}$, the maximum proportion that will move out of a layer in one day under maximum water flow conditions (site-specific) (FPPRP), and a piecewise function of the amount of water in each of the layers (f_{water}). This function increased from 0 at field capacity to a maximum of 1 at a specific volumetric water content and decreased to 0 again in water-logged soil.

$$\text{Flow} = \text{FPPRP} \cdot X_{100-102} \cdot f_{water}$$

3. Nutrient uptake from the soil

The uptake of nutrients by roots from the three soil layers is dependent on several factors. Each plant type was expected to have a maximum daily nutrient uptake (FPMXP) when all modifying factors were at an optimum. The root biomass (X_{33-36}) was important and so was the distribution of roots in the three soil layers (PROOT_{1-3}). The uptake was said to be proportional to the nutrient concentration

(FPECP) in ppm in the three soil layers up to a limit, to growth rate of green biomass (FPEGP) and to the soil temperature in the three layers (FPETP). The soil water (FPEWP) in these layers modified the uptake by a piecewise linear function.

$$\text{Flow} = \text{FPMXP}\cdot X_{33-36}\cdot \text{PROOT}_{1-3}\cdot \text{FPECP}\cdot \text{FPEGP}\cdot \text{FPETP}\cdot \text{FPEWP}$$

The dependency on environmental factors for nutrient uptake by cryptogams was, in the model, said to be similar to that for vascular plants. Above ground reserves (X_{28}) were said to replace roots as absorbing organs and only the conditions in a very thin upper soil layer were of course taken into account.

4. Nitrogen fixation

The biomass of the vascular plants and cryptogam nitrogen fixers was made a fixed proportion ($FNPNF_{1-4}$) of the total biomass, and their nitrogen/carbon ratios ($FNPER_{1-4}$) were made a constant for each plant type. The initial amount of N in the four compartments of each plant type was calculated allowing for their higher concentration of N. The N fixers were assumed to take up any available N that is in the soil layers in the same manner as the non-fixers. To maintain the correct N concentration in these plants a proportion of the daily carbon flow was allocated to the N fixers using $FNPNF_{1-4}$, and sufficient N in addition to that taken up by their roots was fixed from the atmosphere (X_{298}) to make the N/C ratio of the daily flow equal to the predetermined constant ($FNPER_{1-4}$). Thus, the flow for additional N was:

$$\text{Flow} = \text{FNPNF}_{1-4}\cdot \text{P121}_{1-4}\cdot \text{FNPER}_{1-4} - \text{FNPNF}_{1-4}\cdot \text{FN200}_{1-4}$$

= amount of C flow·N/C ratio minus amount of N taken from the soil

where $FN200_{1-4}$ = total N flows from the soil for each of the four plant types

and $P121_{1-4}$ = carbon flows to the four plant types

5. Nutrient translocation in live plants

The translocation of nutrients from the roots to tops and to above and below ground reserves was said to take place in such a way that the nutrient/carbon ratio for each compartment within a plant type, relative to that for roots, was maintained at the initial values given. For example, $FPTRR_{I,J}$ was the P/C ratio for compartment I within the plant type J relative to the P/C ratio in the roots of plant type J. It was expected that FPTRR for tops/root generally is above unity in the growing season. Similar relative ratios were given for other compartments of the various plant types, but will in practice often be close to unity.

The translocation within the plant was dependent on the amount of nutrient taken up by the roots, the current nutrient status, and the current carbon assimilation. The nutrient flow within the plant was expressed by an equation of the form

$$\frac{X(K) + FK}{X(I)} = FPTRR \cdot \frac{X(L) + FP100-F1-F2-F3}{X(J)}$$

where X(I) and X(K) = the amounts of carbon and nutrient, respectively, in the compartment accepting the nutrient, and X(J) and X(L) = the amounts of carbon and nutrient, respectively, in the roots. F1, F2, and F3 = the flows of nutrient uptake to the tops, above ground reserves, and below ground reserves, respectively, FK = either F1,

F2, or F3 depending on the accepting compartment, and FP100 = the total amount of nutrient taken up by the roots from the three soil layers.

This generates a set of simultaneous equations in the unknown flows F1, F2 and F3 which were solved in a subroutine. The term (FP100-F1-F2-F3) was, therefore, the amount of the nutrient uptake (FP100) which remains in the roots.

6. Transfer of nutrients from live to dead compartments, humus and soil pool.

The transfer of nutrients from living to dead plant compartments and between dead compartments was made very simple in that the nutrient transfer was proportional to the transfer of carbon by decomposition between the compartments, but adjusted by specific proportionality factors for each plant type for relative nutrient/carbon transfers. These proportionality factors, however, were set to unity for all transfers between dead material. The general transfer will be

$$\text{Flow} = \text{DFlow} \cdot (\text{XF}/\text{XP}) \cdot \text{FPR}$$

where Flow = nutrient flow between various compartments, DFlow = death rates of living compartments according to the carbon submodel or transfer rates of dead material or decomposition rates of dead material according to the decomposer submodel. XF = nutrient amount in g/m^2 in donor compartment, XP = amount of carbon in donor compartment, and FPR = relative nutrient/carbon transfer ratio.

The transfer of nutrients from humus back to unavailable pool in the soil followed the same pattern as the transfer of dead plant material, i.e. was proportional to the decomposition rate of humus and the nutrient/carbon ratio in humus.

RESULTS

The model was tuned for two Norwegian plant communities, a reindeer lichen community and a perennial grass community. Reliable information was available on site and soil characteristics, biomass changes, and plant nutrient content, but there was limited information on many other aspects, e.g. on aspects of decomposer activity. As might be expected, many of the actual process parameters on which the model runs were unknown at the outset, and their final values were arrived at by testing a range of possible values considered to represent biologically reasonable limits. The modelled variation in biomass of the plants as well as in the plant constituents was mostly of the same order as validated from observed values.

The relationship between plant biomass and productivity formed an interesting point of comparison between the two Norwegian tundra plant communities. It can be seen in Figs. 4 and 5 that biomass for lichens was of the order of trice that for perennial grasses. However, assimilation rates for lichens were much lower (by a factor of 0.1) both in terms of maximum assimilation rates and the actual modified rates following the assessment of plant and environmental factors. In addition, a great difference existed between the decomposition processes in the two communities, and this also contributed to the differences in biomass. Although it was felt that this sort of relationship must hold, there was doubt as to the actual magnitude of these processes. The maximum decomposition rates used here (up to 5-6% per day) exceeded the values considered reasonable on the basis

FIGURE 4

Simulation of changes in live lichen biomass at lichen heath through one "normal" year without grazing (g/m^2). B = green, C = above ground reserves, D = total live biomass.

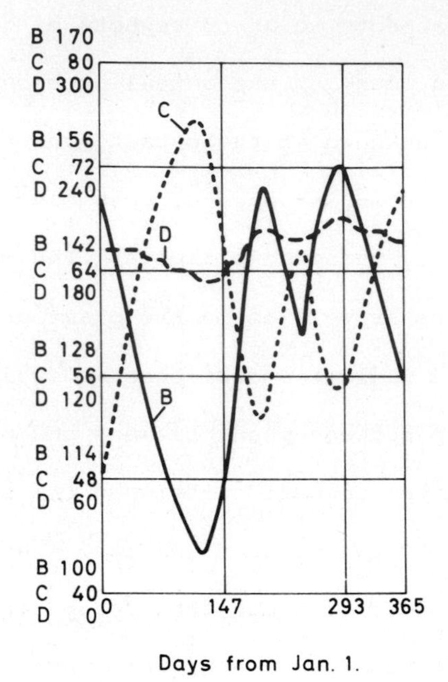

FIGURE 5

Simulation of changes in live above ground perennial herbaceous plant biomass at dry meadow through one "normal" year without grazing (g/m^2). B = green, D = above ground reserves.

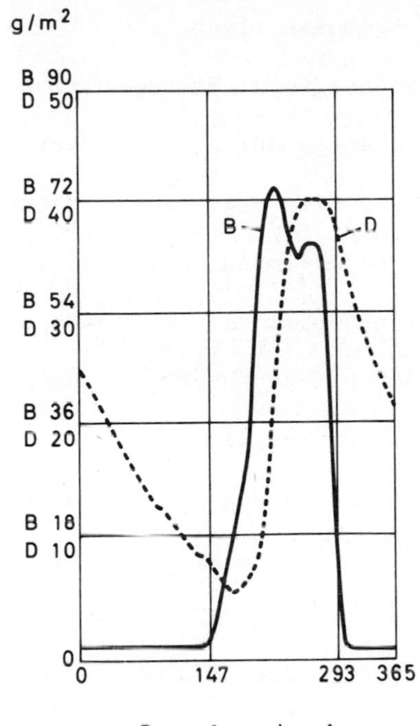

of observation of decomposition using litter bag weight loss technique. However, since the assimilation rates used (Wielgolaski et. al. [17]) were considered to be realistic, it was clear that high decomposition rates were needed to explain why a considerable biomass and/or litter accumulation did not occur. Dead lichens have a much higher percentage of "difficult" decomposed material compared to herbaceous litter of vascular plants.

This also modified the total decomposition of the two plant groups. The biomass changes during two years of dead lichens from the lichen heath are shown in Fig. 6 and of herbaceous litter of vascular plants from the dry meadow in Fig. 7. As an example of the output from the nutrient submodel flow rates are given of phosphorus from dead lichens at the lichen heath to humus and the amount of phosphorus in the humus at the same site during two years (Fig. 8). Generally this part of the present modelling study has proved itself to be of considerable value as a means of synthesizing available data on many aspects of the Norwegian IBP effort.

In all the runs of the model discussed earlier, average or "normal" climatic conditions were used and no grazing activity was included. The soil at the lichen heath was very coarse, and there was no other water supply than precipitation. After two--three days without precipitation the soil water in the upper layers was greatly reduced. In late July--early August, when there was no effect left from snow melting water, drought therefore even in "normal" years often caused a reduced growth (shown in the model as lower biomass), both in lichens (Fig. 4) and in green parts of the shrubs at the

FIGURE 6

Simulation of changes in dead lichen biomass at lichen heath through one "normal" year without grazing (g/m^2). U = Easy soluble material, V = Medium soluble material, W = Difficult soluble material, and $X = \sum (U, V, W)$.

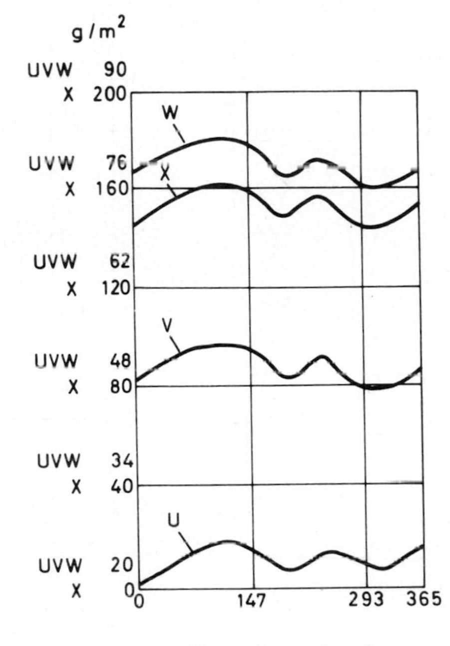

FIGURE 7

Simulation of changes in herbaceous litter of vascular plants at dry meadow through one "normal" year without grazing (g/m^2). J = Easy soluble material, K = Medium soluble material, and L = Difficult soluble material.

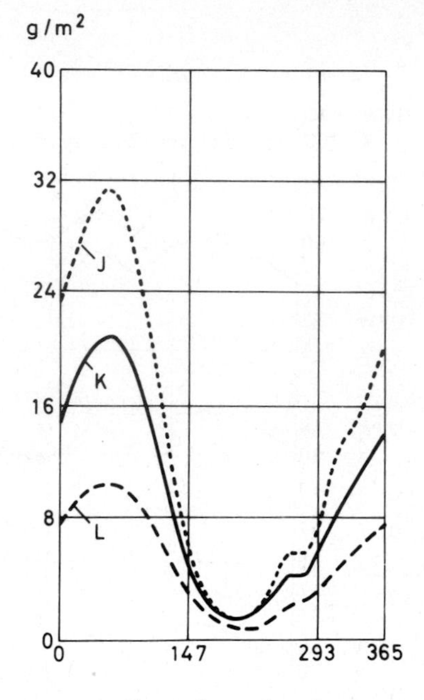

FIGURE 8

Simulation of changes in humus (O) at the lichen heath during two "normal" years (g/m^2) and in transfer rates of phosphorous (g/g/day) from dead lichens to humus (N) in the same years.

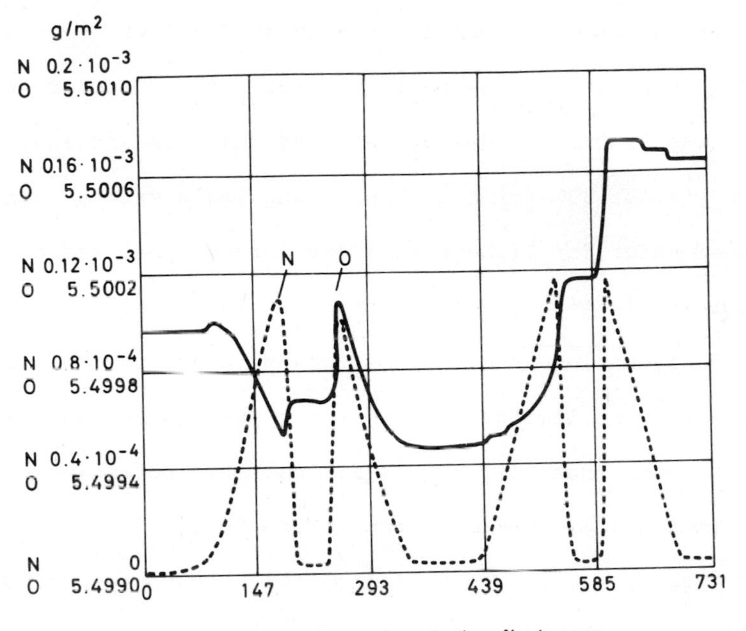

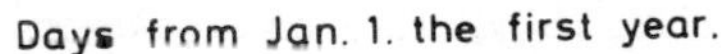

lichen heath studied. The lichens, however, showed strong growth again in the late summer, even stronger than in spring before the drier period. This is in accordance with results found in the field at the site.

If water did not limit the growth of lichens in any period, the biomass might be expected to increase during the whole summer. In a five-year run of the present model this was also found as seen in Fig. 9 in the second year when soil water was said to be non-limiting (temperature conditions "normal" throughout all five years). The growth of shrubs also increased in the wet year. However, while the maximum biomass increased nearly 50% that year for lichens, it increased only about 33% for the shrubs, showing a stronger influence of increased water for lichens than for shrubs. One reason for this is a difference in growth pattern of shrubs and lichens. The maximum increase in green biomass of shrubs occurs in spring (Fig. 9), when there is usually no limitation of water. In this period temperature in tundra areas is normally more limiting. By increasing the temperature (maximum and minimum air temperature by 10°, soil temperature somewhat less) the second year of simulation, the green biomass of shrubs increased about 70%, but only about 20% for lichens (Fig. 10), as temperature was not so limiting later in the season.

It is worth noticing that the increased biomass in the model after one favourable year (increase of water or temperature) did not drop to the previous level in any of (Figs. 9 & 10) the perennials studied because of high production of perennial plant material in the favourable year. One extremely unfavourable year (dry or cold)

FIGURE 9

Five year simulation of changes in green biomass (g/m^2) of shrubs (B) and in live lichens (D) at lichen heath when water was said to be "non-limiting" in the second year and temperature "normal". No grazing.

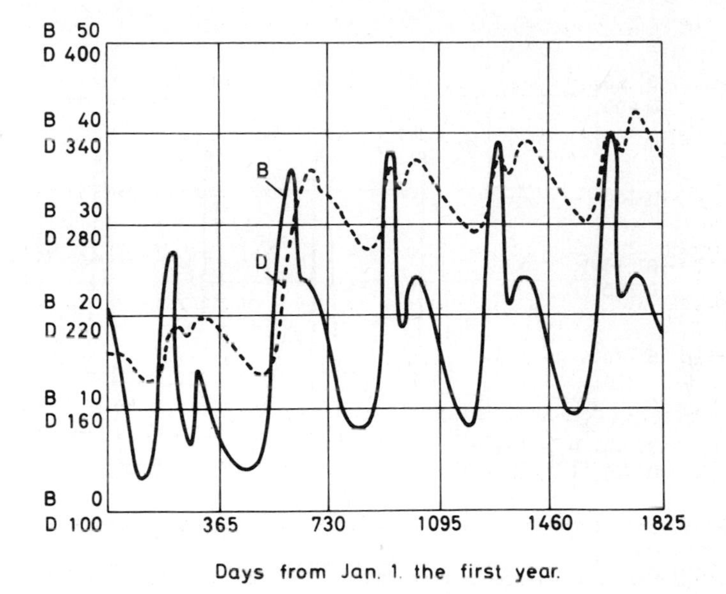

FIGURE 10

Five year simulation of changes in green biomass (g/m^2) of shrubs (B) and in live lichens (D) at lichen heath when the temperature was said to be 10° above "normal" in the second year and water "normal".
No grazing.

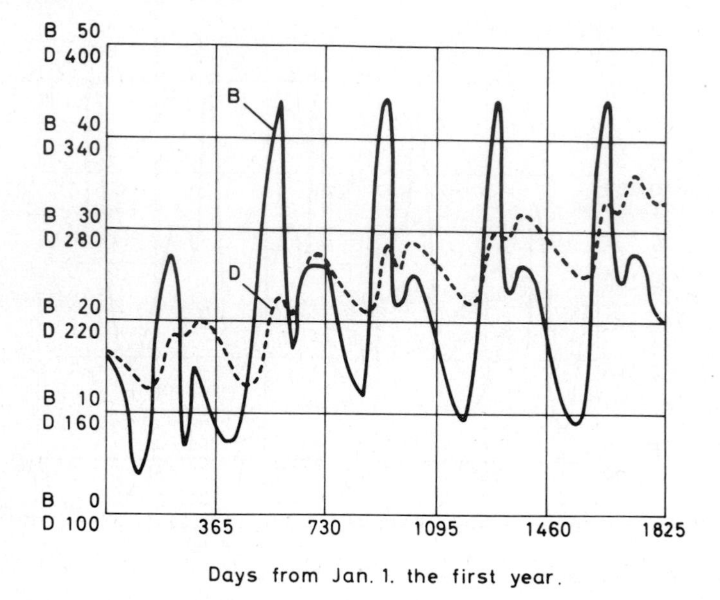

reduces the biomass for that one year--but does not affect the plant's ability to produce shoots and therefore biomass in the following year so that the biomass level is already readjusted to normal in the year following the unfavourable one. This was also found for lichens after one extremely dry year. However, in one cold year (minimum air temperature 10° and maximum air temperature 5° lower than "normal") more lichens die according to the model, and three years were necessary to arrive back at the same biomass level as before the catastrophic year. If the lower temperature limits for death had been lowered, less sensitivity to low temperature would, of course, have been found.

Different levels of summer grazing (mostly invertebrates) and winter grazing (reindeer) were included in the lichen heath model, both in "normal" and "abnormal" years. In normal years only very light winter grazing (0.01% of living lichens grazed per day) by reindeer was possible each year without destroying the ecosystem. The summer grazing could be about 0.02% per day each summer in normal years without reducing the biomass significantly. This shows the higher fragility of the slowgrowing lichens than of dwarf shrubs which was also found in the field. By even very light grazing, summer and/or winter, in one climatically unfavourable year, the green biomass of shrubs and living lichens was reduced for the rest of the five years modelled. One warm or wet year was enough to keep the green biomass of shrubs at a nearly constant level for the rest of the period modelled (Fig. 11) at 0.05% summer grazing during all years. At the same level of winter grazing the biomass of living

FIGURE 11

Five year simulation of changes in green biomass (g/m^2) of shrubs (B) and in live lichens (D) at lichen heath when water was said to be "non-limiting" in the second year and temperature "normal". Grazing at 0.05% of the live biomass per day both summer and winter.

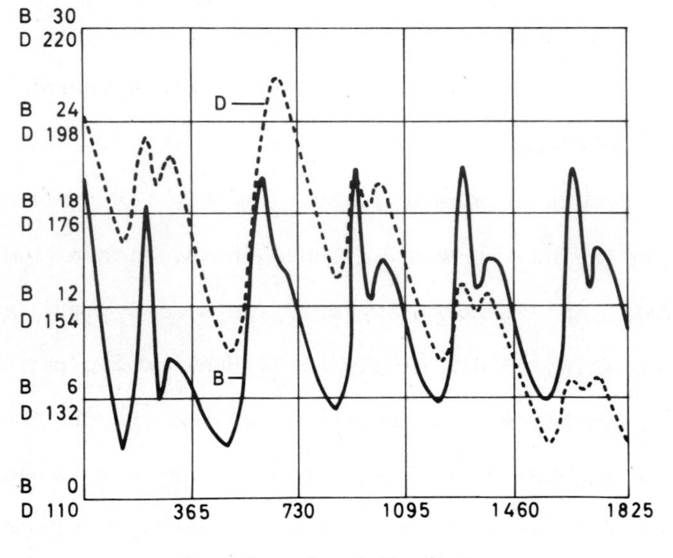

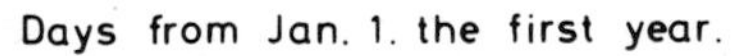

lichens could be kept at a steady state if all years were extremely wet. One wet year increased the lichen biomass enough to cause the maximum biomass to be at the same level in the last year as in the first year modelled at yearly 0.025% winter grazing and 0.05% summer grazing, while this was found in the model output after one extremely warm year at the same level of winter grazing, but only 0.01% summer grazing.

The results show that the lichen heath community in the model is highly sensitive to grazing as well as to climatic changes, the shrubs mostly to summer grazing and temperature changes, the lichens mostly to winter grazing and to moisture changes. This is also in good accordance with results observed in the field. By changing the various parameters in the model it is thus possible to find out which of them are most sensitive e.g. to dynamic validation (Van Dyne [ ]). The ecologist will learn which factors have to be measured most accurately in the field, and may even find indications that parameters not studied earlier could be important in a process. The present model has some shortcomings e.g. absence of any real competitive relations between plant types, and there is no direct possibility of using the model for optimization. However, it is felt that the approach used to construct simple functional relationships to describe flows between compartments has considerable value in the construction of low resolution models for intersite (international) comparisons.

REFERENCES

1. Anway, J.C., Brittain, E.G., Hunt, H.W., Innis, G.S., Parton, W.J., Rodell, C.F. & Sauer, R.H. ELM Version 1.0 Technical Report No. 156, U.S. IBP Grassland Biome, (1972).

2. Bunnell, F. Modelling decomposition and nutrient flux in Proceedings U.S. IBP 1972 Tundra Biome symposium, Seattle, (1972).

3. Dahl, E. & Gore, A.J.P. Proceedings of working meeting on analysis of ecosystems: Tundra Zone, Ustaoset, Norway, (September 1968).

4. de Wit, E.T., Brouwer, R. & Penning de Vries, F.W.T. The simulation of photosynthetic systems in Prediction and measurement of photosynthetic productivity, IBP/PP Tech. Meeting (Trebon, Czechoslovakia), Proc. Center for Agr. Publ. Docu., Wageningen, The Netherlands, (1970).

5. Goodall, D. Computer simulation of changes in vegetation subject to grazing, J. Ind. Bot. Soc. 46, 356 (1967).

6. Goodall, D. Simulating the grazing situation in Concept and models of biomathematics: Simulation Techniques and methods Biomathematics 1, New York, (1969).

7. Gore, A.J.P. & Olson, J.S. Preliminary models for accumulation of organic matter in an Eriophorum/Calluna ecosystem, Aquilo, Ser. Botanica 6, 297-313, (1967).

8. Gustafson, J.D. & Innis, G.S. SIMCOMP. A simulation compiler for biological modelling in Summer Computer Simulation Conf. Proc., San Diego, California, (June 14-16, 1972).

9. Gustafson, J.D. & Innis, G.S. SIMCOMP version 2.0 user's manual U.S. IBP Grassland Biome Tech. Rep. No. 138, Colorado State Univ., Fort Collins, (1972).

10. Kelley, J.M., Ostrup, P.A., Olson, J.S., Auerbach, S.I. & Van Dyne, G.M. Models of seasonal primary productivity in Eastern Tennessee Festuca and Andropogon ecosystems, Oak Ridge National Laboratory Report 1310, (1969).

11. Miller, P.C. A model to incorporate minerals into tundra plant production in Proceedings U.S. IBP 1972 Tundra Biome Symposium, Seattle, 51-54 (1972).

12. Milner, C. The use of computer simulation in conservation management in Jeffers J.N.R. (Ed.) Mathematical models in ecology, 12th Symposium of the British Ecological Soc. (1971).

13. Patten, B. C. _Systems analysis and simulation in ecology_ I, Academic Press, New York & London (1971)

13a. Timin, M. E., Collier, B. D., Zich, J., and Walters, D. "Computer Simulation of the Arctic Tundra Ecosystem near Barrow, Alaska," Proc. U.S. IBP 1972 Tundra Biome Symposium, Seattle, 71-79.

14. Van Dyne, G. M. "Grassland management, research, and training viewed in a systems context," Range Science Series No. 3, Colorado State University (1969).

15. Van Dyne, G. M. _Problems in sensitivity analyses of ecological simulation and optimization models_ (1973)

16. Walters, C. J. and Bunnell, F. "A computer management game of land use in British Columbia," _J. Wildl. Mgmt_ 35, 644-675 (1972).

17. Wielgolaski, F. E., Haydock, K. P. and Connor, D. J. A Grazing lands plant-decomposition, carbon-mineral simulation model, Technical report No. 203, U. S. IBP Grassland Biome (1972).

B.

AIR, NOISE AND WATER POLLUTION CONTROL

GAME THEORETIC APPROACH TO EQUITABLE REGIONAL ENVIRONMENTAL QUALITY MANAGEMENT

James P. Heaney
Hasan Sheikh
Department of Environmental Engineering Sciences
University of Florida

INTRODUCTION

There has been sustained interest over the past decade in determining optimal regional environmental quality management strategies. Numerous investigators have demonstrated that coordinated wastewater treatment strategies are more efficient, in an economic sense, than decentralized individual treatment plants, e.g., Heaney *et. al.*[8]. Similar results have been obtained for air pollution, e.g., Teller [17] and Seinfield and Kyan [15]. However, there has been little success in implementing such proposals due partially to the nonexistence of a real-world regional authority with necessary power to shift decisions in this direction. Lacking such a regional authority, Hass [9] investigated the possibility of setting up a system wherein price guides could be used to direct the activities of the individual waste dischargers toward the regional optimum. He structured the problem using the decomposition principle (Dantzig and Wolfe, [6]). Briefly, the decomposition principle partitions the total regional problem into a series of subproblems--one for each waste discharger, and a regional master problem. Each waste discharger submits a provisional control plan to the regional authority who runs the master problem to see if a regional optimum has been achieved. If not, he transmits a revised set of criterion elements, cost coefficients in this case, to the individual waste dischargers. They resolve their problem and may decide to submit an additional solution for consideration. Each such solution represents an extreme point from their feasible region. Since there are only a finite number of such extreme points, the algorithm eventually converges. The resultant optimal regional solution is actually a weighted average of the extreme points of the solutions submitted by the individual waste dischargers. However, it may occur that the optimal solution for an individual is a convex combination of two adjacent extreme points in which case the notion of decentralization by price guides alone breaks down (Baumol and Fabian, [1]). The reason is that the individual is now indifferent among solutions along this edge connecting the adjacent extreme points while the regional authority knows precisely where along the edge the individual should act in the interest of regional efficiency. Thus, in general, more than price guides are necessary to achieve the regional optimum. Charnes, Clower and Kortanek [4] suggest the inclusion of preemptive goals as a device for providing the requisite amount of information to attain a stable solution.

Dorfman and Jacoby [7] have argued that the optimization models might be used to screen the number of alternatives down to a reasonable number (say 5 to 10) and then submit these Pareto-admissible solutions to further scrutiny by employing a simple political simulation model. The essence of this approach is to assign weights based on the relative importance of each decision making group. One can then generate various weighting schemes and examine how sensitive the solution is to the assumed weights. Burke et. al.[3] have devised a more formalized political simulation model based on work by Bulkley and McLaughlin [2] and applied it to the Dorfman-Jacoby example. This effort describes the relative power of several interest groups and simulates the bargaining process and coalition formation among these groups.

An essential component of any workable regional program is the notion that the resultant solution is not only efficient but also fair to every one of the participants. But the regional optimization models do not even address this question of fairness. Dorfman and Jacoby attempt to incorporate notions of fairness by examining several solutions which are Pareto admissible. We shall present procedures for defining precisely the subset of possible solutions over which negotiations might take place. Furthermore, a unique solution can be promulgated which has some desirable equity properties. A hypothetical example will serve to illustrate the approach.

STATEMENT OF OBJECTIVES

The objective of this analysis is to devise a decision-making model for the specific problem of regional environmental quality management. A fundamental question is to determine the optimal level of pollution control. There are basically two approaches. One is to assess the damages caused by the pollutant, convert them into an equivalent monetary value, and thereby determine an aggregate receptor damage function. Knowing this function and the cost of controlling the pollution, one can determine the optimal degree of control. The primary problem with this approach is that it is extremely difficult, if not impossible, to quantify the receptor damage function. The other approach is based on specifying performance standards which are based on a review by experts of the technical, economic, financial, social and political aspects of the problem. Given this set of standards, the pollutors are assumed to select the least costly way of satisfying that standard. This is the approach we shall use.

Thus, the stated objective for this analysis is to determine the least cost solution subject to meeting a prescribed set of performance standards. This least cost solution might comprise a wide variety of on-site controls as well as off-site controls. However, the solution to the optimization problem is not the final answer. In addition, we need to analyze whether the solution is also fair to everyone. This question will be examined using recent developments from cooperative N-person game theory.

There are numerous ways to control pollution using on-site control and/or off-site control. Thus, a general framework is needed for addressing the problem. The selected approach is based on the planning theory of zoning (Herzog,[10]). Using the planning theory of zoning, each source of pollution calculates the cost of handling the problem on-site as a function of the allowable rate of release from his area. The cost function could look like the curve shown in Figure 1.

Figure 1. Cost function for on-site control.

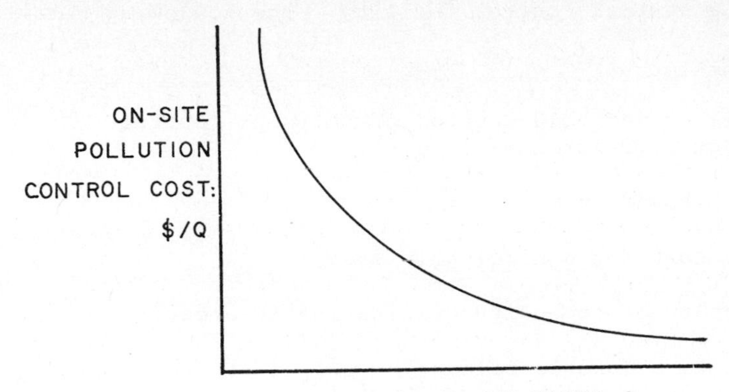

Costs decrease as the allowable release increases. In fact, the cost falls to zero as he is allowed to release more and more pollution. Assume that such a curve exists for each area. Note that the abscissa could be air pollutants, water pollutants, noise or any other "nuisance" which is normally covered by zoning regulations.

In this case, the policy question is to determine how much pollution can be released from each of these areas. Using the planning theory of zoning, the answer depends on a determination of "assimilative capacity." Assume that the required level of control can be specified.

EXAMPLE

The following example is presented in order to illustrate many of the concepts to be discussed. Assume the region under consideration has been partitioned into three study areas. Each study area has two options: (1) on-site control and/or (2) off-site control at a central control facility. The following notation is used. Let

x_{ij} = decision variable: number of units of control j selected for area i

$\bar{x}_{ij}$ = upper bound on x_{ij}

c_{ij} = unit cost for control j in area i

D_i = quantity of commodity originating in area i

$\bar{Q}_i$ = maximum allowable release of pollutant from area i

Z_i = total control cost to area i

π_i = reduction in control cost to area i if $\bar{Q}_i$ is increased by one unit

t_i = unit cost of transporting pollutant from area i to the central control location

c = unit cost of central control

$\bar{W}$ = maximum available control at central facility

The example problem is shown in Figure 2 using a network representation. This problem is deliberately oversimplified to permit us to understand the concepts without getting enmeshed in computational difficulties. The results can be extended easily to more realistic cases where multiple central control facilities exist.

The overall objective function seeks to minimize the total cost of on-site and off-site control. The problem facing each study area is to minimize

$$Z_i = \sum_j c_{ij} x_{ij} + (c + t_i) Q_i$$

Figure 2. Network Representation of Example Problem

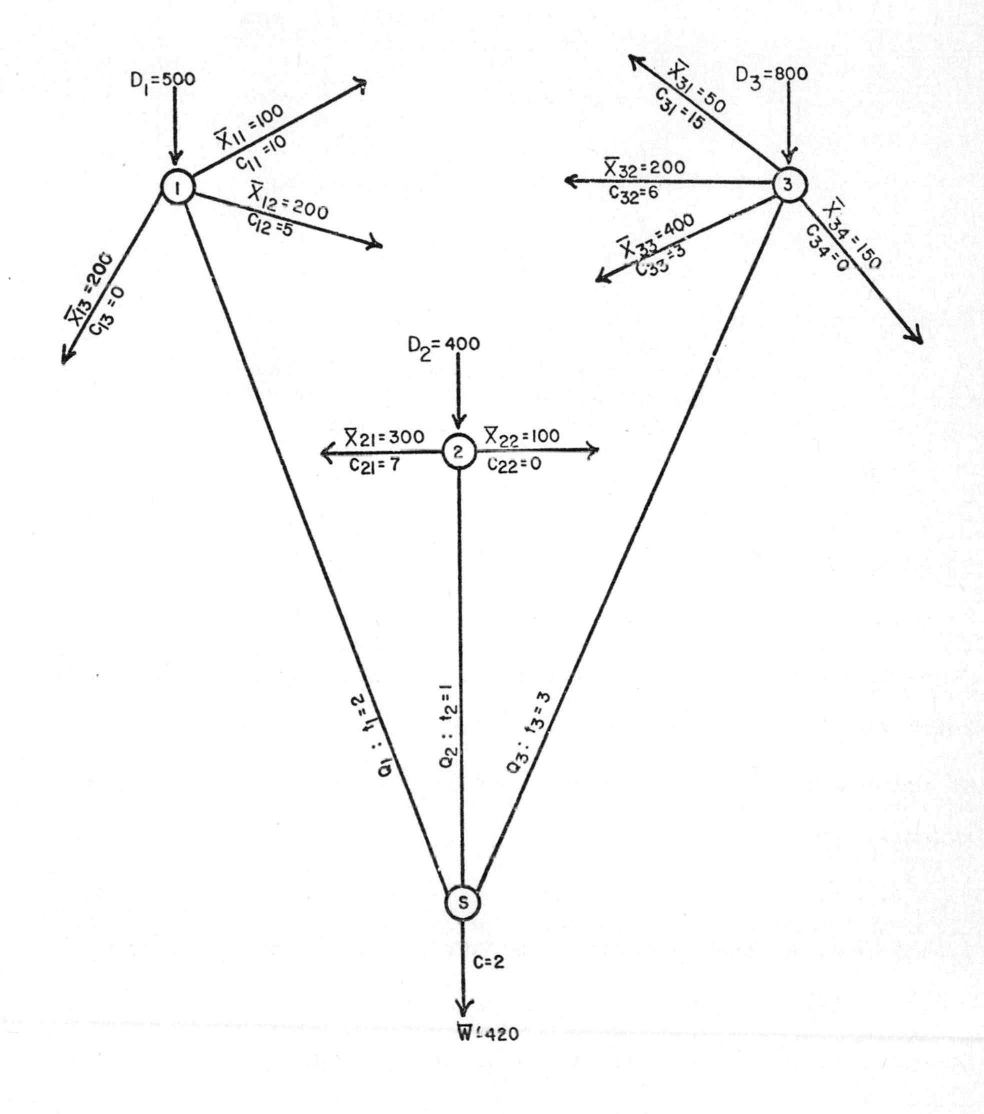

subject to

$$\sum_j x_{ij} + Q_i = D_i$$

$$Q_i \leq \bar{Q}_i \qquad (1)$$

$$x_{ij} \leq \bar{x}_{ij} \quad \text{for all } j,$$

$$x_{ij} \geq 0 \quad \text{for all } j, \text{ and}$$

$$Q_i \geq 0$$

For area 1, his problem is to select x_{11}, x_{12}, and x_{13} so as to

minimize $Z_1 = 10x_{11} + 5x_{12} + 0x_{13} + 4Q_1$

subject to

$$\begin{aligned} x_{11} + x_{12} + x_{13} + Q_1 &= 500 \\ Q_1 &\leq \bar{Q}_1 \\ x_{11} &\leq 100 \\ x_{12} &\leq 200 \\ x_{13} &\leq 200 \\ x_{11}, x_{12}, x_{13}, Q_1 &\geq 0 \end{aligned}$$

This is a very simple problem to solve if $\bar{Q}_1$ is known. As it turns out this is a question of critical importance in some cases. In the context of environmental quality management $\bar{Q}_1$ represents a judgment on the part of the administrator as to the "assimilative capacity" or availability of off-site controls to area 1. Traditionally, the natural system has provided these off-site controls free of charge. For example, the off-site area in this example could be a swamp, river, or the atmosphere.

WILLINGNESS TO PAY FOR OFF-SITE CONTROL

In order to find $\bar{Q}_1$, we need to know how much each area is willing to pay for off-site control <u>based on their alternative on-site control costs</u>. This can be done by solving the above linear program for various values of $\bar{Q}_i$ assuming on-site control is required. Thus, for the moment, $\bar{Q}_i$ is analogous to an effluent standard imposed on area i. The problem can be solved by deleting the last term $(t_i + c)Q_i$, from the objective function and finding the optimal solution for assumed levels of $\bar{Q}_i$. Computationally this can be done quite easily using parametric programming as explained below. Initially, set $\bar{Q}_i = 0$ and solve the linear programming problem. Then, as a post-optimal procedure, one can vary $\bar{Q}_i$ continuously from 0 up to any prescribed upper bound using as the right hand side $\bar{Q}_i = 0 + \theta r$ where θ equals a parameter which will increase continuously from 0 to 1 and r is a scalar, say 1000, in this case. The solution to this problem tells the total cost to area i for any value of $\bar{Q}_i$ which is of interest. This is a very attractive feature of linear programming. Another aspect of linear programming which is of interest is duality theory. The solution to the dual problem is obtained when the above problem (primal problem) is solved. Among other things, it tells, for a given $\bar{Q}_i$, the reduction in cost to area i if $\bar{Q}_i$ is increased by one unit. This unit cost is called the "shadow price" with respect to $\bar{Q}_i$ and will be denoted as π_i.

Let us solve the example problem for area 1 assuming $\bar{Q}_1 = 0$. The answer is simply that on-site control is used. As the constraint on $\bar{Q}_1$ is relaxed, area 1 will substitute off-site control for its most expensive on-site control, x_{11}, in this case. Thus, it is saving \$6 per unit change in $\bar{Q}_i$ in this range. The analysis continues in this manner until all solution possibilities have been identified. The results are shown in Figure 3.

Assume that a similar analysis was done for areas 2 and 3. Then the willingness to pay for each of the three areas would be known. Next assume each area was offered as much discharge as they wanted at its cost of $(t_i + c)$ dollars per unit. In this case, the aggregate demand would be $Q_1 = 300$, $Q_2 = 300$, $Q_3 = 250$, or a total demand of 850 units. But suppose only 420 units are available. How do we allocate the available capacity? If you are the coordinator or planning authority, we assume that you will attempt to maximize the aggregate savings to area 1, 2, and 3 from using the regional facility. Alternatively preference will be given to those areas whose on-site control costs are the highest.

SOLUTION OF THE REGIONAL PROBLEM

This answer to the above question can be obtained by solving one larger optimization model which is formulated on the following page. The constraints for this problem can be divided into two categories: (1) three sets of study area constraints and (2) a coupling constraint. The coupling constraint is the only linkage among the three study areas. If the values of Q_1, Q_2, and Q_3 are prespecified, then the larger problem can be completely

Figure 3. Shadow Price For Area 1 For Assumed Volume of $\bar{Q}_1$

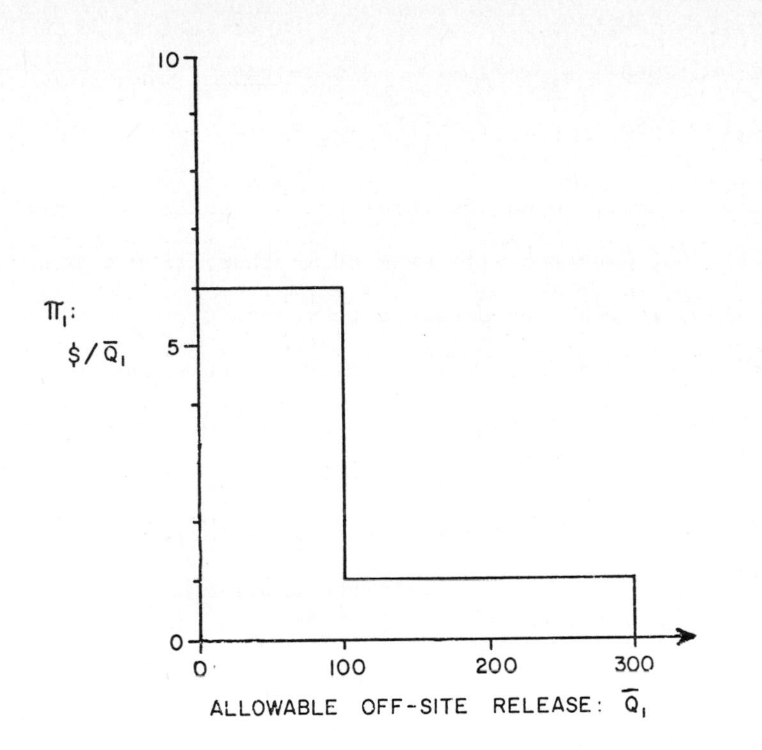

decomposed into three independent subproblems. There are many real world situations in which an a priori apportionment is used, e.g., allocation to each area based on its size, population, etc. For a pollutant, $\bar{W}$ might be the assimilative capacity of the receiving water, which has been apportioned. In the case of a pollutant, the appropriation is equivalent to prescribing effluent standards.

Assume an apportionment is established such that $Q_1 = Q_2 = Q_3 = 420/3$. Given this apportionment, find Z_1, Z_2, and Z_3, the least cost solutions for the three areas. The results are:

$$Z_1 = \$1360 : \pi_1 = \$1 : \qquad Q_1 = 140$$
$$Z_2 = 1540 : \pi_2 = 4 : \qquad Q_2 = 140$$
$$Z_3 = \underline{2560} : \pi_3 = 1 : \qquad Q_3 = \underline{140}$$
$$\sum_{i=1}^{3} Z_i = 5460 \qquad \sum_{i=1}^{3} Q_i = 420$$

Apportioning the capacity among the three areas results in a combined cost of \$5,460. Next we will examine whether, from a least cost point of view, it would be possible to select Q_1, Q_2, and Q_3 such that costs are reduced. This problem can be solved by running the entire linear program with Q_1, Q_2, and Q_3 as decision variables. This approach is equivalent to receiving system standards wherein the coordinator allocates the assimilative capacity in an optimal manner. This is precisely the problem to be addressed here. The optimal solution is:

$$Z_1 = \$1400 \qquad Q_1 = 100$$
$$Z_2 = 1020 \qquad Q_2 = 270$$
$$Z_3 = \underline{2650} \qquad Q_3 = \underline{50}$$
$$\sum_{i=1}^{3} Z_i = \$5070 \qquad \sum_{i=1}^{3} Q_i = 420$$

We see that the solution to the overall optimization problem reduced total costs from \$5,460 to \$5,070 by making more effective use of the available capacity. As can be seen by examining the solutions, the model took capacity from 1 and 3 and allotted it to 2 since his net savings (π_2) were higher. While this latter solution is the least costly from an overall point of view, if the costs are assigned as shown above, areas 1 and 3 are made worse off while area 2 is made significantly better off. Thus, areas 1 and 3 might reasonably object to such a solution. This is pre-

Minimize Z =	$\sum_{j=1}^{3} c_{1j}x_{1j} + (c+t_1)Q_1$	$+ \sum_{j=1}^{2} c_{2j}x_{2j} + (c+t_2)Q_2$	$+ \sum_{j=1}^{4} c_{3j}x_{3j} + (c+t_3)Q_3$	
Subject to				
Area 1's Problem	$\sum_{j=1}^{3} x_{1j} - Q_1$	0	0	$= D_1$
	x_{1j}			$\leq \bar{x}_{1j}$ for $j = 1,2,3$
	x_{1j}			≥ 0 for $j = 1,2,3$
	Q_1			≥ 0
Area 2's Problem	0	$\sum_{j=1}^{2} x_{2j} + Q_2$	0	$= D_2$
		x_{2j}		$\leq \bar{x}_{2j}$ for $j = 1,2$
		x_{2j}		≥ 0 for $j = 1,2$
		Q_2		≥ 0
Area 3's Problem	0	0	$\sum_{j=1}^{4} x_{3j} + Q_3$	$= D_3$
			x_{3j}	$\leq \bar{x}_{3j}$ for $j = 1,2,3,4$
			x_{3j}	≥ 0 for $j = 1,2,3,4$
			Q_3	≥ 0
Coupling Constraints	Q_1	$+ Q_2$	$+ Q_3$	$\leq \bar{W}$

cisely the problem that has thwarted implementation of optimal programs, i.e., while they are the least costly, they do not seem to be "fair" to everyone. We would like to have a procedure which is not only efficient but also is equitable.

SOLUTION USING MARKET PRICE CONCEPT

Perhaps one could use economic theory of demand to determine a better solution. The aggregate demand curve for the three areas is shown in Figure 4. Knowing the demand curve and the supply curve, then it is possible to determine the "market price" for the central facility. Referring to Figure 4, it is \$4/Q. Thus, according to economic theory, to achieve efficient resource allocation, a market price of \$4 per unit of storage should be used. Let π denote the market price (which is the same as the shadow price of the central facility from the overall optimization model); then the assessment to each study area per unit is $(\pi + c + t_i)$.

Economic theory does not address the distributional questions associated with using a market price concept. Thus, until recently, a very fundamental question was left unanswered as indicated below. First, if you actually charge the "market price" then a profit of 420π is realized. What do you do with the money? Traditionally, public services have been priced based on the cost of service. Thus, one could argue that such a price cannot be charged. Economic theory uses the notion of consumer surplus which is defined as the difference between an individual's willingness to pay and the actual assessment levied against him.

Figure 4. Demand For Off-Site Discharge

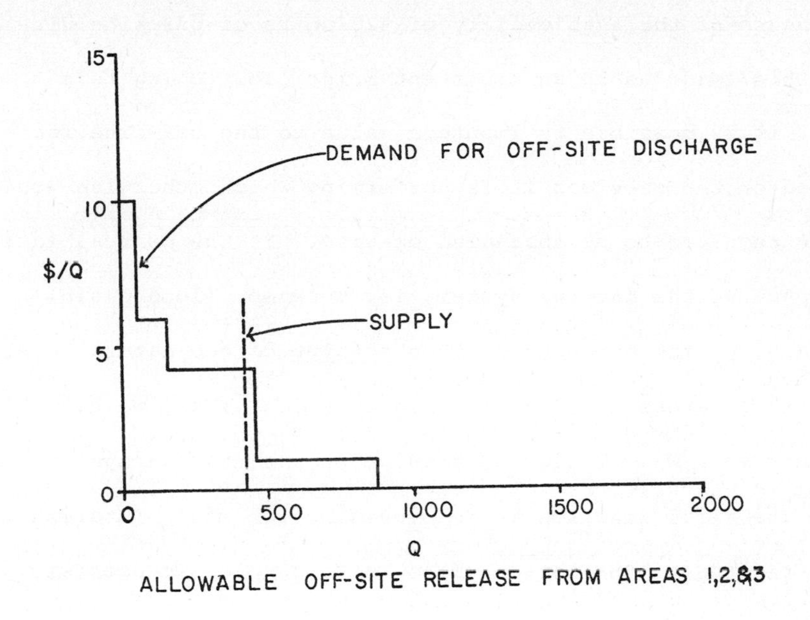

VALUE OF THE OFF-SITE DISPOSAL

Let us determine the aggregate "consumer surplus" in this case if each area pays for off-site discharge it is allotted. As can be seen from this calculation,

Area	Consumer Surplus if $(c + t_i)$ is Charged
1	6(100) = $ 600
2	4(270) = 1080
3	10(50) = 500
	$2180

the aggregate "consumer surplus" is $2180. In this sample, the "consumer surplus" is the savings which accrue to each of the three areas because of the availability of 420 units of off-site disposal. This leads us to an important principle. Using this approach, it is possible to impute a value to the off-site facility based on the services it is performing which otherwise would have been required to be installed on-site. It the central facility is part of the natural system, say a swamp, flood plain, river, lake, or the atmosphere, then a <u>value</u> of the natural system to these three areas is $2180 in this example. Thus, we have been able to derive a way of placing a value on the natural system based on the vital function it is providing for man! Interestingly it is related to the notion of consumer surplus in economic theory.

This ability to place a value on the natural system is of vital importance in resource management. Traditionally, we have attempted to justify open space retention based on its recreation and aesthetic values. This has not provided convincing justification for preserving these areas. Using this technique one can prove, from a functional basis, that these areas are providing other valuable services for man.

IMPLEMENTATION PROBLEMS

Unfortunately, the problem is not yet solved. What happens if we attempt to implement the solution that charges only the actual cost but retains the optimal solution? If we do so, then areas 1 and 3 will probably object since area 2 derives most of the savings. To avoid this possibility one might charge the market price, $c + t_i + \pi$. If this is done, then the consumer surplus is as shown below:

Area	Consumer Surplus if $(c + t_i + \pi)$ is Charged
1	2(100) = \$ 200
2	0(270) = 0
3	6(50) = 300
Net Revenue to "Off-Site Facility"	4(420) = 1680
	\$2180

This solution has at least two problems: (1) what to do with the net revenue and (2) the fact that any positive incentive to area 2 to participate has been eliminated since he is now indifferent between on-site control and off-site control.

COOPERATIVE N-PERSON GAME THEORY

The above problem is amenable to attack using N-person game theory. Luce and Raiffa [12], Owen [13], and Rapoport [14] provide an introduction to the notions of N-person game theory. A key difference between two-person zero-sum game theory and cooperative N-person game theory is the idea of coalition formation and the allocation of associated transfer payments among the players. The balance of this section provides an axiomatic approach to the problem. A related procedure for cost-sharing has recently been proposed by Loehman and Whinston [11].

Let $x\{i\}$ denote the cost assigned to the ith waste discharger if a cooperative arrangement is adopted and $c(N)$ denote the total cost for the coordinated solution. Then, the solutions which are admissible satisfy the following:

$$x\{i\} \geqq 0 \quad \text{for all } i \varepsilon N \qquad (4)$$

$$\sum_{i \varepsilon N} x\{i\} = c(N) \qquad (5)$$

In game theory jargon all such solutions are group rational since the entire cost is partitioned among all of the waste dischargers. The question to be examined is: How should the costs of this central facility be apportioned among the three areas?

It is reasonable to require that

$$x\{i\} \leqq c\{i\} \quad \text{for all } i \varepsilon N \qquad (6)$$

where $c\{i\}$ is the cost of the ith area if he acts independently. Thus, we simply require that the coordinated strategy cannot assess anyone more than their cost if they act independently. This is called the principle of individual rationality. The set of cost shares satisfying (4), (5), and (6) are called <u>imputations</u>. Furthermore, we may wish to extend (6) by requiring the following for all subsets, S, of the master set, N:

$$\sum_{i \varepsilon S} x\{i\} \leqq c(S) \quad \text{for all } S \subset N \qquad (7)$$

All solutions satisfying (4), (5), (6), and (7) belong to the <u>core</u> of the game.

Thus, the first major extension of existing procedures that is revealed using the game theoretic approach is that the notion of the core of an N-person game permits one to delimit more sharply the subset of solutions which are efficient and equitable. It does so by examining only those solutions which are rational for each individual or subset of individuals.

Returning to the example, there are $2^n - 1$ or 7 possible coalitions to be analyzed:

$$\{1\}, \{2\}, \{3\}$$
$$\{12\} \ \{13\} \ \{23\}$$
$$\{123\}$$

The next step in the analysis is to find $c(S)$ for the above combinations. $c(S)$ is called the <u>characteristic</u> <u>function</u> for coalition S and represents the total cost to that coalition. Usually, in game theory, it is assumed that the areas not in the coalition under consideration would form a directly competitive counter coalition which would lead to a minimax solution of a two-person game. Recently, S. Sorenson [16], a student of A. Charnes, generalized some of these notions of cooperative N-person game theory to the cases where alternative definitions of the characteristic function might be used. He describes four possible definitions which are listed below:

$c_1(S)$ = value to coalition if S is given preference over N-S

$c_2(S)$ = value of coalition to S if N-S is not present

$c_3(S)$ = value of coalition in a strictly competitive game between coalition S and N-S

$c_4(S)$ = value of coalition to S if N-S is given preference

As we shall see, alternative definitions can be used depending on how the problem is defined.

We would like to have a technique which permits us to obtain a unique solution to the problem which we can argue is equitable, or fair, to everyone.

The notion of the core reduces the set of feasible solutions considerably. However, there is still room for negotiation among the players as to how the costs are to be divided. Each area seeks to minimize its costs. Assume that the most recent entrant to a coalition pays the incremental cost. Therefore, the cost for the ith area is defined as

$$[c(S) - c(S - \{i\})] \quad \text{for all } S \subseteq N$$

Lacking prior knowledge regarding the likelihood of alternative coalition formation sequences, assume that all sequences may occur and are equilikely. Thus, there are 3! ways of forming the grand coalition as listed:

123	213	312
132	231	321

Under these circumstances, we contend that a fair cost-sharing procedure is to assign to each area the expected value of the cost summed over the above coalition formation sequences. The general relationship is shown below. Let ϕ_i denote the cost assigned to the ith area where

$$\phi_i = \sum_{S \subseteq N} \frac{(S - 1)!(n - S)!}{n!} [c(S) - c(S - \{i\})]$$

The assigned cost, ϕ_i, is called the Shapley value for N-person games (Shapley, 1953). Note that,

$$\sum_{i \in N} \phi_i = c(N)$$

so that the entire cost is apportioned. The Shapley value is a very useful solution notion. Unfortunately, it may happen that the Shapley value can fall outside the core of the game (Sorenson 1972). We want to avoid such cases.

We have examined several problems in cost-sharing and resource allocation. It works out that different definitions of the characteristic function are appropriate depending on how the problem is defined. The easier case is the cost-sharing problem wherein an ownership has been established and the decision needs to be made whether to control on-site only or to form a coalition with others. The more difficult case is to determine a solution when ownership of the central facilities is unspecified. These two cases are considered below.

From a computational point-of-view, $c_1(S)$ can be determined by solving a partition of the overall optimization problem. This procedure will be used in solving the example for various cases. Recall that the overall objective function is to minimize the costs for all three areas. Assume there are n goals over which we want to optimize, i.e.,

$$f_1(x),\ f_2(x)\ .\ .\ .\ f_n(x)$$

In our case the n separate goals represent the goals of all coalition combinations. Then the values of the characteristic function for a given coalition is found by determining the optimal solution to the linear program for the specified objective function. The following cases are considered.

Case I. Ownership of off-site control is specified. Each area "owns" 140 units of the central facility which they may use or sell to other areas.

In this case each coalition seeks to minimize its costs given a specified allocation of the off-site discharge $\bar{W}$. Let $f_i(x) : Q_i \leq \bar{W}'$ denote the objective function for the ith coalition with a specified allocation $\bar{W}'$ of the total of $\bar{W}$ units. The results of this calculation are shown in Table I. Note that the coalition, 13 is inessential, i.e., c(13) = c(1) + c(3).

TABLE I

Optimal Solution with Ownership of Off-Site Facility Prespecified

Coalition	Objective Function	Resource Allocation: $\bar{W}'$	Optimal Solution
c(1)	$f_1(x) = Z_1$	140	$1360
c(2)	$f_2(x) = Z_2$	140	1540
c(3)	$f_3(x) = Z_3$	140	2560
$c(\overline{12})$	$f_4(x) = Z_1 + Z_2$	280	2780
$c(\overline{13})$	$f_5(x) = Z_1 + Z_3$	280	3920
$c(\overline{23})$	$f_6(x) = Z_2 + Z_3$	280	3830
$c(\overline{123})$	$f_7(x) = Z_1 + Z_2 + Z_3$	420	5070

The core of this game is delimited by

$$\begin{aligned} x_1 \quad\quad\quad\quad &\leq 1360 \\ x_2 \quad\quad &\leq 1540 \\ x_3 &\leq 2560 \\ x_1 + x_2 \quad\quad &\leq 2780 \\ x_1 \quad\quad + x_3 &\leq 3920 \\ x_2 + x_3 &\leq 3830 \\ x_1 + x_2 + x_3 &= 5070 \end{aligned}$$

The Shapley values are

$$\phi_i = \frac{1}{3}(1360) + \frac{1}{6}(2780 - 1540) + \frac{1}{6}(3920 - 2560) + \frac{1}{3}(5070 - 3830)$$

$$\phi_1 = \$1300$$

$$\phi_2 = \frac{1}{3}(1540) + \frac{1}{6}(2780 - 1360) + \frac{1}{6}(3830 - 2560) + \frac{1}{3}(5070 - 3920)$$

$$\phi_2 = \$1345$$

$$\phi_3 = \frac{1}{3}(2560) + \frac{1}{6}(3920 - 1360) + \frac{1}{6}(3830 - 1540) + \frac{1}{3}(5070 - 2780)$$

$$\phi_3 = \$2425$$

Note that $\sum_{i=1}^{3} \phi_i = \$5070$ and that the Shapley value does fall within the core of the game. In this case the consumer surplus or savings from having ownership of 140 units is the difference between the cost of on-site control and the Shapley value or \$700, \$755, and \$725 for areas 1, 2, and 3, respectively. This is a much more uniform distribution of savings than existed before.

Case 2. Ownership of off-site control is unspecified. Coordinators own the central facility and allocate it in an efficient and equitable manner.

It has been demonstrated earlier that the optimal solution allocates 100 units to area 1, 270 units to area 2, and 50 units to area 3. The unit costs of providing this central facility are $4, $3, and $5, respectively. At this cost, a larger savings accrues to area 2 than to either areas 1 and 3. Furthermore, areas 1 and 3 would be willing to pay more than $4 to obtain additional units from 2. Thus, they might feel that the solution is unfair.

Let us explore the possibility of using cooperative N-person game theory to solve this problem. In Case I, ownership of the central facility was prespecified. Thus, the values of c(S) were unambiguously defined. This corresponded to the second definition of the characteristic function, $c_2(S)$, wherein the other coalition (N - S) could be viewed as not being present.

Let us examine definition one wherein $c_1(S)$ is defined as the value if S is given preference over N - S. In this case S is allotted as much of the off-site facility as he desires. The resulting values of the characteristic function are shown below:

$$c_1(1) = \$1200; \quad c_1(2) = \$900; \quad c_1(3) = \$2450$$

$$c_1(12) = \$2280; \quad c_1(13) = \$3780; \quad c_1(23) = \$3480$$

$$c_1(123) = \$5070$$

Inspection reveals that this game is not subadditive since coalitions are never advantageous, e.g., $c(12) > c(1) + c(2)$. Thus, this definition of the characteristics function would not result in cooperation in this case since coalitions never leave the players better off.

Let us try the opposite definition wherein $N - S$ is given preference over S. The resulting characteristic function values are shown below:

$$c_4(1) = \$2000; \quad c_4(2) = \$2100; \quad c_4(3) = \$3150$$
$$c_4(12) = \$3220; \quad c_4(13) = \$4230; \quad c_4(23) = \$4470$$
$$c_4(123) = \$5070$$

Unlike the previous situation this game is subadditive for all coalitions. Thus, there is a basis for cooperation.

COOPERATION OR COMPETITION

Indeed one can generalize that in this example coalitions are never advantageous using $c_1(S)$ and are never disadvantageous using $c_4(S)$. The reason is simple. In the former case, the individual area is allotted up to the total available storage, $\bar{W}$, in determining $c(1)$, $c(2)$, and $c(3)$. However, as coalitions form he is forced to share $\bar{W}$ with other areas. Thus, he is never made better off if an additional area joins the coalition since he cannot (due to the nonnegativity restrictions on Q_i) attain more of the central facility than he had before. At best, he is indifferent. On the other hand, using $c_4(S)$, the area will always consider coalitions since they may thereby be able to bid for a larger amount of the capacity of the central facility. Note that $c_1(123) = c_4(123)$ so that the value of the grand coalition is the same. The major difference is in the value of the subcoalitions. This analysis illustrates the importance of a fundamental aspect of resource allocation--the question of ownership. In Case I ownership was specified a priori. This case leads to a pervasively cooperative situation since larger coalitions can never be harmful. Each area has the option of using its allotted share of the off-site facility or selling it to others. Similarly, in the situation where (N - S) is given preference, i.e., $c_4(S)$, a cooperative situation results. However, in the case where S is given priority, a pervasively competitive situation arises.

This discussion can be summarized by introducing the notion of security level which is defined as the value the player can assure himself if he acts independently. We see that the security level depends on how the characteristic function is defined. For the case where ownership is not specified a priori, only definition 4 of the characteristic function guarantees a cooperative solution. Definitions 1 and 3 lead to strict competition and definition 2 is irrelevant since one cannot assume the other N-S players are absent in the case where ownership of the common resource is unspecified. Thus, we feel that definition 4 provides a reasonable basis for an equitable cost sharing arrangement. The Shapley values for this case are:

$$\phi_1 = \frac{1}{3}(2000) + \frac{1}{6}(3220 - 2100) + \frac{1}{6}(4230 - 3150) + \frac{1}{3}(5070 - 4470)$$

$$\phi_1 = \$1233$$

$$\phi_2 = \frac{1}{3}(2100) + \frac{1}{6}(3220 - 2000) + \frac{1}{6}(4470 - 3150) + \frac{1}{3}(5070 - 3220)$$

$$\phi_2 = \$1400$$

$$\phi_3 = \frac{1}{3}(3150) + \frac{1}{6}(4230 - 2000) + \frac{1}{6}(4470 - 2100) + \frac{1}{3}(5070 - 3220)$$

$$\phi_3 = \$2437$$

Note the $\sum_{i=1}^{3} {}_i = \$5070$ and the solution is in the core.

The consumer surplus for this case is \$767, \$700, and \$713, respectively. One could argue that, as long as the calculated consumer surpluses are almost the same, wouldn't it be fairer to simply equalize the consumer surplus so that each area saved an equal amount--about \$727 per study area in this case? Table II shows the charges to each study area for the above two cases.

TABLE II

Charges for Central Facility

Area	Allocation of $\bar{W}$	Total Cost		On-Site Cost	Charge: $\$/Q_i$	
		Shapley	Equal Savings		Shapley	Equal
1	100	\$1233	\$1273	\$1000	\$2.33	\$2.73
2	270	1400	1373	210	4.41	4.32
3	50	2437	2424	2400	0.74	0.48
Totals	420	\$5070	\$5070	\$3610	----	----

In an actual situation one might decide that every area should pay at least a minimum amount or impose other restrictions. It is straightforward to formulate this problem using goal programming wherein one can minimize either the maximum excess savings accruing to an area or minimize the sum of the positive and negative deviations of the excesses relative to a uniform savings solution (see Charnes and Cooper [5] for a description of goal programming).

CONCLUSIONS

The problem of efficient and equitable regional environmental quality management has been addressed using recently developed notions from cooperative N-person game theory. Numerous investigators have demonstrated the value of optimization techniques for solving the problem of determining the pollution control strategy which is in the best overall interest. Unfortunately, such procedures often prescribed solutions which did not seem fair to everyone. Whether or not the solution was fair depends on the cost sharing arrangements which in turn depend on who is entitled to off-site disposal of pollution.

Using a simple example, we have shown how the efficient regional solution is determined using linear programming or a simple market price determination. Included in this analysis is an explicit procedure for placing a value on off-site disposal which is based on savings in alternative on-site control costs. This procedure provides a way of quantifying a value for the natural system which is the recipient of these wastes. Difficulties in achieving an equitable solution were outlined.

Next, introductory concepts from cooperative N-person game theory were presented. This section indicated how it is possible to devise management strategies which retain the efficient solution and permit an equitable cost sharing arrangement to be made. Then two cases were examined. In the first case ownership of the off-site facility was prespecified. Polluters in this situation could utilize their right or exchange it with others. In the second case, ownership was not specified. Depending on how one defined the characteristic function, a competitive or cooperative situation would result. The results of the analysis provide a way of solving this problem.

ACKNOWLEDGMENTS

This research was partially supported by the U. S. Environmental Protection Agency as part of studies of urban stormwater management. Other support comes from the Jacksonville District, U. S. Army Corps of Engineers, who are supporting research on socioeconomic studies of the Upper St. John's River Basin in Florida.

Professor Abraham Charnes of the University of Texas introduced the senior author to the area of cooperative N-person game theory. His personal advice and the writings of his students have been very helpful to us in attempting to apply these techniques to environmental problems.

REFERENCES

1. Baumol, W. J. and Fabian, T. "Decomposition, Pricing for Decentralization and External Economies," Management Science, Vol. 11, No. 1 (1964).

2. Bulkley, S. W. and McLaughlin, R. T. Simulation of Political Interactions in Multiple Purpose River Basin Development, Hydrodynamics Lab Report No. 11, Dept. of Civil Engineering, MIT, Cambridge, Mass. (1966).

3. Burke, R., III, Heaney, J. P. and Pyatt, E. "Political Interaction and Water Resources Management: A Quantitative Approach," presented at the National Symposium on Watershed in Transition, Colorado State University, Fort Collins, Colo. (1972).

4. Charnes, A., Clower, R. W., and Kortanek, K. O. "Effective Control through Coherent Decentralization with Preemptive Goals," Econometrica, Vol. 35, No. 2 (1967).

5. Charnes, A. and Cooper, A. W. Management Models and Industrial Applications of Linear Programming, Vol. I, John Wiley and Sons, Inc., New York (1961).

6. Dantzig, G. B. and Wolfe, P. "Decomposition Principle for Linear Programs," Operations Research, Vol. 8, No. 1 (1960).

7. Dorfman, R. and Jacoby, H. D. "A Model of Public Decisions Illustrated by a Water Pollution Policy Problem," in the Analysis and Evaluation of Public Expenditures: The PPB System. Submitted to the Subcommittee on Economy in Government, U. S. Congress, Joint Economic Committee, Vol 1, Washington, D. C., U. S. Government Printing Office, 1969.

8. Heaney, J. P., Carter, B. J. and Pyatt, E. E. "Equivalent Prices for Upstream BOD Removal, Water Resources Research, Vol. 1, No. 3 (1971).

9. Hass, J. E. "Optimal Taxing for the Abatement of Water Pollution," Water Resources Research,Vol. 6, No. 2 (1970).

10. Herzog, H. W. "The Air Diffusion Model as an Urban Planning Tool," Socio-Economic Planning Sciences, Vol. 3 (1969).

11. Loehman, E. and Whinston, A. "A New Theory of Pricing and Decision Making for Public Investment," The Bell Journal of Economics and Management Science, Vol. 2, No. 2 (1971).

12. Luce, R. and Raiffa, H. Games and Decision, John Wiley and Sons, Inc., New York (1957).

13. Owne, G. Game Theory, W. B. Saunders Co., Philadelphia, Pa. (1968).

14. Rapoport, A. N-Person Game Theory, U. of Michigan Press (1970).

15. Seinfield, J. H. and Kyan, C. P. "Determination of Optimal Air Pollution Control Strategies," Socio-Economic Planning Sciences, Vol. 5 (1971).

16. Sorenson, S. W. A Mathematical Theory of Coalitions and Competition in Resource Development, Ph.D. Dissertation, U. of Texas at Austin (1972).

17. Teller, A. "The Use of Linear Programming to Estimate the Cost of Some Alternative Air Pollution Abatement Policies," Proceedings IBM Scientific Computer Symposium, New York, Data Processing Division (1968).

OPTIMIZATION MODELS FOR REGIONAL AIR POLLUTION CONTROL

Robert Abrams*
Department of Industrial Engineering and Management Sciences
The Technological Institute
Northwestern University

1. INTRODUCTION

Under the Clean Air Act of 1970, the United States government has required the Environmental Protection Agency to set primary and secondary standards** for air pollutants. Standards have been set for six pollutants and these are shown in the Table. Also, under the Act, each state is required to establish regulations on pollutant emissions which will enable the standards to be met by 1975. Thus, in the United States air quality standards have been set by the federal government, but (with certain exceptions such as automobile emissions) methods of meeting the standards must be set by the state and local governments. Past and present air pollution regulations have been established with little or no attempt at any overall optimization. In this paper methods of obtaining optimal emission control strategies will be described and some aspects of the implementation problem will be discussed.

As can be seen from the Table, some standards are given in terms of annual arithmetic mean and others in terms of averaging times which vary between one and 24 hours. Larsen, in [24] and

---

*Partly supported by NSF grant GK27876.

**Primary standards are for the protection of health and secondary standards are for the protection of the public welfare, e.g., property and vegetation.

[25] gives experimental results which establish simple relations between the frequency with which a pollutant concentration is reached and the averaging time over which it was measured. When setting up a model for air pollution control, these transformations are frequently used to reduce all air quality standards to a single averaging time. An alternate approach is the use of chance-constraints (Charnes and Cooper, [4]). One assumption implicit in the use of Larsen's transformation is that abatement procedures will be in effect at all times. In (Teller, [31]), it is suggested that by predicting weather conditions which will cause high pollution levels and by controlling emissions only during these periods, it may be possible to greatly reduce the cost of meeting air quality standards. Two major problems are that weather prediction is not sufficiently accurate, and it is doubtful that large changes in emissions can be made for a short period of time. Switching from coal to natural gas is one change which can be made quickly but availability of natural gas is now extremely limited especially in winter months when air pollution episodes are most likely to occur. The "constant abatement" approach rather than the "forecasting abatement" approach is taken in all models to be considered below.

Almost all methods of controlling air pollution can be divided into four categories, namely, reducing the level of the polluting activity, e.g., reducing the output of a steel mill perhaps by relocation; changing the polluting process, e.g., replacing a more polluting type of steel furnace with a less polluting type;

substituting a less polluting fuel for a more polluting fuel, e.g., low sulfur coal for high sulfur coal; and removal of the pollutants from the effluent, e.g., the use of a precipitator to remove particulates from a coal burning power plant. For details of the many varieties of control methods, especially in the latter two categories, see any of the recent texts in air pollution control e.g., Faith and Atkisson [10], Williamson [33]. The models to be described are sufficiently general to include all the categories of control methods.

Although many of the models described in Section 2 are relatively simple, the air pollution control problem is extremely complex. There are many hidden factors which must be included in the model, explicitly or implicitly, if it is to have any correspondence to the real world. Some of these will be mentioned now.

All of the models to be considered are static. They allow for possible changes in emissions over the years only be basing strategies on some point in the future. However the present best choice of control methods may be significantly influenced by future changes in (uncontrolled) emission levels. Such changes may in part be due to changes in consumption resulting from the costs of pollution control.

Some control methods require large capital investment and have relatively small operating costs while others require little initial investment and have high operating costs. Comparisons of controls will thus depend heavily on interest rate and time horizons assumed.

In many cities automobiles are a principle source of carbon monoxide, nitrous oxides and hydrocarbons, but automobile air pollution standards are set by the federal government and are uniform throughout the country. Therefore, automobile pollution, cannot be taken as a controllable variable (except by limiting traffic) in local air pollution models. For an interesting discussion of automobile air pollution which indicates that current regulations are far from optimal, see Gouse [14].

Finally in section 2, no explicit distinction is made between area, line and point sources. It is assumed that all area and line sources have been approximated by point sources.

2. THE OPTIMIZATION MODELS

Most optimization models for control of air pollution fall into three classes. In models in the first class, it is assumed that air quality standards are given (as indeed they have been by the Environmental Protection Agency). The objective is then to minimize a "cost" of meeting these standards. An alternative model is to put a limit on resources available for control of pollution and to minimize pollutant emissions or pollutant concentrations. Since in general several pollutants are involved in determining any control strategy, models of this class may have vector-valued criteria. In models of the third class a utility function is defined so that costs of control and costs of pollutant damage may be added and then minimized. In this approach optimal concentrations (or emissions) and controls are determined simultaneously. Once a model of one class is set up it is straightforward to formulate a model of one of the other classes using the same basic structure. This, in fact, is done by Kohn [21] and Gorr _et_. _al_.[12].

2.1 Linear Programming Models

Probably the simplest optimization models applied to the air pollution problem are the linear programming models for the St. Louis area studied by Kohn [20], [21], [22], [23]. Kohn assumes that each pollution producing activity, e.g., burning of high sulfur coal, is operated at a given yearly output level which will be maintained in the presence of controls. Every feasible control (including alternate fuel type and alternate production processes) and feasible combination of controls is enumerated and, corresponding to each, a variable is defined to equal the output using that control (or control combination). For example, a variable could be defined to equal the amount of energy produced by low sulfur coal, another would equal the amount produced by high sulfur coal with a precipitator to remove 90% of particulate emissions, etc. Using this structure, only a finite number of control levels may be considered explicitly, e.g., it is not possible to let the fraction of particulates removed be a variable in the problem. The model is formulated as follows:

For each pollution producing activity i, let $J(i) = 1, 2, \ldots k_i$ index the set of feasible controls and combinations of controls. Let x_{ij} equal the output level of activity i using control combination j. Also define:

s_i = the total level of output i to be produced.

c_{ij} = the cost of producing each unit of output i using control method j.

$a_{ij,\ell}$ = output of pollutant ℓ produced by each unit of production of output i with control method j.

E_ℓ = maximum allowable emission of pollutant ℓ.

The model is then

$$\min \sum_{i,j} c_{ij} x_{ij}$$

$$\text{subject to} \sum_{j \in J(i)} x_{ij} = s_i \qquad i = 1, \ldots, n \qquad (1)$$

$$\sum_{i,j} a_{ij,\ell} x_{ij} \leq E_\ell \qquad \ell = 1, 2, \ldots, m.$$

This model is thus of the exact form needed for the Generalized Upper Bounding variant of the simplex method and so significant savings in computational time are possible. (It is not clear whether Generalized Upper Bounding was used in Kohn's work.) If a linear relation is assumed between emissions and the averaging atmospheric pollutant concentrations (after taking natural background into account), the second set of constraints can be converted to a constraint on maximum pollutant concentration simply by multiplying the $[a_{ij,\ell}]$ matrix by an appropriate diagonal matrix. Extension to several receptor points or to any point on a grid is theoretically straightforward assuming it is possible to predict the effect of each source on each receptor.

If resources are fixed and it is desired to minimize pollution, the vector minimization problem is:

$$\min \{\sum_{i,j} a_{ij,\ell} x_{ij}\}$$

$$\text{subject to } \sum_{i,j} c_{ij} x_{ij} \leq R \qquad (2)$$

$$\sum_{j \in J(i)} x_{ij} = s_i \qquad i = 1, \ldots, m$$

where R = resources available for pollution control

One would choose an efficient point which minimizes a damage function of pollution. If such a damage function F is given, a problem of third class which minimizes costs due to damage and control are minimized may be formulated as

$$\min F\,(\sum_{i,j} a_{ij,\ell} x_{ij}) + \sum_{i,j} c_{ij} x_{ij} \qquad (3)$$

$$\text{subject to } \sum_{j \in J(i)} x_{ij} = s_i$$

Taking into account suspended particulates, sulfur dioxide, carbon monoxide, hydrocarbons and oxides of nitrogen, Kohn has calculated minimal cost strategies for the St. Louis area. Although his solutions lump all emissions of each pollutant together, it would clearly be possible to write a set of constraints for each of many points in the area. One disadvantage to the model formulation is that it requires enumeration of all control devices and the efficiency at which each operates. For example in [20], dust collectors with efficiencies of 80% and 95% are considered as controls, but one might want to let the efficiency be a variable in the problem. The problem is, that costs are generally nonlinear functions of efficiency. If one is willing to assume a linear relation, then by including a control with the lowest level of efficiency and one with the highest level to be considered, any convex combination of the two points may be represented in a solution and hence a continuum of controls can be represented with no change in the model. Enumeration of control methods provides the model with two of its principle advantages. One is that by specifying the control method a realistic estimate of the emissions of each pollutant considered may be given. Interactions, such as low sulfur coal affecting the efficiency of a precipitator, can be easily taken into account. The second advantage arising from the fact that at most m (the number of pollutants) of the first n constraints can have more than one positive variable in the optimal solution. This follows because at most m + n variables can be positive in a basic optimal solution and at least one variable must be positive in each of the first m equations. Since m is much less than n, the cost estimates can realistically be based on the assumption that only one method of control will be used for each activity.

Letting the control efficiencies be variables leads to a second linear programming formulation. Define E_i to be the uncontrolled emission and x_i to be the fractional reduction in emission from source i. Let $C_i x_i$ be the cost of removing x_i of the emissions. The problem for one pollutant can then be written

$$\min \sum_{i=1}^{n} C_i x_i \tag{4}$$

subject to $\sum_{i=1}^{n} E_i x_i \geq P$

$$\alpha_i \leq x_i \leq \beta$$

where P is the minimum reduction in emissions and $0 \leq \alpha_i$ and $\beta \leq 1$ are technological bounds in the capabilities of available control equipment. Just as in (1) the constraint can be made on air quality at one or a finite number of points, if a function relating emissions to air quality is known. Cho [5] studies models of this kind for the Chicago area, using both linear and quadratic objective functions. When working with models of this kind, it must be realized that optimal values of the control variable may not correspond to any actual control methods or combination of actual control methods.

2.2 Semi-infinite Programming Models

The air quality standards shown in the Table are supposed to hold at every point in a region. This requirement has been included in the models of Gorr, Gustafson and Kortanek [11], [12], [15], [16]. Thus instead of placing a limit on pollutant concentrations at a finite number of points or on average concentrations, they require that the standards be met throughout the region. This gives rise to an infinite number of constraints--one for each point in the region. Let $U_j(x,y)$ be the average concentration (of a single pollutant) at the point (x,y) in the region, caused by an emission rate of 1 at source j. If there are n sources, then the total concentration at point (x,y) is given by

$$X(x,y) = \sum_{j=1}^{N} Q_j U_j(x,y)$$

where Q_j is average emission rate of source j. Letting x_j be the fraction of the pollutant removed at the source and $c_j(x_j)$ be the cost of removing the fraction x_j of Q_j. The semi-infinite programming problem for the pollution control problem then becomes

$$\min \sum_{j=1}^{N} c_j(x_j) \qquad (5)$$

$$\text{subject to } \sum_{j=1}^{N} (1 - x_j) Q_j U_j(x,y) \leq S \qquad \text{for all } (x,y) \text{ in}$$

where S is the air quality standard.

In addition to giving the optimal strategy, which is guaranteed to meet standards at all points, a solution to (5) will also indicate the points at which the maximum concentration will occur. These are the natural points at which to place air pollution monitoring stations. The primary difficulty is that the function $U_j(x,y)$ must be determined with a reasonable degree of accuracy. Although there has been a large amount of work on diffusion modeling (see the references noted in [12]), the accuracy of diffusion models remains questionable.

Gorr, Gustafson and Kortanek show that if the objective function is assumed to be linear then (5) is the dual of the "moment problem". Computational methods based on this correspondence are given and also several alternate forms of the problem in which objective functions other than cost minimization are mentioned.

2.3 Geometric Programming Models:

For regions with large numbers of sources and many constraints on the controls, it is likely that only linear models will be computationally feasible. However, for regions with smaller numbers of significant sources, nonlinear models which more accurately reflect the real situation will be useful. The models presented below include two nonlinear effects inherent in many situations. The fractional reduction in pollutant emissions due to a control method is taken as a variable, and the cost which is a nonlinear increasing function of the reduction is included in the objective function. The second nonlinearity included is that which will occur when two control methods are used in series. For example, effluent gas from coal burned in a power plant may have particulates removed by a precipitator and then be exhausted from a smokestack. The effective height (actual stack height plus plume rise above the stack) of the stack will greatly influence ground level pollutant concentration. (Frequently, e.g. in Williamson [33], maximum gound level concentration is assumed proportional to h^{-2} where h is the effective stack height.) The combined reduction of the precipitator and the smokestack can be represented as the product of the individual reductions. This leads to the following formulation which for simplicity is formulated for a single pollutant.

Assume that we have a finite number of sources and a finite number of receptor points at which the maximum average concentration is not to exceed s. Let b_{ik} be the concentration at receptor i due to source k. Assume that at source k, ℓ_k controls $\{(k,1)\ (k,2),\ \ldots,\ (k,\ell_k)\}$ are used in series and the resulting effluent is discharged through a stack to be denoted as control $(k,\ \ell_k + 1)$. Let $t_{k,j}$ be equal to 1 minus the fraction of pollutant removed by control (k,j), i.e., $t_{k,j}$ is the fraction of the entering pollutant which passes through control (k,j). It will be assumed that the fractional reduction due to stack height is the same at each receptor point. This assumption can be dropped with no changes in the structure of the model if the t_{k,ℓ_k+1} for each receptor point can be expressed as some other variable raised to a power. We now adopt a device of Charnes and Gemmell in their paper "A Method of Solution of Some Nonlinear Problems in Abatement of Stream Pollution" [4a]. Thus it is further assumed that the cost of operating control devices (k, j) is $d_{kj}\ [\ t_{kj}^{-a_{kj}} - 1\ ]$ for some positive constants c_{kj} and a_{kj}. Note that this function has the desired property of being 0 at $t_{kj} = 1$ (no control) and approaching infinity as t_{kj} approaches 0 (100% control). In the objective function, the constant $-d_{kj}$ is dropped since it will not affect the optimal point.

The problem is then

$$\text{minimize} \sum_{k,j} d_{kj} t_{kj}^{-a_{kj}} \tag{6}$$

$$\text{subject to} \sum_k b_{ik} \prod_{j=1}^{\ell_k+1} t_{kj} \leq S \quad \text{all } i$$

which is a posynomial geometric programming problem (Duffin et.al. [7]), of the same form as the Charnes-Gemmell model for stream pollution.

With the change in variables $t_{kj} = e^{z_{kj}}$, it becomes a convex problem. More important, it has a dual which is a linearly constrained concave maximization problem and which can be easily solved for problems of the order of 100 terms. In many cases, the t_{kj} can be expected to be posynomial functions of design. The parameters can be included in the model.

An extension of a geometric programming model (actually a combination of Kohn's model with the above geometric model) will allow explicit consideration of fuel substitution or alternate production processes together with the efficiency variable for each of the controls acting in series. With the notation of Kohn's model (1) specialized so that x_{ij} is the output level of activity i using fuel j, and the definition of the t_{jk} as above (only controls (j,k) that can be used with fuel j are considered), consider the problem:

$$\text{minimize} \sum_{i,j} c_{ij} x_{ij} + \sum_{j,k} d_{jk} t_{jk}^{-a_{jk}}$$

$$\text{subject to} \sum_{j \in J(i)} x_{ij} \geq s_i$$

$$\sum_{i,j} a_{ij,\ell} x_{ij} \prod_{k=1}^{j_K} t_{jk} \leq E_\ell \qquad \ell = 1, \ldots, m$$

Here the objective function includes the linear cost of fuel and the nonlinear costs of control devices of efficiency $(1-t_{jk})$. The original equality of the first constraint has been replaced by an inequality which is equivalent, since it will be binding. Due to the greater than or equal to constraint, this problem has the term of a "reversed" posynomial programming problem

(Duffin and Peterson, [9]). Such problems are considerably more difficult to solve than those with just "less than or equal to" constraints. One technique (Avriel and Williams [1], Duffin and Peterson [8]) is to approximate the reversed problem with a sequence of prototype posynomial problems. This is computationally time consuming, but feasible for small problems.

2.4 Other Models for Determining Control Strategies

Stochastic optimization models are considered by Goslin and Foes [13] and Blumstein *et. al.*[2]. In [13] the diffusion model of Herzog [17] is coupled with a stochastic linear program. It is similar to the linear programs of Section 2.1 except that the concentrations predicted by the diffusion models are assumed to be the means of normally distributed random variables.

In [2] the effects of breakdowns and maintenance of pollution control equipment are considered and a model for a single source is constructed.

Finally, a number of simulation models have been constructed to display the concentrations that will result from any given set of sources and control strategies. The model of the Chicago area by Argonne National Laboratories, e.g., Croke *et. al.*[6], Kohn [22] and Norco *et. al.*[27] is one of the most extensive studies of this kind.

3. IMPLEMENTATION OF OPTIMAL SOLUTIONS

Early linear programming models for determining optimal air pollution control strategies were proposed at least five years ago, and since that time many more sophisticated models have been developed. However, to the best of this writer's knowledge, there has never been an attempt to implement a control strategy based on an optimization model. Of course, many pollution control regulations do take cost into account, but this has always been done on an ad hoc basis rather than on the basis of an optimization model. The two main reasons for this appear to be the lack of reality in some models and the difficulty in formulating a reasonable set of regulations which would result in adoption of an optimal solution.

Except in the case of a region with only a few polluters or only a few classes of polluters which have similar optimal solutions, it will not be practical to enact legislation specifying different levels of abatement for each polluter or class of polluter. Even if it were possible, it would probably not be considered "fair" to put the burden of cleaning the environment more on one company than another. Thus it appears likely that the only way to implement an optimal solution is by some decentralized means such as taxation based on quantity of emissions or open trading of "air pollution rights".

The idea of using effluent fees to reduce pollution has received much attention recently in the literature, e.g., Solow [30], Schwartz [29], Kneese [19]. In view of some deceptive statements in the literature, it seems worthwhile to review some of the theory connecting the use of effluent fees with optimization models. The situation is analogous to that discussed by

Charnes, Clower and Kortanek [3] for linear models in which decentralization of multidivisional firms through pricing is considered. The results of [3] will be put in the context of the air pollution problem using Kohn's model (1).

If a tax (or effluent fee) proportional to emissions for each pollutant were levied against each polluting activity, the activity would, in order to minimize costs, solve the subproblem

$$\text{minimize} \sum_{i,j} c_{ij}x_{ij} + \sum_{i,j;\ell} a_{ij,\ell}x_{ij}y_{\ell} \tag{7}$$

$$\text{subject to} \sum_{j \in J(i)} x_{ij} = s_i$$

where y_{ℓ} is the tax rate for pollutant ℓ.

It is shown in [3] that the component of a solution to (1) provides a solution to (7) if the y_i are the optimal dual variables. Thus if tax rates were set at levels equal to the optimal value of the dual variables, each polluting activity will find that the solution obtained from (1) is optimal for itself. Unfortunately, the converse is not true, i.e., if the tax rate is set equal to the optimal dual variables, it does <u>not</u> follow that by minimizing its costs, i.e., by solving (7), each polluting activity will adjust its emissions so the composite solutions will solve (1). This phenomena occurs because some of the subproblems (7) will have multiple optimal solutions. Actually, as shown by example in [3], an objective function identically equal to zero may be obtained for the subproblem. It will then follow that the taxed activity is completely indifferent and can pollute in any quantity with no change in its own cost. To avoid this problem, the idea of a "coherent partition" is presented in [3] and it is shown that by adding a constraint to the subproblem, the desired optimal solution will be obtained. But in the present context, the additional constraint is a constraint on pollution emissions which is what we wish to avoid. It is also shown that if the linear objective function is replaced by a strictly convex objective function, then the optimal values of the dual variables will provide tax rates which will result in the composite optimal solutions of the subproblems (7) being optimal for the overall problem (1). Thus it is only in the strictly convex case that a mathematical proof of the equivalence of effluent fees and overall cost minimization can be given.

Although many economists feel that regulation by taxation is the most efficient method of controlling pollution, there is some strong feeling against it. Most of these arguments on both sides are theoretical and it appears that more experimental work based on real data is needed.

A study by Young [35], based on data from 10 sources of particulate pollution, provides some very interesting results. It was found that a system of effluent fees produced a control level which was very close to the optimum and which resulted in considerable savings over present practice. The pollution level was found to be quite insensitive to the level of the fee. In fact, any fee of between ten and ninety dollars would produce close to the optimal emission rate. If similar results can be shown to hold for most of a large number of polluters in a region, we would have at least one practical method of implementing near optimal control strategies.

TABLE
U.S. National Air Quality Standards

Pollutant	Primary Standards	Secondary Standards
Sulfur Oxides	80 micrograms per cubic meter (0.03 ppm) annual arithmetic mean. 365 micrograms per cubic meter (0.13 ppm) as a maximum 24-hour concentration not to be exceeded more than once a year.	60 micrograms per cubic meter (0.02 ppm) annual arithmetic mean. 260 micrograms per cubic meter (0.1 ppm) maximum 24-hour concentration not to be exceeded more than once a year.
Particulate Matter	75 micrograms per cubic meter annual geometric mean. 260 micrograms per cubic meter as maximum 24-hour concentration not to be exceeded more than once a year.	60 micrograms per cubic meter annual geometric mean. 150 micrograms per cubic meter as a maximum 24-hour concentration not to be exceeded more than once a year.
Carbon Monoxide	10 milligrams per cubic meter (9 ppm) as a maximum eight-hour concentration not to be exceeded more than once a year.	Same as Primary Standard
Photochemical Oxidants	160 micrograms per cubic meter (0.08 ppm) as a maximum one-hour concentration not to be exceeded more than once a year.	Same as Primary Standard
Hydrocarbons	160 micrograms per cubic meter (0.24 ppm) as a maximum three-hour concentration (6 to 9 a.m.) not to be exceeded more than once a year.	Same as Primary Standard
Nitrogen Oxides	100 micrograms per cubic meter (0.05 ppm) annual arithmetic mean.	Same as Primary Standard

REFERENCES

1. Avriel, M., Williams, A. C. "Complementary Geometric Programming", Siam J. Appl. Math. 19 (1970).

2. Blumstein, A., Cassidy, R. G., Gorr, W. L., Walter, A. S. "Optimal Specification of Air-Pollution-Emission Regulations Including Reliability Requirements", Operations Research 20, 4 (1972).

3. Charnes, A., Clower, R. W., Kortanek, K. O. "Effective Control Through Coherent Decentralization with Preemptive Goals", Econometrica 35, 2 (1967).

4. Charnes, A., Cooper, W. W. "Deterministic Equivalents for Optimizing and Satisficing Under Chance Constraints", Operations Research 11 (1963).

4a. Charnes, A., Gemmel, R. E. "A Method of Solution of Some Nonlinear Problems in Abatement of Stream Pollution", Systems Research Memorandum #103, The Technological Institute, Northwestern University (1964).

5. Cho, P. "A Scheme for Allocation of Emissions in Urban Areas" Ph.D. Dissertation, Department of Civil Engineering, Northwestern University (1972).

6. Croke, E. J., Roberts, J. J. et. al., "Chicago Air Pollution Systems Analysis Program Final Report", Report prepared by the Center for Environmental Studies, Feb. 1971, Argonne National Laboratory, Argonne, Illinois.

7. Duffin, R. J., Peterson, E. L., Zener, C. Geometric Programming - Theory and Application, Wiley (1967).

8. Duffin, R. J., Peterson, E. L. "Reversed Geometric Programming Treated by Harmonic Means", Indiana University Mathematic Journal 22, 6 (1972).

9. Duffin, R. J., Peterson, E. L. "Geometric Programming with Signomials", J. Optimization Th. Appls. 11, 1 (1973).

10. Faith, W. L., Atkisson, A. A. Air Pollution (second edition), Wiley-Interscience, New York (1972).

11. Gorr, W., Kortanek, K. O. "Numerical Aspects of Pollution Abatement Problems: Constrained Generalized Moment Technique", Institute of Physical Planning Report No. 12, 1970, School of Urban and Public Affairs, Carnegie-Mellon University, Pittsburgh, Penn.

12. Gorr, W., Gustafson, S. A., Kortanek, K. O. "Optimal Control Strategies for Air Quality Standards and Regulatory Policy", Environment and Planning 4 (1972).

13. Goslin, L. N., Foes, C. "An Air Pollution Control Model: Some Comments", paper prepared for the meeting of The Institute of Management Sciences, Washington, D.C. (1971).

14. Gouse, W. S. "Mobile Source Air Pollution - Who Won the War", paper presented at the Joint ORSA/IEEE Meeting, Annaheim, California (1971).

15. Gustafson, S. A., Kortanek, K. O. "Mathematical Models for Air Pollution Control: Numerical Determination of Optimizing Abatement Policies", Report prepared for the Nato Study Institute on Systems Analysis for Environmental Pollution Control, Baiersbrumm, Schwagwald, Germany (1972).

16. Gustafson, S. A., Kortanek, K. O. "On the Numerical Determination of Optimal Control Strategies for Air Quality Standards and Regulatory Policy, Institute of Physical Planning Research Report 22, 1971, School of Urban and Public Affairs, Carnegie-Mellon University, Pittsburgh, Penn.

17. Herzog, H. W. "The Air Diffusion Model as an Urban Planning Tool", Socio. Econ. Plan 3 (1969).

18. Jimeson, R. M. "A Regional Air Pollution Analysis Model", Paper No. FA2.3, prepared for the meeting of the Operations Research Society, Dallas (1971).

19. Kneese, A. V. "Environmental Pollution: Economics and Policy", Amer. Econ. Rev. (Papers Proc.) 61, 153 (1971).

20. Kohn, R. E. "Linear Programming Model for Air Pollution Control", J. Air Pol. Control Assoc. 20 (Feb. 1970).

21. Kohn, R. E. "Abatement Strategy and Air Quality Standards" in Development of Air Quality Standards, Atkinson, A. and Gaines, R. S., eds., Merrill Publishing Co., Columbus,(1970).

22. Kohn, R. E. "Application of Linear Programming to a Controversy in Air Pollution", Management Science 17, 10 (1971).

23. Kohn, R. E. "Optimal Air Quality Standards", Econometrica (Nov. 1971).

24. Larsen, R. I. "A New Mathematical Model of Air Pollution Concentration, Averaging Time and Frequency", J. Air Pol. Control Assoc. 20 (1969).

25. Larsen, R. I., Zimmer, C. E., Lynn, D. A., Blemel, K. A. "Analyzing Air Pollutant Concentrations and Dosage Data", J. Air Pol. Control Assoc. 17 (1967).

26. Ledbetter, J. O. Air Pollution, Marcel Dekker, New York, (1972).

27. Norco, J. E., Snider et. al. "Evaluation of Emission Control Strategies for Sulfur Dioxide and Particulates in the Chicago Metropolitan Air Quality Control Region", Report prepared by the Center for Environmental Studies, Argonne National Laboratory, Argonne, Illinois (Dec. 1970).

28. Passy, U., Wilde, D. J. "Generalized Polynomial Optimization", SIAM J. Appl. Math. 7, 4 (1967).

29. Schwartz, Seymour I. "Models for Decision-Making in Air Quality Management", Technical Report 70-4 (1970), Department of Industrial and Systems Engineering, University of Southern California.

30. Solow, R. M. "The Economist's Approach to Pollution and Its Control", Science 173 (1971).

31. Teller, A. "Air Pollution Abatement: Economic Rationality and Reality", Daedalus (Fall 1967).

32. Teller, A. "The Use of Linear Programming to Estimate the Cost of Some Alternative Air Pollution Abatement Policies", Proceedings, IBM Scientific Computing Symposium, Water and Air Resource Management, IBM Data Processing Division, White Plains, New York (1968).

33. Williamson, S. J. Fundamentals of Air Pollution, Addison-Wesley, Reading (1973).

34. Wolsko, T. D., Matthies, M. T., King, R. F. "A Methodology for Controlling Air Pollution Episodes", Report prepared by the Center for Environmental Studies, Argonne National Laboratory, Argonne, Illinois (Sept. 1970).

35. Young, D. "Optimal Pollution Regulation - A Data Based Study", School of Industrial and Systems Engineering, Georgia Institute of Technology, Atlanta, Georgia.

ENVIRONMENTAL NOISE MANAGEMENT

Daniel P. Loucks
Department of Environmental Engineering
Cornell University

The management of environmental quality has typically included only the control of mass residuals and their effect on the quality of air, land and water resources. With the increased production and use of energy, residuals such as heat and noise are also becoming significant causes of pollution. This paper discusses a quantitative modeling approach for the management of environmental noise, i.e. noise emanating from sources that cannot be controlled by those who hear them. Especially in urban areas, environmental noise is often the most pervasive, annoying and costly of all waste residuals. There exists a number of alternatives that can be used to control both the intensity and duration of environmental noise throughout an urban area. After applying a systematic procedure for identifying the important noise sources and noise receptors, and for defining the costs of alternatives for noise reduction at each source and receptor, both optimization and simulation models can be used for estimating noise attenuation in complex urban areas and for defining and evaluating economically efficient environmental noise management alternatives. Outlined in the appendix of this paper are two simplified separable programming models that can be used as aids in the preliminary screening of alternative noise management policies.

INTRODUCTION

Some sounds add to the quality of the environment, other sounds do not. Sounds that are unwanted or annoying result in what is popularly called noise pollution. This paper is focused primarily on the management of environmental noise, a particular kind of noise that emanates from sources that are not easily controlled by those who hear it. Environmental noise includes, but is not restricted to, what is also termed outdoor noise and community noise. Those who are subjected to environmental noise, whether they are outside or inside of a building, must either endure the noise and the resulting annoyance or damage or incur a cost of protecting themselves from at least a portion of it. For example, outdoor noise entering a room through an open window may interfere with the activities of those in the room. Having to close the window, thereby reducing the circulation of air on a hot day, just to decrease the noise damage is an example of social cost. An economic cost is incurred if an air conditioner is then needed solely because of the closed window. The overall objective of any environmental noise management program is to minimize the sum of the damages resulting from noise and the social and economic costs of any measures taken to reduce noise in the environment.

The approach to environmental noise management proposed in this paper is very similar to the approaches that have been proposed and applied to the management of other waste residuals. Regional air quality management, water quality management and solid waste management are relatively popular and well known terms usually implying an economically efficient or cost-effective approach for planning and maintaining the quality of our airshed, water-ways, and land, and for controlling the residuals that can reduce the value of those natural resources. Among these terms denoting particular aspects of the overall problem of environmental quality management should also be included the term environmental noise management: the management of a residual that in many rural and urban areas reduces the quality of the environment far more than other mass or energy residuals.

Probably most in need of environmental noise management programs are many urban areas, as evidenced by the increasing number

number of public awareness programs and noise abatement and control laws being enacted by the major cities of the United States and by the continued enforcement of noise regulations in many European and Asian cities. While this paper will concentrate on the urban noise management problem, the general rationale and approach to noise management extends to all environmental noise problems. The reason for the concentration here on urban noise stems not only from the fact that it is in these areas where environmental noise and its damage is often the greatest; it is also in these areas where the problem of noise management can be the most complex. The identification of major noise sources and their characteristics, the prediction of noise attenuation from each noise source to each receptor, and the estimation of the effectiveness of noise reduction alternatives at the source and receptor sites and in the transmitting medium between source and receptor sites, are made no easier by the existence of complex configurations of buildings, surfaced with various materials and forming irregular street canyons (which tend to reduce sound attenuation) and barriers (which often increase sound attenuation).

The objective of this paper is not to recommend a set of specific environmental noise levels for various urban, surburban or rural settings, or to suggest a specific means of reducing noise in any given situation. Rather it is to discuss the development and use of quantitative procedures and techniques that can assist in the establishment of specific environmental noise abatement plans and policies. Of interest is an approach to environmental noise management that can define and evaluate numerous noise abatement alternatives based on one or more management objectives. This requires the identification of all significant noise sources and receptors and of the alternative means and costs of reducing the noise at each source, at each receptor, and in the paths between each source and receptor. It requires the prediction of the effectiveness of various noise reduction alternatives in terms of the total decrease (or increase) in the intensity or duration of noise experienced at various locations within the affected area and a knowledge of the effects of the remaining noise on each receptor. The simultaneous consideration of all of these factors, together with information on how changes in various data and assumptions might alter the time distribution and intensity of the noise at various sites within an area, should lead to more effective environmental noise management policies.

This paper will discuss briefly in qualitative terms the proposed approach for defining and evaluating alternative environmental noise management policies. Following a short description of the major urban sources and typical characteristics of environmental noise, various noise reduction alternatives will be reviewed in the context of a hypothetical urban area. Finally a procedure will be outlined for defining and evaluating alternative management policies that satisfy environmental noise standards. These standards specify the maximum noise levels that are allowed at numerous sites within the urban area.

ENVIRONMENTAL NOISE CHARACTERISTICS

It is neither necessary nor practical to identify and characterize all sources of environmental noise simply because all noise sources do not contribute in an equal manner to the total noise level. It is generally acknowledged that the major contributors of environmental noise in most urban environments are transportation, construction, and certain commercial and industrial activities. In addition, noise from heating and air conditioning systems can be locally significant. These are activities over which the typical man on the street has very little control, and yet the noise that these activities create may affect him psychologically, sociologically, physiologically and economically. The effect that each of these types of noise sources may cause will depend on the characteristics of the aggregate noise at various locations as well as on the personal characteristics and the activities of the individuals hearing the noise at those locations.

The annoyance associated with environmental noise is dependent, in part, on a number of its characteristics. Perhaps the most obvious characteristic of noise is its intensity, or amplitude. Sound intensity is usually stated in units called decibels, abbreviated dB. Individuals with unimpaired hearing can detect sounds close to zero decibels. Jet aircraft, on the other hand, can generate over 125 decibels, an intensity that can temporarily if not permanently decrease one's ability to hear and that is painful to all but a very few individuals. Obviously most environmental noises have intensities between these two values. Before describing in any more detail the various noise sources and the intensities associated with urban settings, some more properties of urban noise should be discussed.

The loudness of any noise and the annoyance it may cause is also a function of the frequency of the noise. A higher frequency noise of a given intensity is usually considered more annoying, up to a point, than a lower frequency noise of the same intensity. For this reason most measurements of environmental noise combine in a specific way both frequency and intensity. The net result is called an A-weighted decibel scale, abbreviated dB(A). Two sounds of equal intensity but of different frequencies will have different dB(A) levels, the lower frequency sound having the lower dB(A) level. This method of noise measurement appears to be satisfactory for most environmental noises that contain many frequencies, i.e. broadband noise. For noise containing strong pure frequency tones (e.g. the whine of jet engines), other more complex measurement procedures have been devised.

To complete the description of environmental noise it is necessary to account for the temporal pattern of the changes in the dB(A) levels at various locations. A steady noise is usually less annoying than one whose level is varying, especially if the variations are random. Also, a short duration noise is less annoying than one of a longer duration. While each dB(A) can be weighted depending on its duration and regularity, it is perhaps most useful at this point to describe the temporal distribution of aggregate noise levels, in dB(A), by a continuous graph recording.

As a means of illustrating how noise is characterized, consider a portion of the hypothetical and simplified urban community called Ninevah. This portion of Ninevah, as shown in Figure 1, contains a number of different types of urban settings: manufacturing and industrial areas, commercial and shopping districts, thoroughfares and railroads, school and hospital areas, a residential section, and a park. The major noise sources include the motor vehicle traffic on the thoroughfare, the railroad, aircraft flying on airways V-64 and V-32 that cross over the commercial and industrial portions of Ninevah, industrial and manufacturing operations, construction equipment, and in the residential area, occasional trash collection, dog barking and lawn mowing activities. Obviously there are many other noise sources that contribute to the total noise level at any particular location in the area.

The outdoor noise environment at locations marked A, B, C and D in Figure 1 might be as indicated by the temporal patterns shown in Figure 2. Each pattern shown in Figure 2 represents a different type of urban setting, but each illustrates three features of most urban noise environments.

The first feature of urban noise environments is that the aggregate noise level can vary with time over a range of more than 30 dB(A), which is usually perceived as an eight-fold range of noisiness. This eight-fold range is based on the observation that a change of approximately 10 dB represents a doubling of halving of perceived loudness or noisiness of a sound. Hence a 30 dB(A) increase represents three doublings (2x2x2) or eight times the original noisiness.

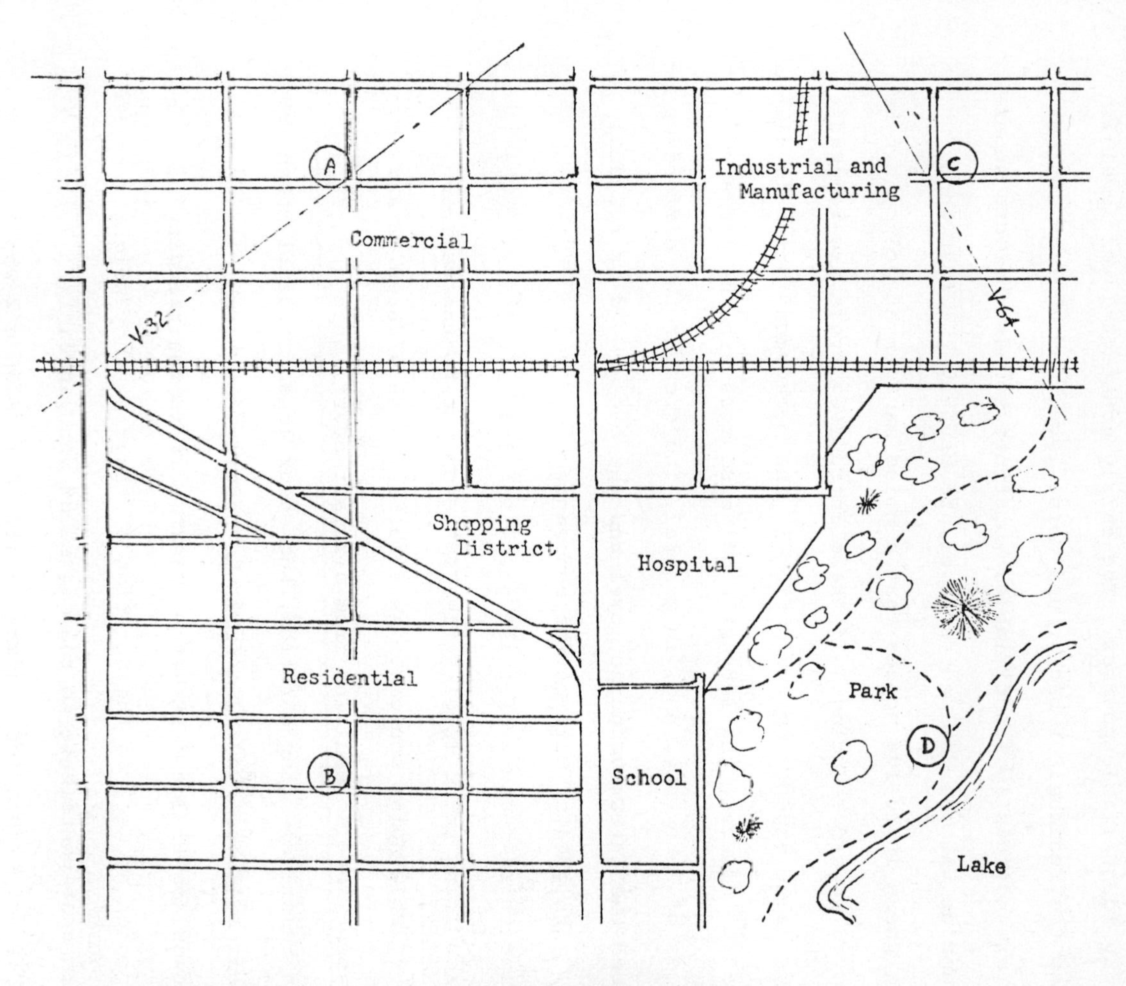

Figure 1: Portion of Urban Community "Ninevah"

The second feature of these typical temporal noise distributions is that there tends to exist a relatively steady lower noise level upon which is superimposed higher levels that originate from clearly identifiable sources. The fairly constant lower level is termed the residual or background noise level coming from sources that are not easily identifiable. Distinct sounds that are superimposed on the residual noise level are usually the most annoying. It is these intrusive noises that a noise management policy can be directed toward reducing. If these intrusive levels are reduced one can also expect some reduction in the background noise level as well. However, the intrusive noise sources are the primary sources of annoyance, and, in general, the greater the difference between the intrusive noise level and the background or residual level, the greater will be the annoyance.

The third feature of these noise samples is the difference in the resulting noise level - time patterns due to the various identifiable noise sources. As illustrated in Figures 2a and 2c, the noise levels resulting from aircraft overflights are heard for a relatively long duration, but occur less frequently than the noise levels resulting from the passing motor vehicles on the thoroughfare or local streets. Clearly it is the rapidity of occurrence as well as the duration of the frequency weighted noise levels that contribute to the degree of annoyance.

The noise-duration patterns depicted in Figure 2 would likely change during different periods of the day or night, and also during different times of the year. Also of importance is the

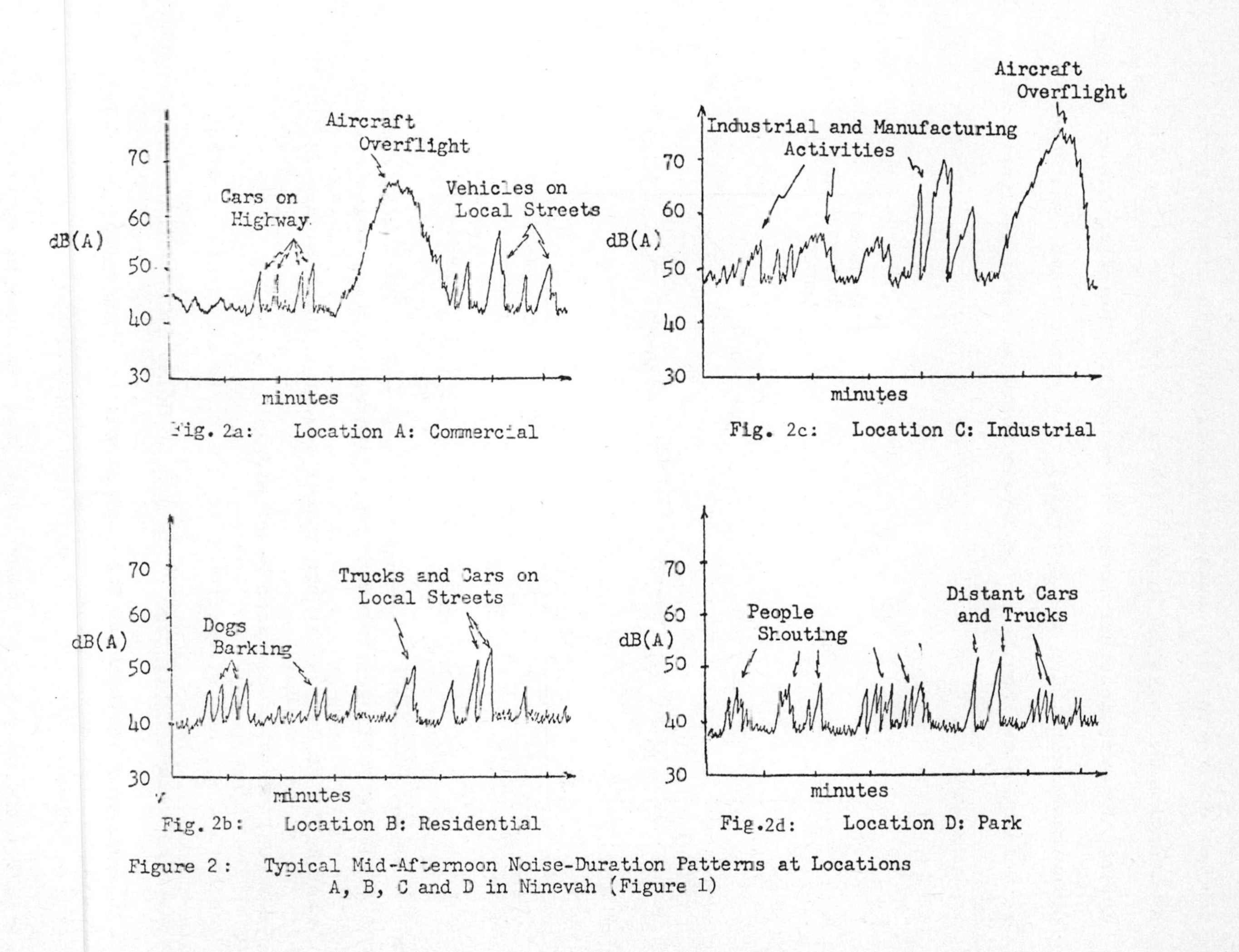

Fig. 2a: Location A: Commercial

Fig. 2c: Location C: Industrial

Fig. 2b: Location B: Residential

Fig. 2d: Location D: Park

Figure 2 : Typical Mid-Afternoon Noise-Duration Patterns at Locations A, B, C and D in Ninevah (Figure 1) noise environment of individuals who travel from one place to another during the day, such as an individual in Ninevah who lives at location B, works in locations A or C, and spends some time during the week in other locations as well. The objective of any noise management plan is to control this environment, which can be done by altering the noise levels heard at specific locations such as at A, B, C or D.

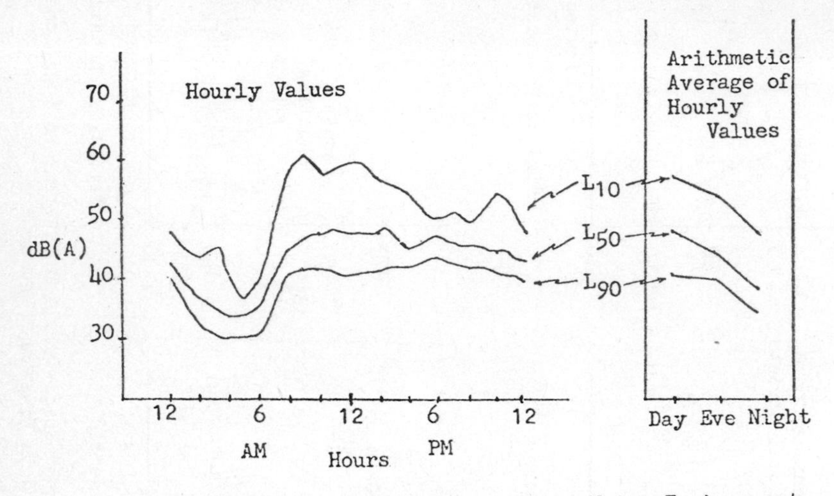

Figure 3: An Individual's Daily Noise Environment

Figure 3 illustrates what an individual's daily noise environment might be during say the summer season. In this example the residual noise level drops after midnight, reaching a minimum at about 5 A.M., and rises from 6 to 8 A.M. to an almost constant daytime value. This time variation of noise is dependent on the location and activity of the individual, and hence this pattern, while typical, may vary for different individuals.

For the purposes of measurement, the residual background noise level is often defined as that noise originating from nonidentifiable sources which is exceeded 90 percent of the time (L_{90}). The median noise level (L_{50}) is a measure of the average noise environment in the sense that half the time it is quieter and noiser than this average value. Finally, the value L_{10} in Figure 3 indicates the noise levels that are exceeded only 10 percent of the time.

To summarize, the noise environment associated at any particular location or with any particular individual can vary substantially. Nevertheless, environmental noise consists of two main parts, an all-pervasive and non-specific background or residual noise and a constant and intermittent readily identifiable intrusive noise. It is the difference between the residual and intrusive noise levels that causes most annoyance, and as a consequence, the management of environmental noise need be concerned mainly with the reduction of the intrusive noise. While the annoyance associated with indoor noise is sometimes reduced by adding to the residual noise, that is not usually considered a reasonable alternative for the management of outdoor environmental noise.

AN APPROACH TO ENVIRONMENTAL NOISE MANAGEMENT

As outlined above, at least one of the goals of any environmental noise management policy is to minimize the sum of the damages that result from the effects of environmental noise and the net costs incurred by any measures taken to reduce the noise, either at the noise source sites, between the source sites and the receptor sites, or at the receptor sites. However, like other mass and energy residuals released into the environment, the effects of noise are difficult to quantify and hence the approach suggested here for the management of noise will be similar to that developed for and applied to the management of other residuals, namely one that efficiently meets predefined environmental quality standards. Numerous noise abatement alternatives exist that can be used to satisfy standards specifying the maximum allowable duration-frequency weighted noise levels at various locations in an urban area. A method for defining these alternatives and evaluating them based on their costs will be discussed in this final section of the paper. The discussion will continue to be qualitative, and introduce to the reader what will be discussed in more quantitative terms in the appendix of this paper.

Decisions regarding the expenditure of funds for the maintenance or improvement of environmental quality are often made on the basis of a mostly qualitative integration of numerous economic, political, social and technological objectives. The explicit tradeoffs between each of these partially complementary and conflicting objectives have not always been clear and therefore the selection and implementation of environmental quality management plans too often have failed to meet these objectives to the extent originally envisioned. In many cases, the current approach to environmental noise management promises to be no exception.

The environmental management decision-making process is further complicated by the uncertainty as to the outcome of any decision due to factors outside the control of the decision maker, and by the limited ability of decision makers to consider simultaneously all information pertinent to the decision even if there were no uncertainty. Even when supplemented with computers, the capacity to plan and implement an efficient environmental quality management policy is limited. To decrease the extent of this limitation with regard to the management of air, water and solid wastes, planners and decision makers have constructed simplified models of their management problem. These models enable them to more effectively process what information they have in order to predict and evaluate the possible outcomes of various policies. These simplified models can range from those that are wholly conceptual and contained within the mind of the planner or decision maker to those that are specified by sets of algebraic equations designed to be solved on high speed computers. It is this latter approach, applied to the management of environmental noise, that is proposed in this paper. The use of this quantitative modeling approach can be of considerable value as an aid, but not as a substitute, to the responsible decision-making process.

Very briefly, the quantitative modeling approach to be discussed involves:

1) the identification of all intrusive noise sources and the measurement of the intensity and duration of each resulting noise;

2) the identification of alternative methods for reducing the noise intensity and/or duration at each noise source

and of the net cost of this reduction as a function of the amount reduced;

3) the identification of a variety of receptor sites, i.e., sites where the control or management of environmental noise is desired and which are located in such a manner that the environmental noise in areas in-between the specific receptor sites will also be adequately controlled;
4) the identification of alternatives that can be implemented at each receptor site for reducing both the intensity and/ or duration of the noise heard at those sites, and of the net cost of these control measures as a function of their effectiveness;
5) the identification of all significant paths that sound can travel between each source and each receptor;
6) the identification of measures that can be used to modify the attenuation of sound as it travels along each path from source to receptor, and of the net cost and the effectiveness of each of these alternatives as a function of some appropriate scale of these alternatives; and
7) the use of both optimization and simulation techniques to aid in the economic evaluation of all combinations of noise abatement alternatives and to assist in defining the particular combinations of noise reduction measures that satisfy various noise standards and management objectives.

This systematic approach for defining and evaluating noise

management alternatives obviously involves a lot of work. However, in the case of gaseous, liquid and solid mass residuals, and thermal energy, the economic benefits (costs saved) achieved from the use of such an approach for defining and evaluating management alternatives have proven to be substantially more than the costs of data collection and analysis, model development, and computation. There is no reason to believe that the application of these techniques to the management of environmental noise will be any less beneficial.

Three of the steps mentioned above warrant further discussion, namely, the types of alternatives that are available for reducing noise intensities and durations, the prediction of noise attenuation in urban areas, and the structure of models that can be used to aid in the definition and evaluation of efficient noise management policies.

Noise Reduction Alternatives

Noise can be reduced at most source sites by improving on the design of the object that emits acoustical energy, by shielding the object, or by a combination of design changes and shielding. It may also be possible to relocate the noise source or change its time and/or duration of operation. In some cases just changing the method of operating certain kinds of equipment may substantially reduce the resulting noise level. Examples of these alternatives for controlling noise at its source include shielding and improving the muffling of combustion engines and air compressors, restricting the operation of noise producing equipment to certain hours of the day, reducing motor vehicle speed and the number of stops required along city streets, and rerouting both surface and air traffic. Of considerable impact in many areas would be simply the effective enforcement of noise abatement procedures already required by law.

The construction of solid barriers, the planting of belts of trees and other vegetation, and the modification of the material and shape of building exteriors are examples of measures that tend to modify the propagation of sound from its source to any receptor. The effectiveness of these control alternatives for increasing sound attenuation will depend on the location and characteristics of the noise sources as well as on the location of the receptor sites.

At the receptor sites, the insulation of buildings and the wearing of ear protectors are two obvious means of reducing the noise that is heard, but each has its disadvantages. Noise reduction measures such as these applied at receptor sites are usually considered only in special circumstances, but where the social cost of other noise reduction alternatives is excessive, these measures are available and should be considered.

Noise Propagation Prediction

The prediction of noise propagation and attenuation in an urban area is complicated by the existence of numerous obstructions of different shapes, sizes and materials. Nevertheless, it is possible to estimate the attenuation of noise from each noise source to each receptor site along each noise path by the application of a series of appropriate rules for the various situations encountered along each path. These situations include free field or unobstructed conditions, barriers of buildings or trees, surfaces that absorb or reflect sound to various extents, street canyons, intersections, and so on. Local meteorological conditions also affect the attenuation of noise. The noise that each identifiable source contributes to each receptor site can be derived from a knowledge of the intensity of the noise source and the total attenuation along each path from that source to the receptor site. Once this is known, it is relatively simple to predict the effect of any noise abatement measure, applied at any noise source, within the transmission medium, or at the receptor site, on the aggregate noise level heard at any particular receptor site.

Cost-Effective Noise Management Policies

While there are numerous noise management objectives that can and should be identified, perhaps one of the more significant objectives is cost minimization. This single objective can be used as a means of illustrating the application of quantitative modeling techniques to the definition and evaluation of noise management alternatives. Through the use of such models one can estimate how much reduction in noise intensity and/or duration is needed, and the manner in which it can be achieved, in order to satisfy a set of noise standards (incorporating duration, frequency and intensity) at each receptor site at a minimum total cost.

To obtain an estimate of the solution to this cost minimization problem involves the use of both mathematical optimization and computer simulation techniques. As sophisticated as these models and solution techniques may be, they still are considerable simplifications of reality, hence the emphasis in this paper on their use as guides rather than substitutes for the planning and implementation of effective noise management policy.

APPENDIX

Preliminary Noise Management Models

By defining some notation, two models for the preliminary screening of alternative environmental noise management policies can be structured. The first model to be developed will include only alternatives that reduce the frequency weighted dB(A) noise levels. This model will then be modified to include noise duration as well. Of interest in this example will be the identification of cost-effective policies, i.e., noise abatement alternatives that minimize the total cost of meeting standards specifying the maximum allowable noise levels at various receptor sites. Before constructing the management models, it may be useful to review in quantitative terms some of the more elementary properties of environmental noise and how it is measured.

A bel is simply the logarithm (to the base 10) of a ratio of two quantities Q and Q_o. One tenth of a bel is a decibel, hence the number of decibels associated with the quantity Q is

$$dB = 10 \log \frac{Q}{Q_o} \text{ re } Q_o. \tag{1}$$

The quantity, Q_o, is a predetermined reference quantity, and to be precise, decibels should be stated with reference to this quantity. One of three ratios used to measure the level of sound is the ratio of the square of the sound pressures, p^2/p_o^2, where the reference pressure, p_o, equals 20 micro newtons per square meter, $20\mu N/m^2$. Thus a sound pressure of $p = 20\mu N/m^2$ is equivalent to 0 dB re $20\mu N/m^2$, which at 1000 cycles per second is the threshold of hearing for normal young ears. Note that 0 dB is not the absence of sound; indeed some individuals can hear sound levels less than 0 dB.

Denoting dB_i as the sound pressure level one unit distance from a source at site i, Equation (1) can be rewritten

$$dB_i = 10 \log 10^{dB_i/10} \qquad (2)$$

since p^2/p_o^2 equals $10^{dB_i/10}$. In a free, unobstructed field, the square of the sound pressure decreases with the square of the distance. Letting d_{ij} be the number of unit distances from source site i to a receptor site j, the sound pressure level heard at site j due to source i equals

$$dB_{ij} = 10 \log \frac{10^{dB_i/10}}{d_{ij}^2} \qquad (3)$$

Of course in most environments free field conditions do not usually exist and a variety of other equations are required to estimate dB_{ij}.*

---

*(See, for example, Beranek, L. L., Noise and Vibration Control, McGraw Hill, 1971.)

Once each source site i and receptor site j have been identified together with the contributing sound pressure levels dB_{ij} for each pair of source and receptor sites, the total sound pressure levels, dB_j, at each receptor site j can be estimated by summing over all source sites i the square of the corresponding sound pressure ratios

$$dB_j = 10 \log \sum_i 10^{dB_{ij}/10} \qquad (4)$$

The application of the last two equations can be illustrated by considering two noise sources, each producing 80 dB as measured one unit distance away, hence $dB_i = 80$ for $i = 1,2$. From Equation (3), the sound pressure level at a receptor site 10 unit distances away is 60 dB if just one of the two sound sources is operating.

$$dB_{ij} = 10 \log 10^{80/10} - 10 \log 10^2 = 60$$

If both sources are heard simultaneously, Equation (4) can be used to calculate the total sound pressure level at site j.

$$dB_j = 10 \log [10^{60/10} + 10^{60/10}] = 10 \log 2(10^{60/10})$$

$$= 10 \log 2 + 10 \log 10^{60/10} = 63$$

(Note that this 3 dB increase when two equal sound levels are heard simultaneously will apply regardless of the original levels. For example, 0 dB + 0 dB = 3 dB).

When frequency is included in the calculations by using the dB(A) weighted decibel measure, Equation (4) may be regarded only as an approximation. Recall from the previous qualitative discussion that the dB(A) scale is simply a summation over various bands of frequencies of the corrected squares of the sound pressure ratios associated with those frequency bands. Defining the correction factor associated with the A weighting scale as C_f for each frequency band f,

$$dB(A) = 10 \log \sum_f 10^{(dB_f + C_f)/10} \qquad (5)$$

To accurately define $dB(A)_j$ it would be necessary to identify the contributing sound pressure levels for each frequency band f, dB_{ijf}, for use in Equation (6).

$$dB(A)_j = 10 \log \sum_f \sum_i 10^{(dB_{ijf} + C_f)/10} \qquad (6)$$

For the purposes of preliminary screening prior to a more detailed analysis of noise abatement policies, the following approximation of Equation (6) can be used:

$$dB(A) \cong 10 \log \sum_i 10^{dB(A)_{ij}/10} \qquad (7)$$

At this point one can begin to structure a management model for estimating the dB(A) reductions at each source and receptor site, and the scale of structures placed between various source and receptor sites, that minimize the total cost of meeting a set of environmental noise standards, $dB(A)_j^{max}$. These standards specify the maximum allowable dB(A) level at each receptor site j.

$$dB(A)_j \leq dB(A)_j^{max} \qquad (8)$$

Let the variables X_i^S and X_j^R be the dB(A) reductions at each source site i and receptor site j costing $C_i^S(X_i^S)$ and $C_j^R(X_j^R)$ respectively. The scale of any alternative h used to modify the noise attenuation between a source and receptor site can be defined by the variable S_h which costs $C_h(S_h)$ and results in an attenuation of $A^h_{ij}(S_h)$ between source i and receptor j.

Combining these costs into an objective function and incorporating the decision variables X^S_i, X^R_j and S_h into Equations (7) and (8) completes the nonlinear but separable cost-effective noise management model.

Model I:

$$\text{Minimize} \sum_i C^S_i(X^S_i) + \sum_j C^R_j(X^R_j) + \sum_h C_h(S_h) \qquad (9)$$

subject to environmental noise standards

$$10 \log \{ \sum_i 10^{[dB(A)_{ij} - X^S_i - \sum_h A^h_{ij}(S_h)]/10} \} - X^R_j \leq dB(A)^{max}_j \qquad \forall j \qquad (10)$$

and non-negativity conditions:

$$X^S_i,\ X^R_j,\ S_h \geq 0 \qquad \forall i,j,h \qquad (11)$$

The above model can be transformed for solution by separable programming algorithms which are generally available at most scientific computing facilities. Letting the parameter e_{ij} equal the known constant $10^{dB(A)_{ij}/10}$, Equation (10) can be expressed as

$$\sum_i e_{ij}\ 10^{-(X^S_i + X^R_j + \sum_h A^h_{ij}(S_h))/10} \leq 10^{dB(A)^{max}_j/10} \qquad \forall j \qquad (12)$$

If the variable P_{ij} equals $10^{-(X^S_i + X^R_j + \sum_h A^h_{ij}(S_h))/10}$, then the environmental noise standard, Equations (10) or (12), can be written,

$$\sum_i e_{ij}\ P_{ij} \leq 10^{dB(A)^{max}_j/10} \qquad \forall j \qquad (13)$$

Since $0 < P_{ij} \leq 1$, the logarithm of P_{ij} exists. Hence

$$-10 \log P_{ij} = X_i^S + X_j^R + \sum_h A_{ij}^h(S_h) \qquad \forall\ i,j \quad (14)$$

and by definition

$$P_{ij} = 10^{\log P_{ij}} \qquad \forall\ i,j \quad (15)$$

Each variable "$\log P_{ij}$" can be approximated by piecewise linear segments k having slopes L_k as illustrated below.

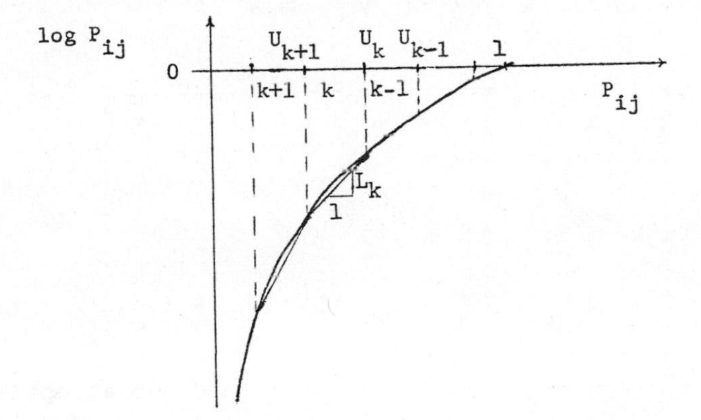

Figure 1. Piecewise Linear Approximation of Logarithmic Function.

Defining variables P_{ijk} such that

$$1 - \sum_k P_{ijk}/L_k = P_{ij} \qquad \forall\ i,j \quad (16)$$

$$0 \leq P_{ijk} \leq (U_k - U_{k+1})L_k \qquad \forall\ i,j,k \quad (17)$$

permits a linear approximation of Equation (15)

$$\log P_{ij} \cong -\sum_k P_{ijk} \qquad \forall\ i,j,k \quad (18)$$

Model I, in a form suitable for solution by separable programming algorithms, can now be summarized. The objective is to estimate the minimum cost combination of dB(A) reductions X_i^S and X_j^R at source sites i and receptor sites j, and the scale S_h of alternatives h for modifying the attenuation of sound between source and receptor sites, required to satisfy environmental noise standards expressed as a function of the dB(A) levels at each receptor site.

Model I:

$$\text{Minimize} \sum_i C_i^S(X_i^S) + \sum_j C_j^R(X_j^R) + \sum_h C_h(S_h) \quad (9)$$

subject to:

$$\sum_i e_{ij} \sum_k P_{ijk}/L_k \geq \sum_i e_{ij} - 10^{dB(A)_j^{max}/10} \quad \forall\, j \quad (13,16)$$

$$10 \sum_k P_{ijk} = X_i^S + X_j^R + \sum_h A_{ij}^h(S_h) \quad \forall\, i,j \quad (14,18)$$

$$0 \leq P_{ijk} \leq (U_k - U_{k+1})L_k \quad \forall\, i,j,k \quad (17)$$

Experience in solving this model indicates that global optimality (in the mathematical sense) can be achieved if the slopes L_k and segment bounds U_k are the same for each P_{ij}, if the attenuation functions $A_{ij}^h(S_h)$ are concave and if the cost functions in Equation (9) are convex, as they appear to be based on the limited data currently available.*

The above model can be modified to include the duration of noise as well as its intensity and frequency. In fact, both

---

*(National Bureau of Standards, The Economic Impact of Noise, Report NT1D300.14 of Environmental Protection Agency, December, 1971).

occupational and nonoccupational noise standards have been expressed in a manner that includes the duration of noise. The Walsh-Healy Act, which applies to individuals working in activities involved in interstate commerce, specifies the maximum time that any sound pressure level above 90 dB(A) can be heard in an 8-hour working day if the remaining levels are less than 90 dB(A). These times are listed in the following table. (Federal Register, Volume 34, No. 96, Part 2, pp. 7948-7949, May 20, 1969; also in Volume 36, No. 105, paragraph 1910.95, May 29, 1971).

Table 1. Walsh-Healy Noise Level Standards

Sound Level Index s	Sound Level $dB(A)_s$	Max. Permissible Duration Hours T_s
1	90	8
2	92	6
3	95	4
4	97	3
5	100	2
6	102	1.5
7	105	1.0
8	110	0.5
9	115	0.25

A similar set of noise levels and maximum permissible durations have been recommended for non-occupational environments for a 16-24 hour day.* Both sets of standards state that for all discrete sound levels $dB(A)_s$, the actual duration of those levels, t_s, must be such that the sum of the ratios t_s/T_s, where T_s is

*(Cohen, Articaglia and Jones, "Sciocusis-Hearing Loss from Non-Occupational Noise Exposure," Sound and Vibration, pp. 12-20, November 1970).

the maximum permissible duration, cannot exceed unity.

$$\sum_{s} \frac{t_s}{T_s} \leq 1 \qquad (19)$$

This form of environmental noise standard can be incorporated into Model I by first noting that for both the occupational and non-occupational noise standards just discussed, the product of the square of the pressure ratios, $10^{dB(A)_s/10}$, raised to the power 0.602, times the maximum permissible duration, T_s, is a constant. For the occupational standards this constant equals $T_s\,(10^{dB(A)_s/10})^{0.602} = 2.095 \times 10^6$. The corresponding constant for the recommended non-occupational standards equals 0.262×10^6.

Next, define $10^{dB(A)_{sj}/10}$ as the square of the sound pressure ratio s heard at a receptor site j. Let the constant α_j be the appropriate weighting exponent for that receptor site (e.g. α_j = 0.602). Multiplying each term of the numerator and denominator of Equation (19) by $(10^{dB(A)_{sj}/10})^{\alpha_j}$ does not change its value.

$$\sum_{s} \frac{t_s(10^{dB(A)_{sj}/10})^{\alpha_j}}{T_s(10^{dB(A)_{sj}/10})^{\alpha_j}} \leq 1 \qquad (20)$$

If each term in Equation (20) has a constant denominator, as it does for the occupational and non-occupational standards just described, that constant can be labeled E_j^{max}, the maximum allowable sound duration-pressure ratio permitted by site j. Multiplying each term in Equation (20) by E_j^{max} yields an environmental noise standard equivalent to the original standard expressed by Equation (19).

$$\sum_s t_s (10^{dB(A)_{sj}/10})^{\alpha_j} \leq E_j^{max} \qquad \forall j \qquad (21)$$

Finally, consider a set N_s of simultaneously operating noise sources i, each individually contributing a sound pressure level of $dB(A)_{ij}$ at site j for a duration of t_s hours. The total squared sound pressure ratio heard at each receptor site j during this period of t_s hours follows from portions of Equations (7), (12) and (13).

$$10^{dB(A)_{sj}/10} = \sum_{i \varepsilon N_s} 10^{[dB(A)_{ij} - X_i^S - X_j^R - \sum_h A_{ij}^h(S_h)]/10}$$

$$= \sum_{i \varepsilon N_s} e_{ij} P_{ij} \qquad \forall j,s \qquad (22)$$

Substituting Equation (22) into Equation (21) yields the environmental noise standard that, in addition to frequency and intensity, includes duration.

$$\sum_s [\sum_{i \varepsilon N_s} e_{ij} P_{ij}]^{\alpha_j} t_s \leq E_j^{max} \qquad \forall j \qquad (23)$$

This standard can replace the previous standard, Equation (13), that considered only intensity and frequency.

To render Equation (23) amenable to solution by separable programming algorithms requires some more logarithmic transformations. Let the variable

$$Q_{js} = \sum_{i \varepsilon N_s} e_{ij} P_{ij} \qquad \forall j,s \qquad (24)$$

Since $0 < Q_{js} \leq \sum_{i \varepsilon N_s} e_{ij}$, the logarithm of Q_{js} can be defined and approximated by a piecewise linear function

$$\log Q_{js} \cong \log \sum_{i \varepsilon N_s} e_{ij} - \sum_k Q_{jsk} \qquad \forall j,s \qquad (25)$$

where

$$\sum_{i \varepsilon N_s} e_{ij} - \sum_k Q_{jsk}/L'_k = Q_{js} \qquad \forall\, j,s \qquad (26)$$

$$0 \leq Q_{jsk} \leq (U'_k - U'_{k+1})L'_k \qquad \forall\, j,s,k \qquad (27)$$

Next let

$$R_{js} = Q_{js}^{\alpha_j}\, t_s \qquad \forall\, j,s \qquad (28)$$

since Q_{js}, α_j and $t_s > 0$, $0 < R_{js} \leq [\sum_{i \varepsilon N_s} e_{ij}]^{\alpha_j} t_s$ and its logarithm exists.

$$\log R_{js} = \log t_s + \alpha_j \log Q_{js} \qquad \forall\, j,s \qquad (29)$$

Again approximating $\log R_{js}$ by a piecewise linear function,

$$\log R_{js} \cong \log \{[\sum_{i \varepsilon N_s} e_{ij}]^{\alpha_j t_s}\} - \sum_k R_{jsk} \qquad \forall\, j,s \qquad (30)$$

where

$$t_s[\sum_{i \varepsilon N_s} e_{ij}]^{\alpha_j} - \sum_k R_{jsk}/L''_k = R_{js} \qquad \forall\, j,s \qquad (31)$$

$$0 \leq R_{jsk} \leq (U''_k - U''_{k+1})L''_k \qquad \forall\, j,k,s \qquad (32)$$

The environmental noise standard, Equation (23), can now be written in a linear form:

$$\sum_s R_{js} \leq E_j^{max} \qquad \forall\, j \qquad (33)$$

To summarize, the problem is to estimate the minimum cost combination of dB(A) reductions, X_i^S and X_j^R, at each noise source and receptor site, and the scale S_h of alternatives h that will modify the noise abatement between each source and receptor site, required to meet environment noise standards expressed as a function of dB(A) levels and durations.

Model II:

$$\text{Minimize: } \sum_i C_i^S(X_i^S) + \sum_j C_j^R(X_j^R) + \sum_h C_h(S_h)$$

subject to environmental noise standards:

$$\sum_s \sum_k R_{jsk}/L''_k \geq \sum_s t_s [\sum_{i\in N_s} e_{ij}]^{\alpha_j} - E_j^{max} \qquad \forall\, j \qquad (31,33)$$

and definitions relating the decision variables $\bar{X}$ (dB(A) reductions) and $\bar{S}$ (scale of alternatives for modifying noise transmission) to the variables R_{jsk}:

$$10 \sum_k P_{ijk} = X_i^S + X_j^R + \sum_h A_{ij}^h(S_h) \qquad \forall\, i,j \qquad (14,18)$$

$$0 \leq P_{ijk} \leq (U_k - U_{k+1})L_k \qquad \forall\, i,j,k \qquad (17)$$

$$\sum_k Q_{jsk}/L'_k = \sum_{i\in N_s} e_{ij} \sum_k P_{ijk}/L_k \qquad \forall\, j,s \qquad (16,24,26)$$

$$0 \leq Q_{jsk} \leq (U'_k - U'_{k+1})L'_k \qquad \forall\, j,s,k \qquad (27)$$

$$\alpha_j \sum_k Q_{jsk} = \sum_k R_{jsk} \qquad \forall\, j,s \qquad (25,29,30)$$

$$0 \leq R_{jsk} \leq (U''_k - U''_{k+1})L''_k \qquad \forall\, j,s,k \qquad (32)$$

Additional piecewise linear functions may be needed to define the costs in Equation (9) and the attenuation functions $A_{ij}^h(S_h)$ in Equation (14,18) if they are nonlinear. Also note that the upper bound U_1 for the variables P_{ij1}, Q_{js1} and R_{js1} are the maximum values of P_{ij}, Q_{js} and R_{js}, namely 1, $\sum_{i\in N_s} e_{ij}$ and $[\sum_{i\in N_s} e_{ij}]^{\alpha_j} t_s$ respectively, and hence the segment index k will begin at different values for these three sets of variables. As illustrated in Figure 4, the segments are ordered so that $U_k > U_{k+1}$ for each segment k.

It is clear that some noise control measures could modify the duration of the noise with or without a concurrent dB(A) reduction. In these situations the variables t_s would be unknowns. While the algebra involved to include changes in duration among the decision variables is relatively straightforward, it is not a practical exercise at this point for a variety of reasons. One reason is that limited data exist on the costs as well as on the effects of noise duration control measures. In addition, very often the duration of noise generated from each source is stochastic, and this important feature has not been included in either of the above models.

Some Numerical Examples

This section ends with some simple numerical examples to illustrate the application of these models. Consider an area having five identifiable noise sources and two receptor sites. The table below indicates the sets of sources, N_s, that operate simultaneously for the specified times, t_s, and also the A-weighted sound pressure levels, $dB(A)_{ij}$, that each source i contributes to each of two receptor sites j.

Given the information presented in Table 2, each parameter $e_{ij} = 10^{dB(A)_{ij}/10}$ can be computed. A variety of solutions were obtained using Model I in which the duration of each noise level was not considered. Also assumed for simplicity were linear cost functions and no alternatives for increasing noise abatement in the transmitting medium. Some of these solutions are included in

Table 2. Durations, Combinations and Levels of Noise Sources

Source No. i	Source Level $dB(A)_i$	Combinations of Sources Operating Source Set Index s					$dB(A)_{ij}$ Levels Receptor Site j	
		1	2	3	4	5	1	2
		Source Sets N_s						
1	80	*	*		*	*	40	60
2	85	*	*	*		*	73	60
3	72	*	*	*	*	*	40	60
4	68		*		*	*	41	60
5	110				*	*	97	60
Expected Duration, t_s hours:		3	2	1	1	1		

Table 3 as a means of illustrating how the model works in situations where the optimal solution is relatively obvious. The example problems and solutions shown in Table 3 are clearly not intended to be realistic with respect to costs or standards, but they do illustrate better than would more realistic and complex examples the types of solutions that can be expected and the reasonableness of the solutions that result from these simplified problems.

Without any noise abatement measures, the total sound pressure level at receptor sites 1 and 2 with all sources operating simultaneously, as they do for 1 hour, equals 97 and 67 dB(A) respectively. Noise standards in the first two problems constrain the sound level at receptor site 2 to be no greater than 55 dB(A), requiring a reduction of 12 dB(A) at each source site, or a similar reduction at the receptor site, or a combination of equivalent reductions at various source and receptor sites. In the first problem the total cost is less if each source is reduced

Table 3. Solutions From Model I

Problem No.	Standards $dB(A)_j^{max}$ Receptor Site j 1	2	Unit Cost and dB(A) Reduction (Solution) Source Sites i 1	2	3	4	5	Vectors Receptor Sites j 1	2
1	100	55	C: 1	1	1	1	1	0	6
			X: 12	12	12	12	12	0	0
2	100	55	C: 1	1	1	1	1	0	4
			X: 0	0	0	0	0	0	12
3	70	70	C: 1	1	1	1	1	3	0
			X: 0	6	0	0	30	0	0
4	35	70	C: 1	1	1	1	1	3	0
			X: 0	25	0	0	55	15	0
5	35	70	C: 1	1	1	1	1	6	0
			X: 12	45	12	13	69	0	0
6	35	35	C: 1	1	1	1	1	2	1
			X: 0	11	0	0	40	29	30

by 12 dB(A), and conversely in the second problem, it is cheaper to reduce the sound level heard at the receptor site by 12 dB(A). Problems 3, 4 and 5 include standards that are binding for receptor site 1, and the solutions of the model are again what would be expected given the relative costs of abatement at each source as compared to the abatement costs at the receptor site. Note in Problems 3 and 4 that as the maximum permissible environmental noise level is lowered so that more noise sources are affected, abatement at the receptor site becomes more attractive even though the relative abatement costs have not changed. Increasing the receptor costs, Problem 5, again shifts more of the burden of abatement to the source sites. These are other rather obvious solutions.

Problem 6 requires noise level reductions to meet the standards at both receptor sites, and in this case, as simple as it is, the solution is not as obvious as it is for the other five problems. Nevertheless as in the first five examples the combination of source and receptor reductions indicated in the solution of the model does in fact minimize the total cost of meeting the environmental noise standards.

Finally an example that includes the duration of noise can be illustrated using the information from Table 2 and Model II. The sets of noise sources, N_s, that operate simultaneously for the specified number of hours, t_s, are indicated by the columns of asterisks in Table 2. For example, $N_1 = \{i=1, 2, 3\}$ and $N_4 = \{i=1, 3, 4, 5\}$. Accepting the recommended non-occupational environmental noise standard of $E_j^{max} = 0.262 \times 10^6$ for receptor site j = 1 and a more strict standard of $E_j^{max} = 0.050 \times 10^6$ for receptor site j = 2, Table 4 summarizes the solutions obtained from Model II. In these example problems the exponent α_j equals 0.602 for each receptor site.

CONCLUSIONS

This introductory paper has presented only an outline of an approach to urban noise management. Considerable research is required before quantitative methods similar to those proposed and illustrated in this study can be used with confidence. Particularly necessary is the development of improved techniques for predicting and verifying noise attenuation in urban areas, for

Table 4. Example Solutions From Model II

Solution No.	Standards E_j^{max} Receptor Site j 1	2	Unit Cost and dB(A) Reduction (Solution) Source Sites i 1	2	3	4	5	Vectors Receptor Sites j 1	2
1	0.262×10^6	None	C: 1 X: 0	1 3	1 0	1 0	1 12	3 0	0 0
2	None	0.05×10^6	C: 1 X: 0	1 0	1 0	1 0	1 0	0 0	2 3
3	0.262×10^6	0.05×10^6	C: 1 X: 0	1 3	1 0	1 0	1 12	3 0	2 2

Solution No.	Receptor Site 1 Noise Level $dB(A)_1$	Receptor Site 1 Duration Hours	Receptor Site 2 Noise Level $dB(A)_2$	Receptor Site 2 Duration Hours
1	70 85	6 2	66 65 ≤ 64	3 1 4
2	73 97	6 2	64 63 ≤ 62	1 3 4
3	70 85	6 2	64 63 ≤ 62	3 1 4

estimating costs of noise abatement as a function of the reduction achieved, and for defining and evaluating noise control alternatives based on various economic and social criteria. Without this information and methodology, attempts to achieve desired environmental noise levels may not only be ineffective but needlessly costly.

ACKNOWLEDGMENTS

The research reported in this paper was supported in part by a grant from Resources for the Future, Inc. The writer is indebted also to Blair Bower, Walter Spofford and Jing-Yea Yang whose assistance and guidance made this study possible.

MODELS IN BIOLOGICAL WASTEWATER TREATMENT:

OBJECTIVES, RESULTS, APPLICABILITY.

L. Lijklema
Department of Chemical Engineering
Twente University of Technology
The Netherlands

Many attempts have been made to model different aspects of the processes in biological wastewater purification. Although not always clearly stated, generally the objective is to optimize the performance of a plant or a part of it. Sometimes the goal is a better understanding of the processes. Aspects modeled concentrate especially on the kinetics of waste removel and oxygen consumption.

There is a wide variety fiom simple empirical models to complex conceptual models based on metabolism, growth kinetics, transport mechanisms, mass balances etc. Whereas models based on empirical experience may be applied successfully within the limits set by past experience, an extension to new conditions generally will be impossible. On the other hand the conceptual models will have a wider applicability but the uncertainty of the relations involved and the difficulty of assessing the correct parameters limit the applicability and reliability.

A review of different models will be presented. The development, structure and application of such models will be exemplified by a specific example.

INTRODUCTION

In wastewater treatment plants the settleable solids of industrial or domestic origin are separated from the water by sedimentation in a primary clarification step. Colloidal and soluble contaminants however can only be removed after conversion into a form suitable for phase separation. In biological treatment processes micro organisms are instrumental in this conversion by adsorption and metabolic transformations. Both aerobic and anaerobic processes are employed. The discussion here will be limited to the activated sludge process and its modifications, an aerobic process in which the active micro organisms grow in flocs suspended in the wastewater. In the activated sludge process the micro organisms and the wastewater are mixed in the aeration tank (Figure 1). The air provides the turbulence preventing the sludge flocs from settling and introduces the oxygen required by the metabolic reactions. In the secondary clarification the sludge is separated from the effluent and recycled. Excess sludge can be wasted.

In mathematical models of the activated sludge process biological, physical, chemical and technological aspects are combined into a meaningful set of relations between measurable quantities of the process. The complexity of a model varies with its intended use and the information available. In the following section a short review of the most important aspects is presented.

PROCESS ANALYSIS

Population Dynamics

The activated sludge process is the largest industrial application of the continuous culture technique (Pipes [54]). Probably it is also the least understood technical application in microbiology due to its complex ecology.

The interactions between the physical and chemical nature of the wastewater and the organisms in the activated sludge flocs are not well known and the existing food chains received only modest attention.

The lack of quantitative information on growth and death rates of mixed cultures in complex media, the complexity of selection mechanisms in a daily and seasonally varying environment and the unknown

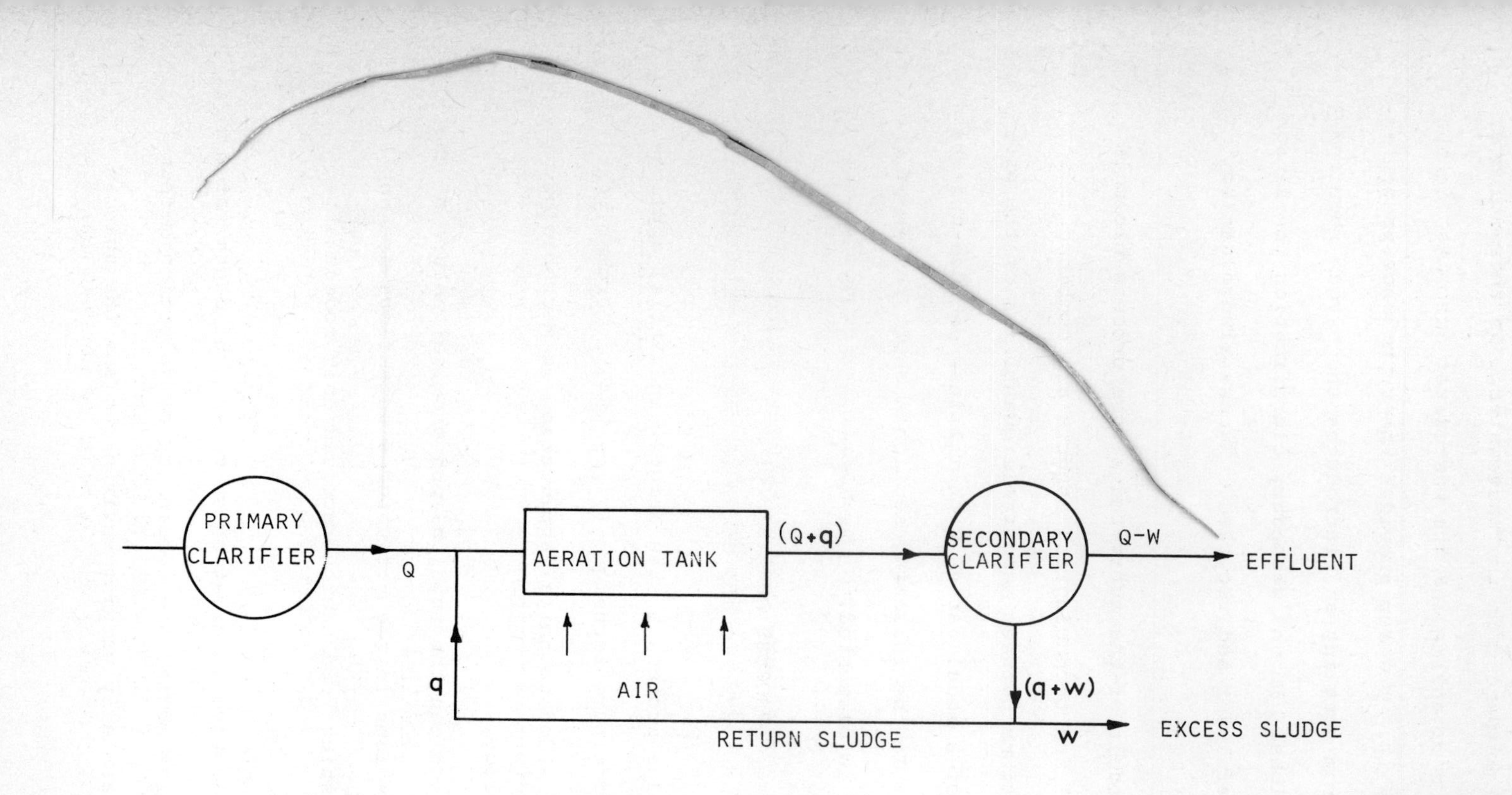
PRIMARY CLARIFIER
Q
q
AERATION TANK
AIR
(Q+q)
SECONDARY CLARIFIER
Q-W
EFFLUENT
(q+w)
w
EXCESS SLUDGE
RETURN SLUDGE

FIGURE 1.

effects of environmental shocks such as the anaerobic periods during settling and recycling of the sludge are a few of the problems inhibiting detailed modeling of the community structure (Fredrickson et. al.[21], Bella [5]).

A basic simplification used in most models is the neglect of differences in organisms and of age or size distribution within a population (Fredrickson [21]). Generally the active mass is represented by only one parameter; the biomass, representing all bacterial species including the protozoa preying on bacteria. In some recent studies the role of protozoa has been investigated (Curds [11], Curds et. al. [12, 13], Sudo et. al.[65]) and attempts are made to imply their role in models (Curds [11], Lijklema [37]). Especially, their role in the removal of single cells is important for a high effluent quality (Curds et. al. [12], Pipes [54]).

The shift in protozoan population with decreasing organic loading of the sludge is a well-known phenomenon (Hawkes [27]).

Models neglecting the variations in microbial composition etc. can still be successful in predicting purification performance or in other respects. In a "steady state" continuous mixed reactor a highly variable community structure caused only slight variations in effluent quality (Cassel et. al.[10]).

In the comparatively simple situation of nitrification with a single substrate (ammonia) and one organism (Nitrosomonas) reasonable successful models were developed (Lijklema [37], Downing et. al.[16]). The fact that the substrate concentration can be measured readily is of considerable interest in the testing of these models.

A poorly understood phenomenon is the relation between the

community structure of the activated sludge and its settleability. The concentration of the biomass in the aeration tank depends on the degree of compacting attained in the secondary clarifier. Several sophisticated models with elaborate kinetic formulations containing terms for biomass concentrations completely ignore this aspect. A special situation is the nuisance caused by filamentous growth resulting in "bulking" sludge (Pipes [55], Pasveer [53], Forster [20]).

Energetics

The thermodynamic aspects of microbial growth received increasing interest in recent years (McCarty [43], Servizi [59], Stouthamer [63], Senez [58], Lehninger [35]). Though the processes occurring during growth are irreversible, the classical concepts of thermodynamics proved to be useful in relating growth and substrate utilisation. In aerobic microbial processes the major energy producing process is the oxidative phosphorylation. Through the ATP-ADP system the catabolic process is coupled with the synthesis process. The efficiency of the oxidative phosphorylation and of the coupling determine the fraction of the substrate that will be converted into cell material (*Ibid.* cf. also Forrest [19]). The yield can be reduced by adverse temperatures (Stouthamer [63], Senez [58]) or nutrient deficiencies (McCarty [42], Stouthamer [63]).

Since the energy produced in oxidative phosphorylation is proportional to the number of electrons passing through the cytochrome system towards oxygen as the ultimate electron acceptor, there must be a close relation between the amount of cell material produced and the oxygen equivalence (COD) of the substrate (Servizi [59], McCarty

[42], Pipes [56], Hadjipetron [26], Hetling et. al.[28]). The fraction of the substrate oxidised thus depends on the oxidation level of the substrate (Servizi [59], Lijklema [39]).

It has been observed frequently in the activated sludge process and other microbial cultures that the yield decreases with decreasing organic loading of the biomass or with increasing mean cells residence time (Hetling et. al.[28], Hopwood et. al.[29], Schulze [57], Sherrard et. al.[60], Dawes et. al.[14]).

Concurrently the oxygen consumption increases. This can be due to the activity of protozoa (Hopwood et. al.[29]), lysis of cells (Hadipetron [26], Bucksteeg et. al.[9], Gandy et. al.[22]), maintenance energy (Hetling et. al.[28], Schulze [57]) or combinations. In models generally one term in introduced to represent this decay of biomass as a first order process.

Though for several substrates including sewage for normal operating conditions the yield as a function of loading is reasonably well established, extrapolations to more extreme conditions must be considered carefully. In particular transient conditions and ecological changes may cause variations in yield (Gandy et. al.[23], Monod [41], Storen et. al.[62]).

Reactors

Batch reactors are seldom used in wastewater treatment. The chemical engineering literature has provided ample information on mixing and residence time distributions in various types continuous flow reactors (Levenspiel [36]). Most aeration tanks in waste water treatment plants have mixing conditions intermediate between plug flow and complete mixing and can probably be described by the axial dispersion model slightly better than by the completely mixed tanks in series model (Murphy et. al.[46]). The axial dispersion depends on the applied air flow rate causing the spiral flow condition (Murphy et. al. [47]) but is not sensitive to the water flow rate.

Treatment efficiency models based on Monod kinetics or modifications have been formulated for steady state situations for different reactors or combinations of reactors (Andrews [1], Lawrence et. al. [34], Erickson et. al.[17], Fan et. al. [18], Naito et. al.[48]). The mathematical analysis predicts optimal substrate removal when longitudinal dispersion is minimal; this is especially true for high feed concentrations (Fan et. al.[18]).

The equalisation of toxic or otherwise interfering substances is not included in such studies. Consideration of inhibition (Andrews [1]) or repression might influence the selection of the optimal reactor type.

The distribution of sewage along the length of the aeration tank practised in the step loading modification reduces the total tank volume needed (Erickson et. al.[17]). There is a strong interrelation with the kinetic model used in calculation of the volume since each addition of influent raises the substrate concentration and lowers the concentration of micro organisms.

The bottleneck in activated sludge models is the role of the sedimentation tank. In models concerned with finding optimal design criteria, e.g. mean cells residence time (Lawrence et. al.[34]), or the optimal volume ratio for aeration and sedimentation tanks (Naito et. al.[48]), an important variable always is the unpredictable sludge concentration factor of the sedimentation tank. The degree of compacting itself will be a function of another important variable--the sludge recycle ratio q/Q (Figure 1). This variable can be manipulated for process control.

Experimental values for sedimentation tanks measured and modeled in a particular situation (Fan *et. al.* [18], Naito *et. al.* [48]) cannot be used under different conditions.

Kinetics

The kinetic aspects of bacterial growth have been studied extensively and excellent reviews are available (Dean *et. al.*[15]). Though in activated sludge the rate of removal of substrate (waste) is of primary interest, the kinetic models proposed by micro-biologists are models for the rate of bacterial growth (Bengston [6]). The coupling of rates of removal with growth through a constant yield coefficient is questionable, especially under changing conditions (Gandy *et. al.* [22], Mateles [41], Storen *et. al.*[62]).

All kinetic formulations applied assume growth proportional to the bacterial concentration, generally expressed as biomass per volume.

This implicitly assumes constant viability of the cells and neglects the character of storage products as a solid form of substrate rather than active biomass capable of reproduction (Andrews [3]). For the dependency of growth on the substrate concentration various formulations were proposed and applied. All these expressions account for a decline in growth rate with decreasing substrate concentration.

The most common model is that proposed by Monod [44]; an empirical extension of the conceptual Michaelis-Menten equation for enzyme substrate interaction:

$$\frac{dX}{dt} = \frac{\mu_m S}{K_s + S} X \qquad (1)$$

$$\frac{dX}{dt} = -Y \frac{dS}{dt} \qquad (2)$$

where X = bacterial concentration (mass per volume)

S = substrate concentration (mass per volume)

μ_m = maximum specific growth rate (time^{-1})

K_s = saturation constant (mass per volume)

Y = yield coefficient

In activated sludge it is assumed that the carbon source is growth limiting (Wuhrmann [71]). Combination of equations (1) and (2) gives the rate of substrate disappearance:

$$\frac{dS}{dt} = -\frac{\mu_m S X}{Y(K_s + S)} \qquad (3)$$

The saturation constant K_s can be evaluated as the substrate concentration which gives half the maximum growth rate. Its physical meaning may be a saturation phenomenon (Dean <u>et. al.</u>[15]) or a mass transfer limitation (Baillod [4], Mueller [45], Gulevich [25]) or both. Under certain conditions oxygen diffusion limitation might occur (Mueller [45], Kalinske <u>et. al.</u>[30]).

It has been observed that the saturation constant of a heterogenous culture increases with decreasing detention time due to population selection (Ghosh [24]). Selection induced by variation of the dilution rate or substrate concentration for various microbial systems demonstrates the variability of the kinetic parameters in heterogenous systems with operating conditions (Stumm-Zollinger [64]).

The long cell residence times as applied usually in the activated sludge process (Lawrence _et. al._ [34]) cause an appreciable auto-oxidation or organism decay (Andrews [3]). Equation (1) is usually modified to include a death rate:

$$\frac{dX}{dt} = \frac{\mu_m S}{K_s + S} X - bX \qquad (4)$$

The first order decay rate constant b is not a true constant since it decreases with organism age (Lijklema [39], Andrews [3]). A death rate inversely proportional to the substrate concentration has been suggested (Westberg [68]).

Since the decay products are in part available as substrate equation (3) should be modified to include redissolution (Naito _et. al._ [48], Westberg [68]);

$$\frac{dS}{dt} = \frac{\mu_m S X}{Y(k_s + S)} + b \beta X \qquad (5)$$

in which β is a proportionality constant.

Other modifications of the Monod equation are the use of an inhibition function (Andrews [1, 2]) and a proportionality of K_s with the substrate concentration in the influent (Koga _et. al._ [33]).

An alternative expression is the exponential relation (Teissier [66], Schulze [57]):

$$\mu = \mu_m(1 - \exp(-cS)) \qquad (6)$$

in which μ is the actual specific growth rate and c a system constant.

A selection of one particular model and assessment of the kinetic constants on the basis of laboratory experiments with artificial substrates should always be considered with suspicion when the results are extrapolated to different conditions (Pipes [54], Fredrickson [21], Ghosh [24], Stumm-Zollinger [64]). For the individual components in a mixed substrate zero order removal rates have been observed (Tischler [67], Wuhrmann [71]) whereas the overall rate approached Monod type kinetics (Tischler [67]).

An important phenomenon in the activated sludge process is the rapid initial removal of substrate. Whether the underlying mechanism is absorption on the surface of the activated sludge (Naito _et. al._ [49]) or assimilation (Bengston [6]) is not clear.

Introduction of an activity parameter defining the fraction of active sites on the sludge (Naito _et. al._ [49]) seems promising although the number of variables and parameters to be evaluated is higher. Expecially for the description and optimisation of the contact stabilisation and the step aeration modifications of the activated sludge process such models might prove useful.

Optimisation of design for substrate removal (Erickson _et. al._ [17], Fan _et. al._ [18], Naito _et. al._ [48]) should be coupled with the optimisation of aeration (Brouzes [8]).

The dependence of substrate removal on the interrelation of reactor type and kinetics stresses the need to assess the mechanism of substrate removal.

The theoretical advantage of plug flow reactors over well-mixed reactors would disappear when individual components of the sewage are removed at zero order rates. Other factors which can influence the design of the aeration tank are the occurrence of inhibition, repression, diphasic growth and the variability in flow and loading of real plants.

In recent years theoretical dynamic kinetic models for the activated sludge process were developed (Lijklema [37, 40], Storen et. al. [62], Klei et. al. [33], Koga et. al. [32], Westberg [69], Brett et. al. [7], Ott et. al. [51]). These studies partly focus on the feasibility of process control through the sludge recycle ratio (Westberg [69], Brett et. al.[7], Ott et. al.[51]). In this important area experimental data are scarce.

OBJECTIVES AND APPLICATION

Although not always clearly stated, generally the objective of mathematical models developed for the activated sludge process falls in one or more of the following categories:

- optimisation of design of wastewater treatment plants
- optimisation of operation; process control
- improved understanding of the process; research tool.

Models based on empirical relations from practical experience with existing plants are often applied successfully by engineers to find design criteria for treatment plants. In the past the tendency was to optimise the individual unit operations, but there is an increasing awareness of the need to consider the system as a whole (Parkin et. al.[52], Smith [61]). Programs for computerised design of complete plants are available nowadays.

Whereas empirical models can be useful within the limits set by past experience, extension of the application to new conditions generally is not justified. An example: a simple method to reduce the cost of aeration is to measure the hourly variation in oxygen consumption is various parts of the aeration tank during dry weather conditions. By means of mechanical or electronic devices the air flow can be regulated to supply the required oxygen, thus reducing excess aeration.

Such an empirical system will fail during storm conditions or during holidays, strikes, other seasons etc.

Conceptual models on the other hand, based on mass balances, kinetics etc. may have a wider applicability, but the uncertainty of the relations involved and the difficulty of assessing the correct parameters limit the reliability. Such models are generally developed by scientists, but are seldom applied by engineers in existing plants.

There is certainly a permanent need for exchange of ideas and experience between these two groups.

A MODEL FOR pH IN ACTIVATED SLUDGE

A practical problem with the treatment of strongly alkaline wastewater from the textile industry serves as an example to demonstrate the development of a mathematical model. The question raised was: to what extent is neutralisation or equalisation of this alkaline wastewater necessary prior to discharge in the municipal sewerage system connected with a biological wastewater treatment plant. The adverse effects of a high pH on the organisms in the activated sludge are well-known (Kiefer and Meisel [31]). The steps involved in developing this model can be summarised as:

- Formulation of the variables governing pH in activated sludge; assessment of their relations.
- inventory of the factors causing changes in these variables; assessment of their magnitude.
- model building; testing of the model.
- experimental verification.
- application in practical situation.

pH Formulation

The pH determining compounds in sewage are the weak acids and their salts, chiefly the carbonic acid system, and to a lesser extent the ionic strength and the temperature of the water. The weak acid carbon dioxide can be present in three forms: free carbon dioxide, bicarbonate and carbonate (CO_2, HCO_3^-, CO_3^{--}). The distribution over these species depends on the relative quantity of hydroxyl ions (OH^-) present. For a normal temperature and ionic strength the relation with pH is plotted in Figure 2.

Factors affecting pH

Since the pH is governed by the ratio of total carbon dioxide and total hydroxyl ions present $\frac{CO_2 + HCO_3^- + CO_3^{--}}{HCO_3^- + 2CO_3^{--} + OH^-}$,

the pH can be affected by production or removal of alkalinity and carbon dioxide.

Several physical and chemical processes are of importance during passage of the sewage through the aeration tank. The quantity of

FIGURE 2
Relation between ratio of total quantities of CO_2 and OH^- in liquid and pH. Ionic strength 0.027, Temp. $20^{o}C$

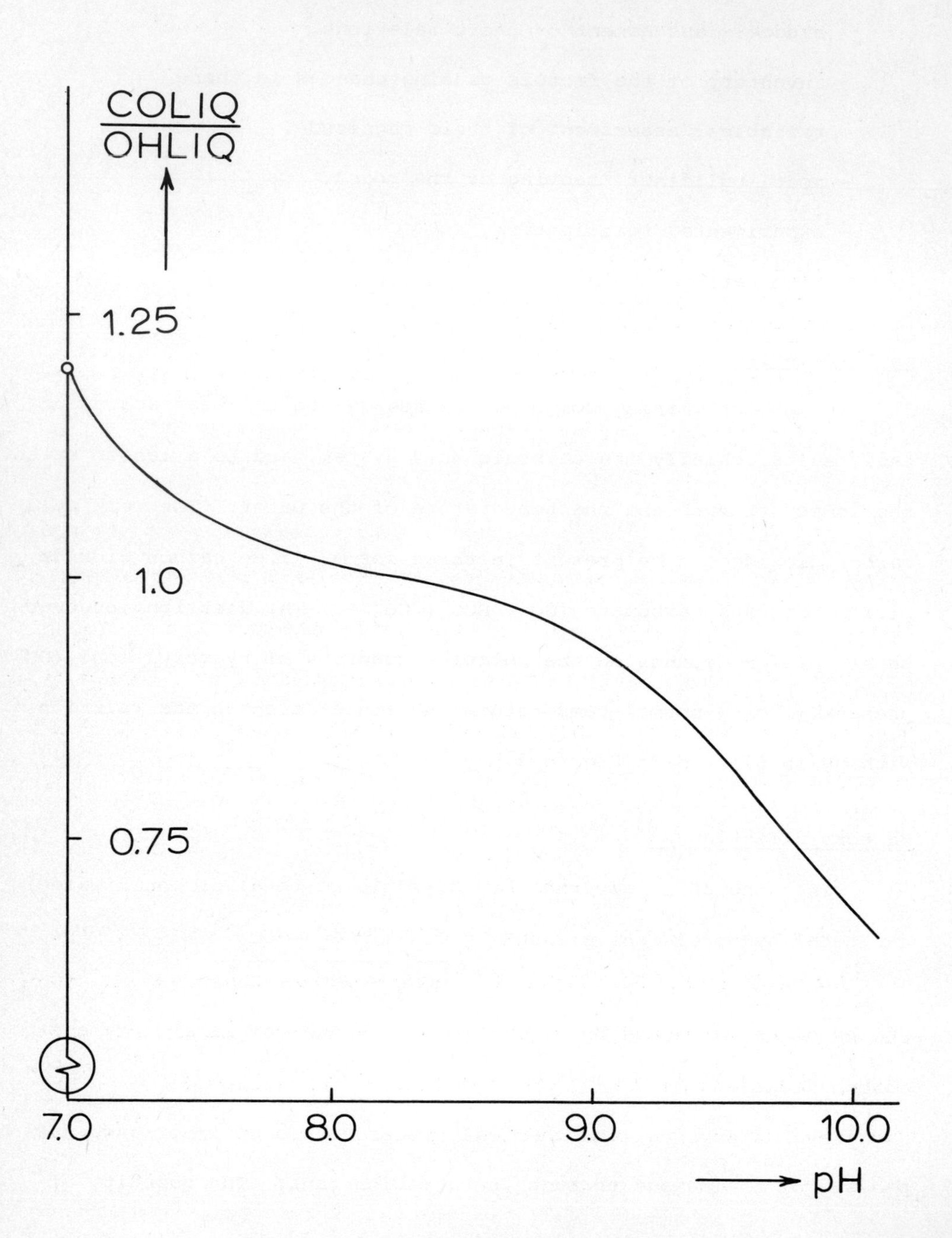

carbon dioxide is affected by:

- input and output of the reactor
- desorption with air
- production by biochemical oxidation of organic wastes
- hydrolysis of urea.

The total alkalinity is affected by:

- input and output of the reactor
- nitrification
- hydrolysis of urea
- buffering in the sludge.

The reactor input and output are calculated from data on flow and composition of influent and reactor content.

The hydrolysis of urea:

$$CO(NH_2)_2 + 2H_2O = 2NH_4^+ + CO_3^{--}$$

is often completed before the sewage enters the aeration tank. The desorption of carbon dioxide was found to be proportional to the air flow and the concentration of free carbon dioxide. Hence the rate of removal is a function of the system variables. Theoretical and experimental investigations resulted in the conclusion that enhancement of the desorption rate by the chemical reaction

$$H^+ + HCO_3^+ = H_2CO_3 = H_2O + CO_2$$

in the boundary layer of the air bubbles is of little importance.

The carbon dioxide produced by oxidation of organic waste material is a complex function of the COD reduction, the ratio between oxidation and synthesis in cell metabolism, the oxidation level of the substrate and its physical nature (Lijklema [39]). The COD reduction is a function of pH.

COD = Chemical Oxygen Demand, is the waste or waste concentration expressed as oxygen equivalents.

The carbon dioxide production as a function of COD removal can be expressed as the product:

$$\frac{O_2 \text{ consumed}}{\text{COD removed}} \cdot \frac{CO_2 \text{ produced}}{O_2 \text{ consumed}} = \alpha$$

The first term describes the efficiency of the use of substrate energy by the bacteria. The second term is the well-known respiration quotient R.Q., and depends mainly on the oxidation level of the substrate. From a mass balance these terms can be evaluated:

$$C_pH_qO_r + \frac{1}{7} f\, p\, NH_3 + \left(\frac{9}{14} f\, p - \frac{q}{2}\right) H_2O + \left(2p + \frac{q}{2} - r - \frac{29}{14} f\, p\right) O =$$

$$= \frac{1}{7} f\, p\, C_7H_{12}NO_4 + (1 - f)\, p\, CO_2.$$

A fraction f of substrate $C_pH_qO_r$ is converted into cell material of the average composition $C_7H_{12}NO_4$. An efficient use of the oxidation energy will result in a high value for f. Microbiologists often express this efficiency in terms of Y/O; the yield in grams cell material per half mole of oxygen respired. According to the mass balance

$$Y/O = \frac{24.85\, f\, p}{\left(2p + \frac{q}{2} - r - \frac{29}{14} f\, p\right)}$$

the COD of the substrate - the oxygen required for complete oxidation - is $(2p + \frac{q}{2} - r)$ half moles of oxygen. For a given Y/O value it is possible to express f p in terms of COD and hence to calculate the oxygen consumption as a percentage of the COD removed. For the Y/O values in the range reported in the literature (Stouthamer [63]) and higher, some results are summarized in Table I. For substrates of known composition the values of f and the R.Q. can also be evaluated.

A high $\frac{COD}{p}$ ratio (low oxidation level) results in a high value f and a low R.Q.

TABLE I

Y/O	18			22			25		
$\frac{O_2 \text{ consumed}}{\text{COD removed}}$	0.40			0.35			0.325		
	f	RQ	α	f	RQ	α	f	RQ	α
$C_{89}H_{167}N_{5.8}O_{65}$	0.62	0.98	0.39	0.67	0.99	0.35	0.69	0.99	0.32
$C_{89}H_{171}N_{3.7}O_{20}$	0.78	0.42	0.17	0.84	0.35	0.12	0.88	0.29	0.10
$C_7H_{12}NO_4$	0.62	0.97	0.39	0.67	0.97	0.34	0.70	0.97	0.32
$C_6H_{12}O_2$	0.435	1.88	0.75	0.47	2.02	0.71	0.49	2.10	0.68

It can be seen from the table that α is mainly determined by the substrate composition and to a lesser extent by the Y/O value. The substrates in Table I represent respectively the gross composition of the soluble fraction of domestic sewage (Oldham [50]), the suspended material in preclarified sewage, the average substrate composition, caproic acid and aspartic acid. If at high cell residence times a greater fraction of the COD is oxidised than calculated in Table I, it can be assumed that the extra oxidation is due to oxidation of cell material. Hence the overall value for R.Q. can be calculated from the weighted contributions of substrate respiration and endogenous respiration. Calculated and experimental values for caproic acid and aspartic acid agree well (Figure 3).

For sewage a calculation is not possible along these lines. The soluble fraction will be metabolised comparatively fast; at longer detention times an increasing fraction of the less soluble material will be oxidised, thus reducing the overall R.Q. Qualitatively the experimental curve for the sewage under study (Figure 3) can thus be understood. The increasing oxidation at low loads thus is partly counterbalanced by a decreasing R.Q. and the value of α is a fairly constant 0.40 in the loading range where most activated sludge plants operate.

The effect of the boichemical oxidation of ammonia on the alkalinity is a reduction proportional to the amount of ammonia oxidised:

$$NH_4^+ + 2O_2 = NO_3^- + 2H^+ + H_2O$$

The nitrification is mainly a function of temperature and cells residence time and can be predicted fairly accurately (Lijklema [37], Downing and Knowles [16]). The theoretical reduction in alkalinity with 2 mval per mole ammonia oxidised has been verified by experiments (Lijklema [38]).

The buffering of hydroxyl ions by the proteins in the biomass has two important aspects for a pH model: first the buffering capacity, determining the amount of hydroxyl ions necessary to raise the pH of a unit quantity of sludge to a certain value, and secondly the rate of transport of hydroxyl ions from liquid to sludge or vice versa as a function of concentration differences. The empirical relation between pH and buffered hydroxyl ions was measured by titration of elutriated sludge with alkali, allowing the system to equilibrate. The "sludge function" relating pH and buffered alkali is approximately linear. An analysis of titration curves with variable speed of alkali addition (Figure 4) showed that the transport of hydroxyl ions could be described reasonably by assuming that 28% of the sludge equilibrates instantaneously with the liquid while 72% attains equilibrium

FIGURE 3

The respiration quotient as a function of percentage substrate oxidation.

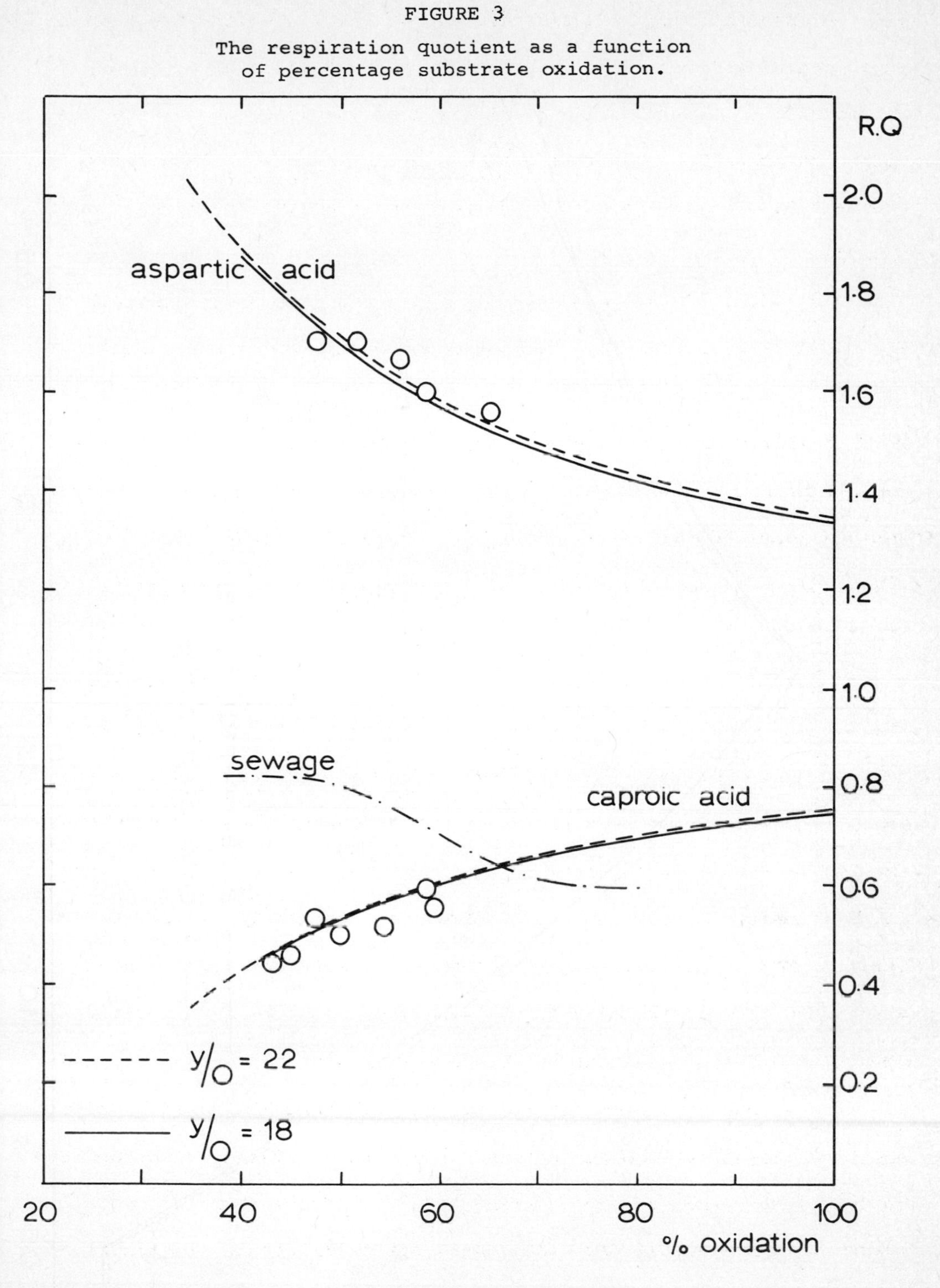

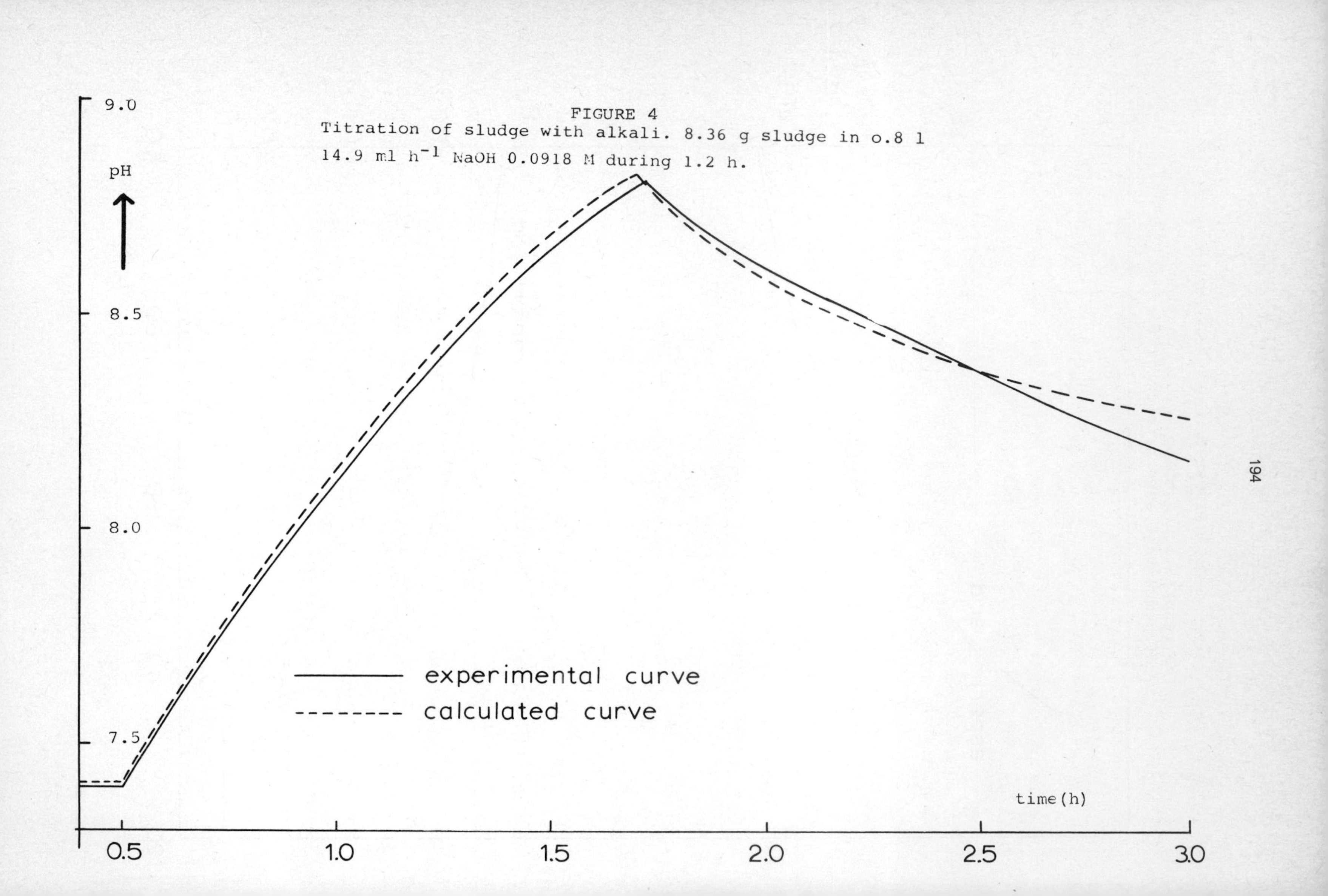

FIGURE 4
Titration of sludge with alkali. 8.36 g sludge in o.8 l
14.9 ml h^{-1} NaOH 0.0918 M during 1.2 h.

through a first order diffusion process. The proportionality coefficient for this diffusion was also evaluated by trial and error.

This empirical sub-model was integrated into a dynamic simulation model describing the pH in activated sludge as a function of the variable inputs (flow, COD and alkalinity of the sewage etc.) Figure 5 depicts the block diagram.

Model building, testing and experimental verification

Details of the model as presented in Figure 5 are described elsewhere (Lijklema [40]). An interesting feature is the use of an implicit function block, a standard iteration procedure to break an algebraic loop in computations. The physical background is the fact that the quantity of alkali buffered in the equilibrated sludge depends on the pH, but otherwise the pH depends on the distribution of alkali between sludge and liquid.

An important task in testing a model is to evaluate the sensitivity to the individual parameters. Those parameters causing an appreciable shift in the calculated results by a small change in their magnitude, deserve special attention. When experimental and calculated data do not fit very well a re-examination of these parameters and if necessary a closer investigation is in order. This is the background of the rather elaborate study of the carbon dioxide production in this research.

Figure 6 shows a result of the experimental verification of the model. It is a dynamic simulation of the effect of a shock load of alkali in the influent. It is desirable to extend the range of experimental verifications of a model as far as possible towards the

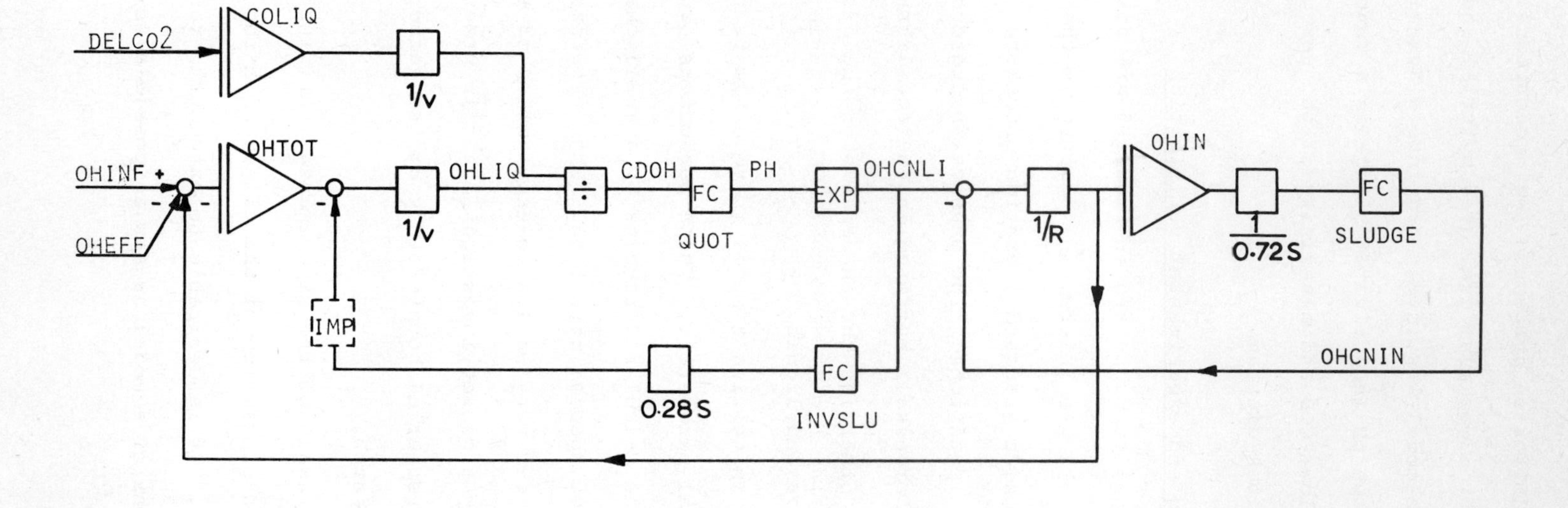

FIGURE 5. BLOCK DIAGRAM - SIMULATION OF PH IN ACTIVATED SLUDGE.

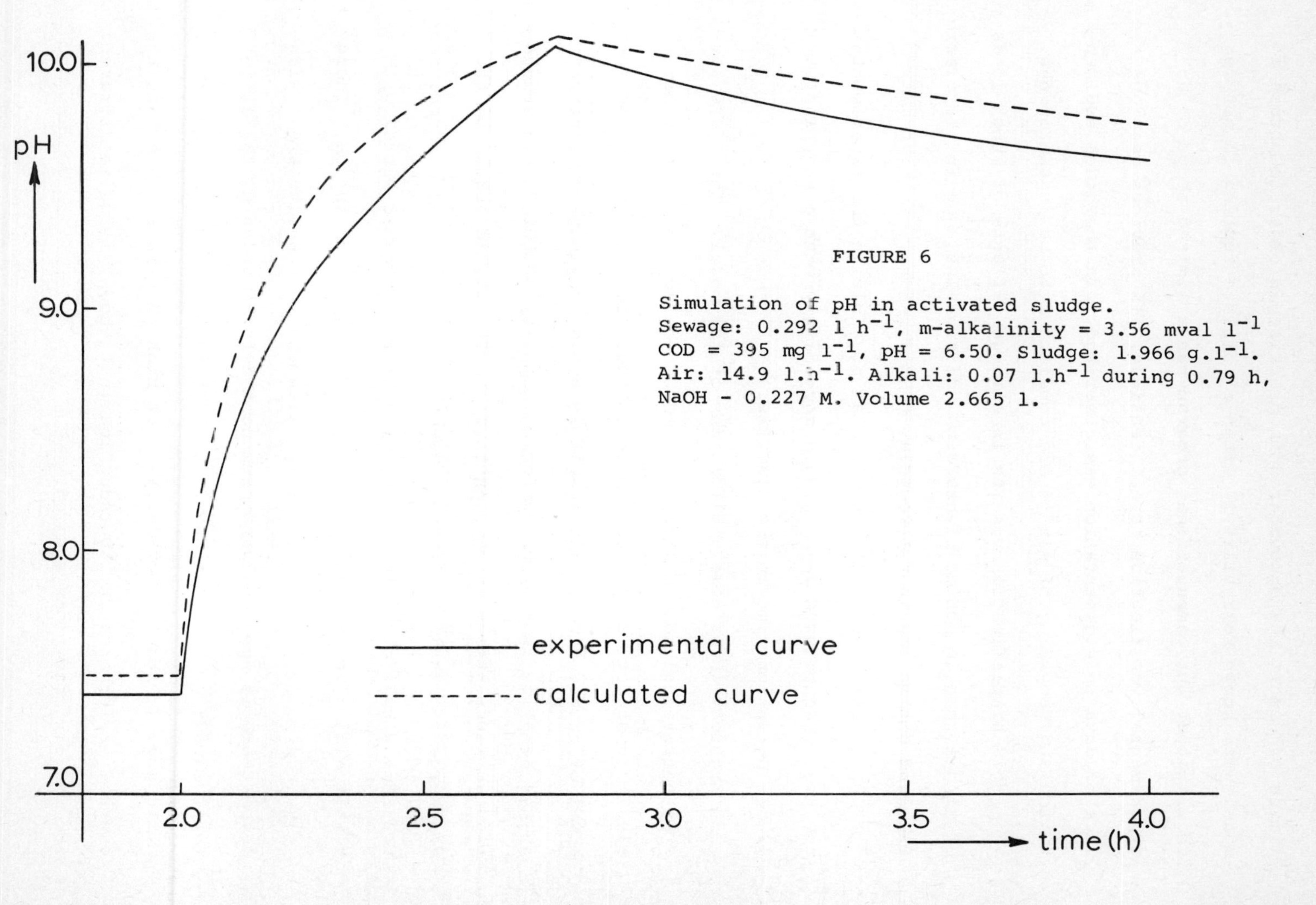

FIGURE 6

Simulation of pH in activated sludge.
Sewage: 0.292 l h^{-1}, m-alkalinity = 3.56 mval l^{-1}
COD = 395 mg l^{-1}, pH = 6.50. Sludge: 1.966 g.l^{-1}.
Air: 14.9 l.h^{-1}. Alkali: 0.07 l.h^{-1} during 0.79 h,
NaOH - 0.227 M. Volume 2.665 l.

limits of conditions encountered in practise in order to obtain a model with known reliability. Often a complete testing is not feasible; either the computational program has a limited scope or an experimental verification is not feasible. In previous sections the lack of data on sedimentation was described. In the model sedimentation and recycling of sludge are considered to be instantaneous.

This simplification results in stronger effects of shock loads of alkali than in actual plants with additional sludge in the clarifier and a time lag in the recycle.

To approach the model as close as possible in the experiments the test reactor used was designed accordingly with a very short sludge detention time in the clarifier.

Figure 7 shows some applications of the model. There are several operational possibilities to reduce the adverse effects of alkali in an activated sludge plant. Reduction of the rate of removal of excess sludge will result in a higher sludge concentration in the aeration tank. The buffering capacity of this extra sludge reduces fluctuations in pH. Another possibility is spreading the discharge of alkali in time. Since the carbon dioxide produced neutralises alkali the simultaneous discharge of organic wastes with alkali will also reduce the effect of the alkali on the pH. These effects are analysed in Figure 7. Although in the situations simulated there is apparently no permanent deterioration of the activated sludge process, it should be kept in mind that the purification efficiency is reduced considerably at high pH. This means that the overall performance in COD reduction (not shown) is quite different for the six conditions studied.

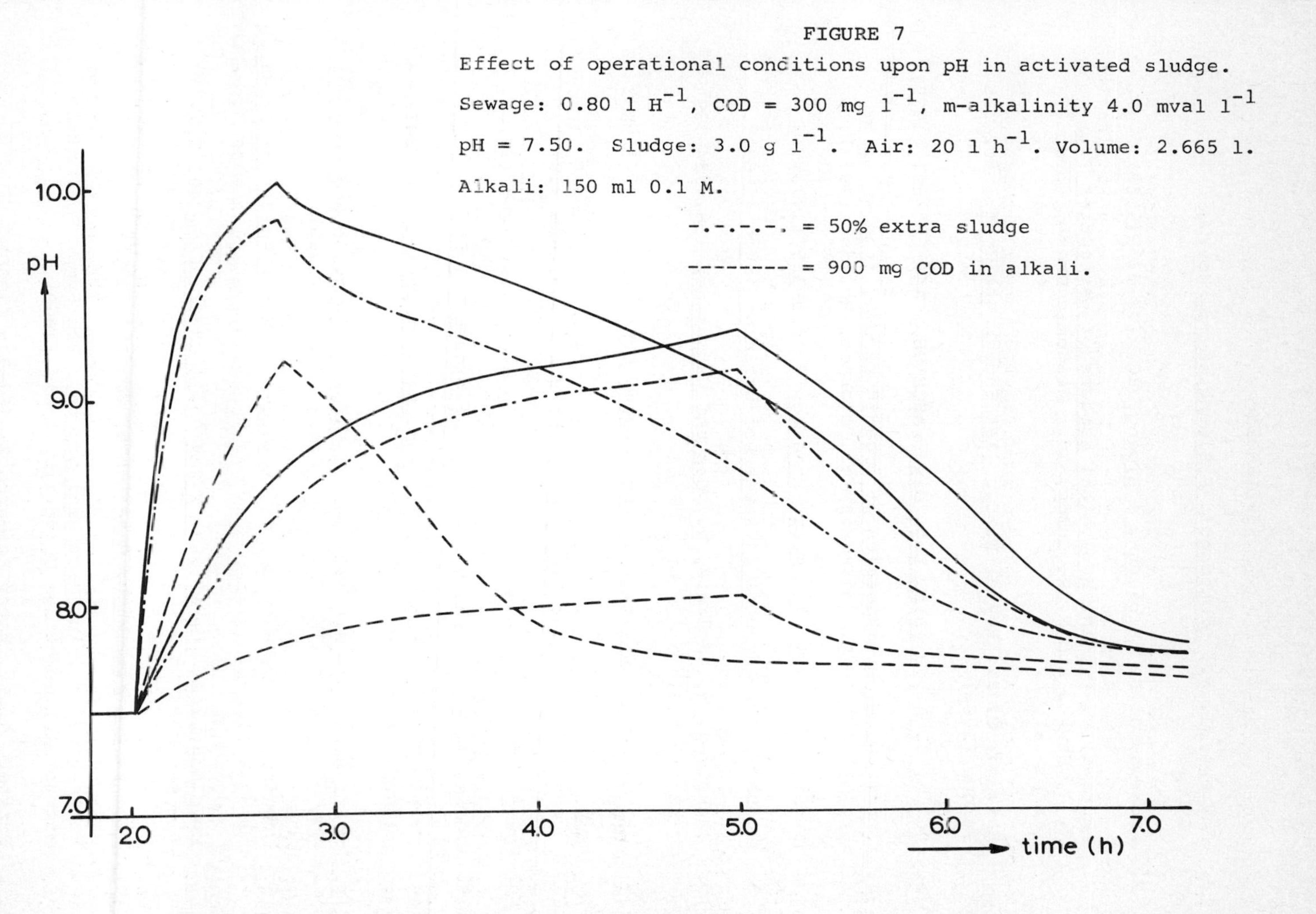

FIGURE 7

Effect of operational conditions upon pH in activated sludge. Sewage: 0.80 l H^{-1}, COD = 300 mg l^{-1}, m-alkalinity 4.0 mval l^{-1} pH = 7.50. Sludge: 3.0 g l^{-1}. Air: 20 l h^{-1}. Volume: 2.665 l. Alkali: 150 ml 0.1 M.

REFERENCES

1. Andrews, J. F., Biotechnol. & Bioeng. Symp. No. 2, 5-33 (1971).

2. Andrews, J. F., Biotechn. & Bioeng. 10, 707-723 (1968).

3. Andrews, J. F., in: Ecological Aspects of Wastewater Treatment H. A. Hawkes and C. R. Curds, Ed., Academic Press, London (in press).

4. Baillod, C. R., Boyle, W. C., 21st Ind. Waste Conf., 302-326 Purdue (1968).

5. Bella, D. A., J. San. Eng. Div., ASCE 98, SA4, 685-691 (1972).

6. Bengtson, H. H., Kinetics of organic removal from water by bacteria, Thesis. University of Colorado, Boulder (1971).

7. Brett, R. W. et. al., Water Research 7, 525-535 (1973).

8. Brouzes, P., Advances in Water Pollution Research (1969).

9. Bucksteeg, W., Wolters, N., Gas und Wasserfach 108, 734-739 (1967).

10. Cassel, E. A. et. al., J. Water Poll. Contr. Fed. 38, 1398-1409 (1966).

11. Curds, C. R., Water Research 5, 793-812; 1049-1066 (1971).

12. Curds, C. R.,Cockburn, A., Water Research 4, 237-249 (1970).

13. Curds, C. R., Cockburn, A., J. Gen. Microbiol. 66, 95-108 (1971).

14. Dawes, E. A., Ribbons, D. W., Ann. Rev. Microbiol. 16, 241-264 (1962).

15. Dean, A. C. R., Hinshelwood, C., Growth, function and regulation in bacterial cells, Clarendon Press, Oxford (1966).

16. Downing, A. L., Knowles, G., Advances in Water Pollution Research 117-142 (1966).

17. Erickson,L. E. et. al., J. Water Poll. Contr. Fed. 40, 717-732 (1968).

18. Fan, L. et. al., Water Research 4, 271-284 (1970).

19. Forrest, W. W., in: Microbial growth, 19th Symp. Soc. gen. Microbiol., Cambridge University Press (1969).

20. Forster, C. F., Water Research 2, 767-776 (1968).

21. Fredrickson, A. G. et. al., Advances in applied microbiology 13, 419-465 (1970).

22. Gandy, A. F. et. al., J. Water Poll. Contr. Fed. 42, 165-179 (1970).

23. Gandy, A. F. et. al., Biotechn. & Bioeng. 9, 387-411 (1967).

24. Ghosh, S., Poland, F. G., Water Research 6, 99-115 (1972).

25. Gulevich, W., The role of diffusion in biological waste treatment, Thesis. John Hopkins Univ.,Baltimore (1967).

26. Hadjipetron, L. P., Relation between energy production and aerobic growth of bacteria, Thesis. Utrecht, The Netherlands (1965).

27. Hawkes, H. A., The ecology of wastewater treatment, Pergamon.

28. Hetling, L. J. et. al., 19th Ind. Waste Confer. Eng. Ext. Series 117, 687-715, Purdue (1964).

29. Hopwood, A. P., Downing, A. L., J. Proc. Inst. Sew. Purif., 435-452 (1965).

30. Kalinske, A. A., J. Water Poll. Control Fed. 43, 73-80 (1971).

31. Keefer, C. E., Meisel, J., Sewage Ind. Wastes 23, 982-991 (1951).

32. Klei, H. E. et. al., CEP Symp. Series 65, no. 97, 232-237 (1969).

33. Koga, S., Humphrey, A. E., Biotechn. & Bioeng. 9, 375-386 (1967).

34. Lawrence, A. W., McCarty, P. L., J. San Eng. Div. ASCE 96, SA3, 757-778 (1970).

35. Lehninger, A. L., Bioenergetics, Benjamin Inc., New York (1965).

36. Levenspiel, O., Chemical Reaction Engineering, J. Wiley Inc. (1962).

37. Lijklema, L., Env. Science and Technology, 428-433 (1973).

38. Lijklema, L., Water Research 3, 913-930 (1969).

39. Lijklema, L., Water Research 5, 123-142 (1971).

40. Lijklema, L., Water Research 6, 165-182 (1972).

41. Mateles, R. I., Chian, S. K., Env. Science and Technology 3, 569-574 (1969).

42. McCarty, P. L., Advances in Water Pollution Research, 169-185 (1964).

43. McCarty, P. L., in: Organic Compounds in Aquatic Environments, M. Dekker, New York (1971).

44. Monod, J., Ann. Rev. Microbiol. 3, 371-394 (1949).

45. Mueller, J. A., Oxygen diffusion through a pure culture floc of Zoogloea Ramigera, Thesis. Univ. of Wisconsin (1967).

46. Murphy, K. L., Tympany, P. L., J. San. Eng. Div. ASCE 93, SA5, 1-15 (1967).

47. Murphy, K. L., Boyko, B. I., J. San. Eng. Div. ASCE 96, SA2, 211-221 (1970)

48. Naito, M. et. al., Water Research 3, 433-443 (1969).

49. Naito, M. et. al., Biotechn. & Bioeng. 11, 731-743 (1969).

50. Oldham, G. F., Effluent Water Treat. J. 8, 234-242 (1968).

51. Ott, C. R., Bogan, R. H., J. San. Eng. Div. ASCE 97, SA1, 1-17 (1971).

52. Parkin, G. F., Dague, R. R., J. San. Eng. Div. ASCE 98, SA6, 833-852 (1972).

53. Pasveer, A., Advances in Water Pollution Research (1969).

54. Pipes, W. O., Advances in applied microbiology 8, 77-103 (1966).

55. Pipes, W. O., Advances in applied microbiology 9, 185-234 (1967).

56. Pipes, W. O., Advances in Water Pollution Research, 437-440 (1962).

57. Schulze, K. L., Water and Sewage Works 111, 526-538 (1964).

58. Senez, J. C., Bact. Rev. 26, 95-107 (1962).

59. Servizi, J. A., Bogan, R. H., J. San. Eng. Div. ASCE 89, SA3, 17-40 (1963).

60. Sherrard, J. H., Schroeder, E. D., Water Research 6, 1039-1049 (1972).

61. Smith, R., Report No. WP 20-9, US Depart. of the Interior, Fed. Water Pollut. Control Admin., Cincinnati (1968).

62. Storen, F. F., Gandy, A. F., Env. Science and Technology 3, 143-149 (1969).

63. Stouthamer, A. H., in: Methods in Microbiology: Vol. I., Academic Press, New York (1969).

64. Stumm-Zollinger, E., in: Organic Compounds in Aquatic Environments, M. Dekker, New York (1971).

65. Sudo, R., Aiba, S., Water Research 7, 615-621 (1973).

66. Teissier, G., Ann. Physiol. Physicochim. Biol. 12, 527-586 (1936).

67. Tischler, L. F., Eckenfelder, W. W., Advances in Water Pollution Research, 361-383 (1969).

68. Westberg, N., Water Research 1, 795-804 (1967).

69. Westberg, N., Water Research 3, 613-621 (1969).

70. Wuhrmann, K., Advances in applied microbiology 6, 119-151 (1964).

71. Wuhrmann, K., in: Biological Waste Treatment, W. W. Eckenfelder J. M. McCabe, Ed., Pergamon Press (1963).

BIOLOGICAL FILTER DESIGN OPTIMIZATION

Harold B. Gotaas
The Technological Institute
Northwestern University
Evanston, Illinois

and

William S. Galler
North Carolina State University
Raleigh, North Carolina

An optimization method is used to determine the most economical design for biological trickling filters based on four different models, the Eckenfelder, the Galler-Gotaas, the National Research Council and the Upper Mississippi and Great Lakes Board. Filter depth has the greatest influence on the most economical design with the Galler-Gotaas model, some influence in the Eckenfelder and NRC models and no influence in the Upper Mississippi and Great Lakes model.

The most economical design is achieved when the depth is maximum to permit adequate filter ventilation, after which recirculation is economical up to about four recirculations. The radius of the filter must then be increased to provide greater volume to achieve the required BOD removal. Design curves are shown for the four models. Because more variables are included in the Galler-Gotaas model, it appears to provide the best fit to treatment observations for different types of wastes.

If pumping to the filters is necessary regardless of the filter depth, deeper filters with forced ventilation are more economical than shallower filters.

INTRODUCTION

Biological filtration is a well established process for the removal of dissolved and non-settleable organic matter from liquid wastes. Studies of the process have led to proposals of a variety of mathematical models* to describe the effect of the physical variables. Different designs, almost infinite in number, can be developed to remove a given amount of biochemical oxygen demand (BOD) from a wastewater in accordance with a model when different values of depth, area (or radius for a circular filter), recirculation, hydraulic application rate, and temperature are assumed. The most effective design should 1) accomplish the desired treatment and 2) should be optimal as to the initial cost and operation for the particular situation.

It has been shown (Lynn <u>et. al.</u> [12]) that the overall cost of a treatment plant could be minimized by choosing the most economical unit operations and processes for a flow path through the plant necessary to obtain the required BOD reduction. Hence, the problem is to determine the optimum design with respect to the unit operation or process. This study presents an optimization analysis using four different models as constraints for optimal design of biological filters. The development of the method for optimizing the design is the same as presented by Galler and Gotaas [5]. Only the overall basic concepts will be presented here in relationship to the four models, since the details of application of the optimization method to one model are in the above reference, and the procedure in relationship to other filter models is similar.

---

*Velz [18], National Research Council [17], Upper Mississippi, River Board [2], Rankin [14], Howland [8], Schulze [16], Eckenfelder [3], and Galler and Gotaas [4].

BIOLOGICAL FILTER MODELS

As noted, there are a variety of mathematical models for biological filters. The four models used for these optimization design studies cover the range of models as to differences in approach.

Velz [18] postulated that the BOD removed per unit depth of filter was proportional to the BOD remaining

$$\frac{dL}{dD} = -kL \tag{1}$$

which integrates to

$$\frac{L_e}{L_o} = e^{-kD} \tag{2}$$

where L_e is the BOD remaining in the effluent, L_o is the BOD applied in the influent, and D is the depth of the filter.

Howland [8] later proposed that the BOD removal was a function of time of contact giving the model

$$\frac{L_e}{L_o} = e^{-kt} \tag{3}$$

$$\text{and} \quad t = k' \frac{D}{Q^n} \tag{4}$$

in which n and k' are constants and Q is the hydraulic rate through the filter. Hence the remaining BOD in the effluent is obtained by substituting Eq. 4 in Eq. 3 yielding

$$\frac{L_e}{L_o} = e^{-k'(\frac{D}{Q^n})} \tag{5}$$

in which k' is a proportionality constant and n was determined to be 2/3. Howland also introduced the effect of temperature on the reaction rate, k, in the BOD reduction equation by introducing the factor ;

$$\theta = 1.035^{(T-20)} \tag{6}$$

in which T is the temperature in degrees Centigrade Eq. 5 becomes

$$\frac{L_e}{L_o} = e^{-k\,\theta\,(\frac{D}{Q^n})} \tag{7}$$

Eckenfelder [3] modified the Howland [8] and Shultz [16] models considering a decreasing amount of BOD removal per unit of depth with increasing filter depth resulting in the model:

$$\frac{L_e}{L_o} = \frac{1}{1 + k\,\frac{D^{(1-m)}}{Q^n}} \tag{8}$$

in which the constants were evaluated as m = 0.33, n = 0.5 and k = 2.85. The temperature effect is considered the same as for the Howland model. The derivation considered no recirculation, and recirculation was considered as a dilution effect on the influent BOD L_i, as described by

$$L_o = \frac{L_i + RL_e}{1 + R} \tag{9}$$

where R is the recirculation ratio.

The National Research Council (NRC) model [17] developed empirically, based on data from plants at military installations during World War II:

$$E = \frac{1}{1 + K(\frac{W}{VF})^{0.5}} \tag{10}$$

in which $E = \frac{L_o - L_e}{L_o}$ is the fraction of BOD removed, W is the loading in pounds of 5-day 20° C BOD, V is the volume of the filter in cu. ft., F is the recirculation factor, and K is 0.0085. The recirculation factor is

$$F = \frac{\frac{r}{i} + 1}{[1 + (1-p)\frac{r}{i}]^2} \qquad (11)$$

in which r is the recirculation rate, i is the influent rate, and p is a weighting factor of approximately 0.9 which reflects the availability of organic material for biological change. In this model recirculation is equivalent to additional passage of sewage through the filter.

The Upper Mississippi River - Great Lakes Board standards [2] specifying a maximum filter loading of 110 pounds of 5-day 20° BOD per 1,000 cu. ft. of filter, a filter depth of not less than 5 ft. nor more than 7 ft., a filter influent BOD concentration not to exceed three times the effluent BOD concentration and a hydraulic loading of not less than 10 mgad nor more than 30 mgad, was formulated into a mathematical model by Rankin [14]:

$$E = \frac{\frac{r}{i} + 1}{1.5 + \frac{r}{i}} \qquad (12)$$

Galler and Gotaas [4] formulated another empirical model for trickling filter efficiency from a multiple regression analysis of data from pilot plants and existing plants with effluent BOD, L_e, as the dependent variable:

$$L_e = \frac{0.464\, L_o^{1.19} \left(\frac{i+r}{i}\right)^{0.28} \left(\frac{Q}{A}\right)^{0.13}}{(1+D)^{0.67}\, T^{0.15}} \qquad (13)$$

which $L_o = \frac{iL_i + rL_e}{i + r}$, r = recirculation rate in mgd, Q = mgd, A = area in acres, D = depth in feet and T is the temperature in degrees centigrade.

It is noted that Moore, et. al.[13] found that the increase in plant efficiency resulting from recirculation was greatest when sludge from the bottom of the secondary tank was returned to the influent to the primary settling tank. However, the increase observed when only the effluent from the secondary tank after settling was recirculated was significantly less. Lumb and Eastwood [11] found that dilution of the influent to the filter with rain water did not improve the efficiency of BOD removal, but when effluent from a "Simplex" aeration unit was used for pseudo-recirculation the efficiency was improved. This indicates that perhaps the most beneficial effect of recirculation is seeding the sewage with organisms predominating in the filter environment.

BIOLOGICAL FILTER DESIGN

Mathematical Basis for Optimization

Problems which can be stated in linear form, i.e. as a linear functional that expresses the objective of the problem and a set of simultaneous linear inequalities that express the limitations or constraining effects can be optimized by the use of linear programming. This may be stated mathematically as

$$\text{Minimize} \quad \sum_{k=1}^{n} C_k x_k$$

$$\text{Subject to} \quad \sum_{k=1}^{n} h_{jk} x_k \leq b_j (j = 1, \ldots, m) \quad x_k \geq 0 \tag{14}$$

in which C_k = cost coefficients, x_k = activity levels, h_{jk} = input - output coefficients, and b_j = stipulations. In the case of the biological filter, neither the functional or the constraints are all in linear form. However the nonlinear functional and the nonlinear constraint equations can be linearized using the cutting plane method of J. E. Kelley, Jr. [9]. After linearization, a solution can be obtained by the dual method developed by C. E. Lemke [10].

Detailed studies of filter design minimizing the cost of construction and operation using the Galler-Gotaas model have been made (Galler [6]) for (a) the trickling filter, (b) the trickling filter with settling basins, (c) the deep trickling filter with forced ventilization and (d) the maximization of BOD removal from a liquid waste subject to limited capital. Only the formulation of the trickling filter alone, using the four model equations described, will be presented here.

Assumptions

The following assumptions and conditions relating to construction and operation of plants which are significant to the cost optimization and hold the problem within reasonable limits while maintaining the accuracy of the solution were made:

1. The labor costs for operating a sewage treatment plant as well as the maintenance costs are relatively constant for a given plant capacity;
2. the cost of recirculation piping is a relatively small fraction of the total costs and hence, if the conditions for determining the piping (flow rate, length, etc.) can be found from the optimization of the filter and plant layout, the pipe sizing can be optimized separately with almost no effect on the overall optimization;

3. the cost of excavation is not included, but can be readily included;
4. the filters are circular; however, the formulation can be adapted to other shapes.

FIGURE 1 TRICKLING FILTER

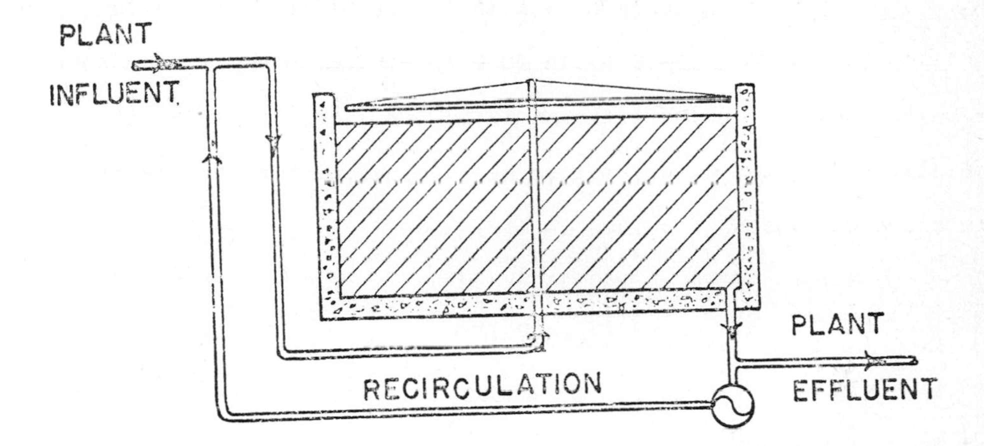

The Trickling Filter

The cost of a trickling filter, Fig. 1, is a function of the radius, a, in feet, the depth, D, in feet, and the recirculation rate, r, in mgd and based on the cost derivations in Appendix I, can be stated as:

$$\text{Cost} = \Big(\frac{2c_1\pi W_a D}{27} + \frac{2c_1\pi W F a}{27} + \frac{c_1 \pi W^2 D}{27}$$

$$+ c_2 \pi a^2 + \frac{c_3 \pi a^2 D}{27} + 2c_4 a + \frac{r}{c_6 + c_7 r}\Big) \lambda$$

$$+ \frac{c_5 i\,(8.34)\,(365)\,(r)\,(D + 1)}{(2.65\,p)} \qquad (15)$$

This cost equation has been used to optimize the design with the Galler-Gotaas, the Eckenfelder, the NRC and the Upper Mississippi - Great Lakes Board models. Since the method is the same only one application illustration will be shown, that of the Galler-Gotaas model, because this model includes more variables which affect filter performance. The model describing BOD reduction may be written (Galler and Gotaas [41]):

$$\ln L_e = \ln K_1 + 1.19 \ln (iL_i + rL_e) - 0.78 \ln (i+r) - 0.67 \ln (D + 1) - 0.25 \ln a \qquad (16)$$

in which L_e = BOD in the effluent in mg per liter, L_i = BOD in the influent, in mg per liter, before being mixed with recirculation, i = the plant influent volume in mgd, and

$$K_1 = \frac{0.47(\frac{43,560}{\pi})^{0.13}}{(i)^{0.28}\ (T)^{0.15}} \qquad (17)$$

in which T = temperature in ^{o}C; 43,560 converts square feet to acres; and 0.47 is a constant for Eq. 16.

A practical maximum for recirculation has been found to be about four times the influent, hence the constraint for recirculation is taken as

$$0 \leq r \leq 4i\ . \qquad (18)$$

The limits of depth are established at a minimum of three feet and a maximum of ten feet. The maximum depth of ten feet is to avoid the possibility of improper ventilation, but the problem is also studied for a maximum filter depth of twenty feet and pumping plant influent to the filter. The depth constraint for a ten foot depth filter is

$$3 \leq D \leq 10 \qquad (19)$$

The radius of the filter is allowed to vary from ten feet to one hundred feet to conform to the size of distributor equipment available, giving the constraint

$$10 \leq a \leq 100 \tag{20}$$

If a maximum hydraulic through-put rate, Q_{max}, is established, the lower boundary condition for the radius becomes

$$\left[\frac{43,560\ (i + r)}{\pi\ Q_{max}}\right]^{0.5} \leq a. \tag{21}$$

Hence, the boundary conditions for the radius are

$$\text{Min}\left[10, \left(\frac{43,560\ (i + r)}{\pi\ Q_{max}}\right)^{0.5}\right] \leq a \leq 100 \tag{22}$$

Neither the functional nor the constraint set is linear, but a logarithmic transformation can be used to put the problem in almost completely linear form. If

$$x_1 = \ln\ (i\ L_i + r\ L_e) \tag{23a}$$

$$x_2 = \ln\ \frac{r + i}{i} \tag{23b}$$

$$x_3 = \ln\ \frac{D + 1}{4} \tag{23c}$$

and

$$x_4 = \ln\ \frac{a}{10} \tag{23d}$$

the following programming problem can be formulated:

Minimize $\quad C_1 e^{x_2} + C_2 e^{x_3} + C_3 e^{x_4} + C_4 e^{x_2 + x_3} + C_5 e^{x_3 + x_4}$

$$+ C_6 e^{x_3 + 2x_4} + C_7 e^{2x_4} + \frac{i\ (e^{x_2} - 1)}{C_6 + iC_7\ (e^{x_2} - 1)} \tag{24}$$

Subject to

$$\ln L_e - \ln K_1' = 1.19x_1 - 0.78x_2 - 0.67x_3 - 0.25x_4 \quad (25a)$$

$$\ln 5 \geq x_2 \quad (25b)$$

$$\ln \frac{11}{4} \geq x_3 \quad (25c)$$

$$\ln 10 \geq x_4 \quad (25d)$$

$$1/2 \ln K_2 \leq -1/2x_2 + x_4 \quad (25e)$$

$$0 = x_1 - \ln[iL_i + iL_e (e^{x_2} - 1)] \quad (25f)$$

in which

$$K_1' = K_1 \left(\frac{1}{10^{0.25}\ 4^{0.67}\ (i)^{0.78}}\right) \quad (26a)$$

and

$$K_2 = \frac{435.6i}{\pi\ Q_{max}} \quad (26b)$$

Although the functional and one constraint are still non linear, this can be solved using the cutting plane method of Kelley [9]. An example of the solution for the Galler-Gotaas is shown in Appendix II.

The same procedure is used for optimizing the Upper Mississippi River-Great Lakes Board, the NRC and the Eckenfelder models. Equation (23a) for x_1 will change for model, but the equations (23b) (23c) and (23d) for x_2, x_3 and x_4 will be the same.

Data Used

The maximum hydraulic rate through the filter is 30 mgad. All models were analyzed for a temperature of 18°C and the Galler-Gotaas model was studied at temperatures of 10°C and 25°C (Galler and Gotaas [4]). The temperature is contained in the Galler-Gotaas model. The effect of temperature in the other models can be determined by changing the 18°C values by $E_t = E_{18} \times 1.035^{(T-18)}$. The BOD influent and effluent ranges studies are as shown in the results. The filter wall thickness used is 0.75 feet and a freeboard of the wall above the filter rock is considered to be one foot.

The cost conditions used in the optimization study vary as follows: cost of wall - \$60 to \$200 per cubic yard; cost of floor - \$2 to \$6 per square ft.; cost of filter media - \$5 to \$20 per cubic yd.; cost of distribution system - \$25 to \$100 per ft. of diameter; power costs - \$0.01 to \$0.05 per kwh; interest rate - 0% to 10%; life of filter - 20 yrs.; and life of pump and blower - 10 yr. and 20 yr. respectively.

RESULTS

Optimized Design Models

Figures 2 and 3 present the results of the design optimization for the Galler-Gotaas model for effluent BOD concentrations related to influent BOD concentrations of 100 mg per liter to 400 mg per liter, filter depth 3 ft. to 10 ft. and influent temperatures of 18°C and 25°C respectively. The 25°C curves are shown since this may be more nearly the sewage influent temperature in tropical regions than 18°C. Figures 4 and 5 present the model for the depth range 10 ft. to a maximum of 20 ft., and influent temperatures of 18°C and 25°C respectively. These are an extension of the depth curves of Figures 2 and 3 and do not include the filter radius which can be computed or determined from the radius curves of Figures 2 and 3.

Figure 6 presents similar results for the NRC model for a maximum depth of 10 ft. and an influent temperature of 18°C. The NRC model also yields a family of curves, since efficiency is a function

FIG 2. TRICKLING FILTER DESIGN CHARACTERISTIC AT 30 M.G.A.D. AND 18°C. MAX. DEPTH 10FT. — PLANT INFLUENT 1M.G.D.

GALLER - GOTAAS MODEL

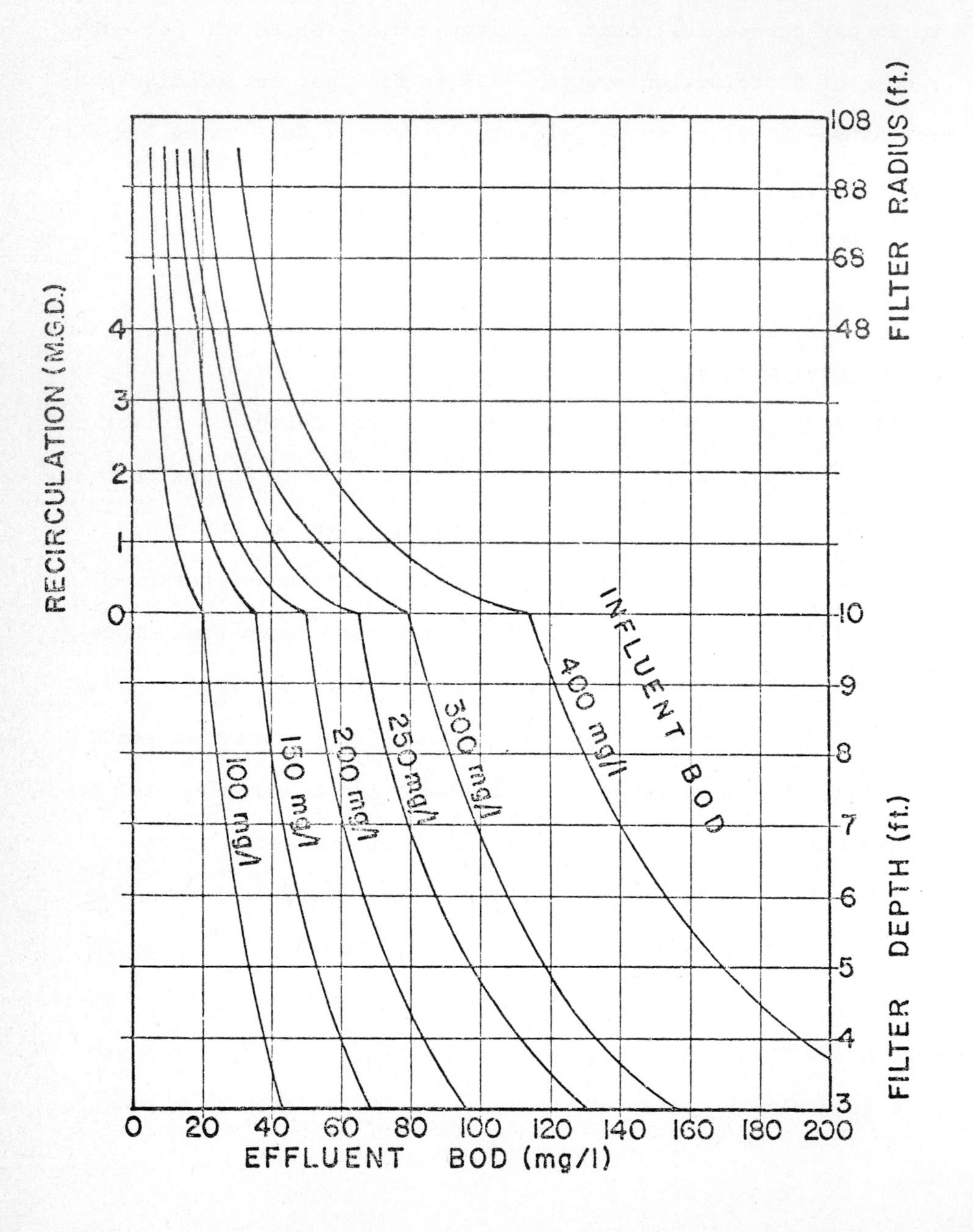

FIG 3. TRICKLING FILTER DESIGN CHARACTERISTIC AT 30 M.G.A.D. AND 25°C. MAX. DEPTH 10 FT. — PLANT INFLUENT 1 M.G.D. GALLER - GOTAAS MODEL

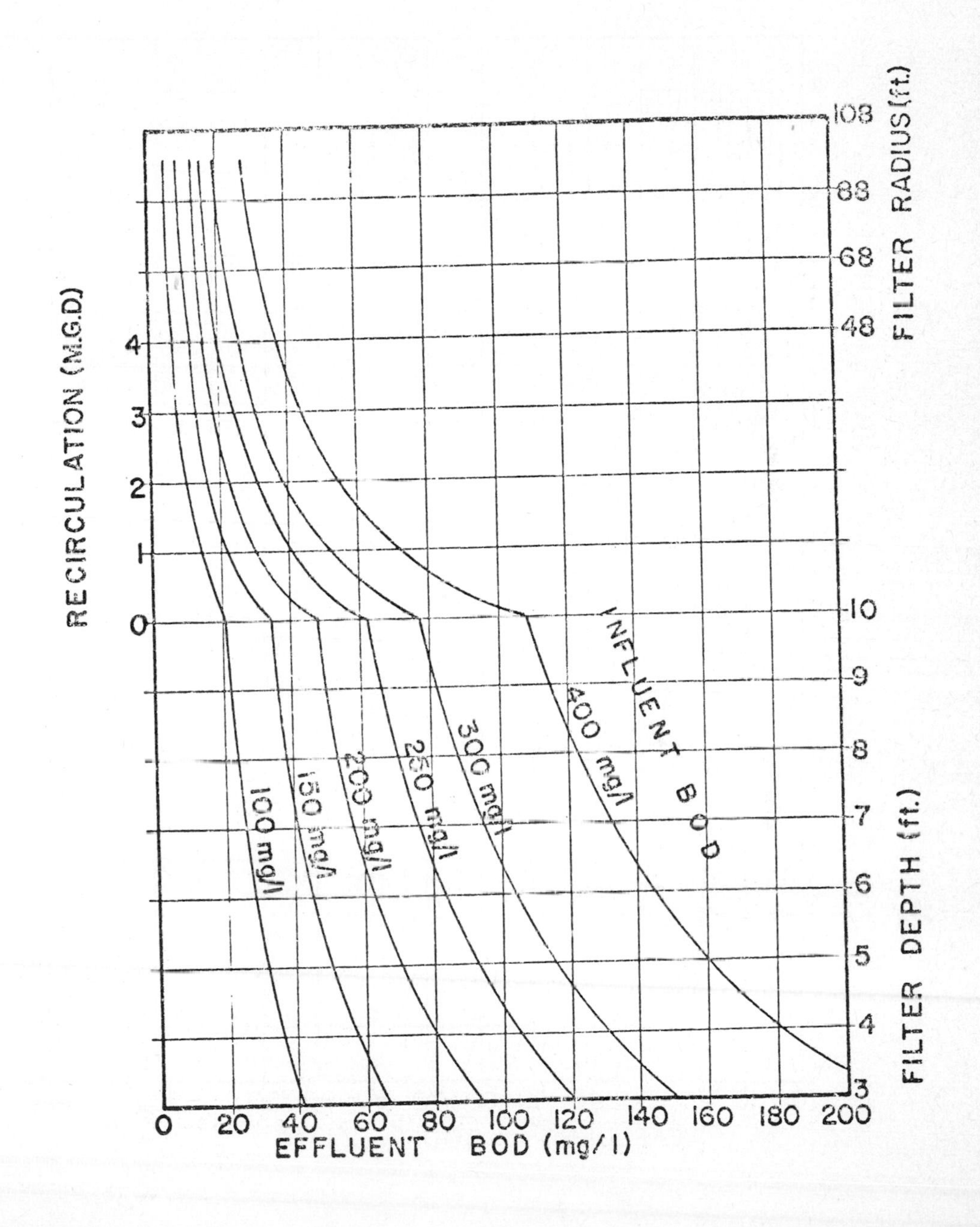

FIG 4. TRICKLING FILTER DESIGN CHARACTERISTIC AT 30 M.G.A.D. AND 18°C. MAX. DEPTH 20 FT. — PLANT INFLUENT 1M.G.D. GALLER - GOTAAS MODEL

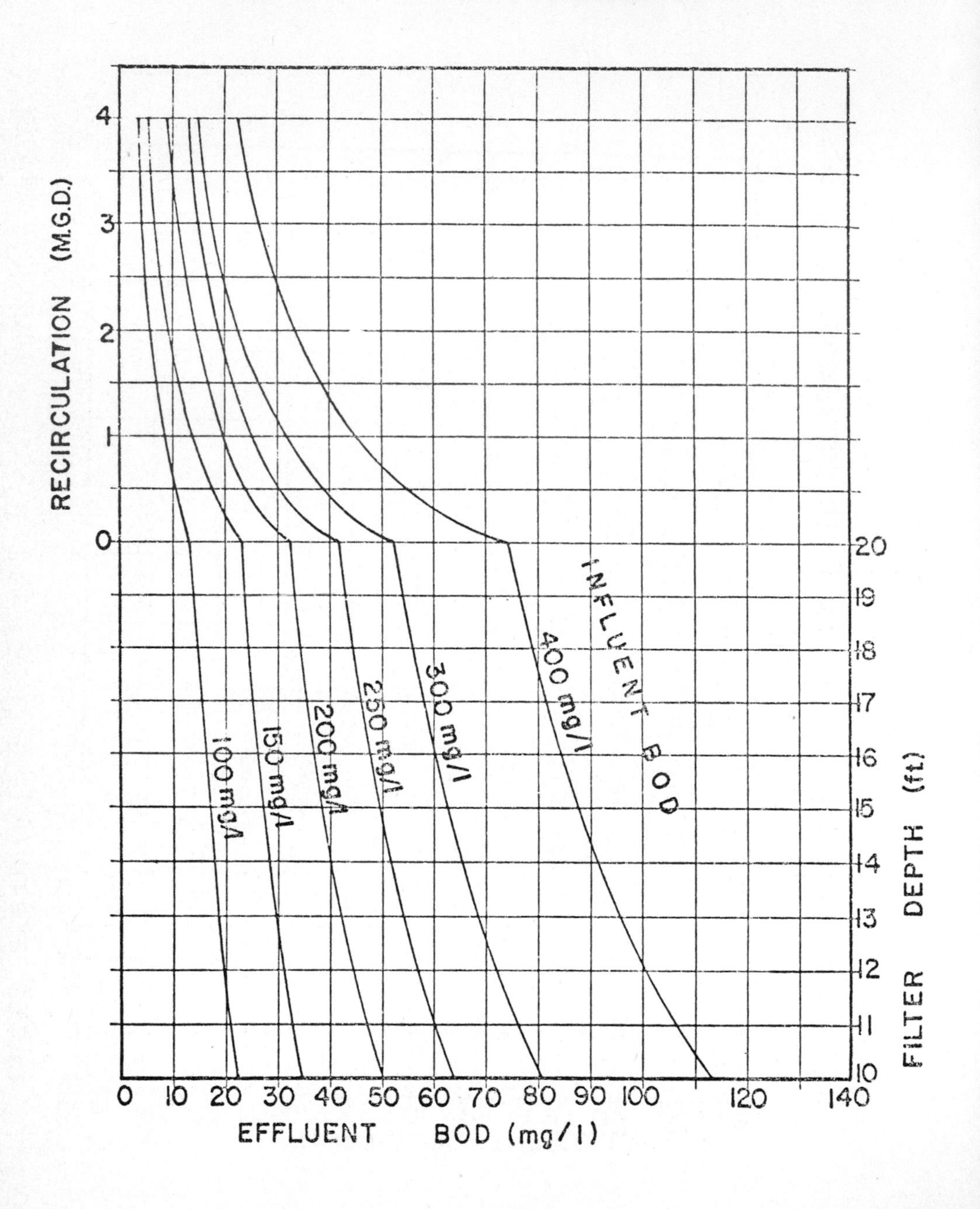

FIG 5. TRICKLING FILTER DESIGN CHARACTERISTIC AT 30 M.G.A.D. AND 25°C. MAX. DEPTH 20 FT. — PLANT INFLUENT 1 M.G.D. GALLER - GOTAAS MODEL

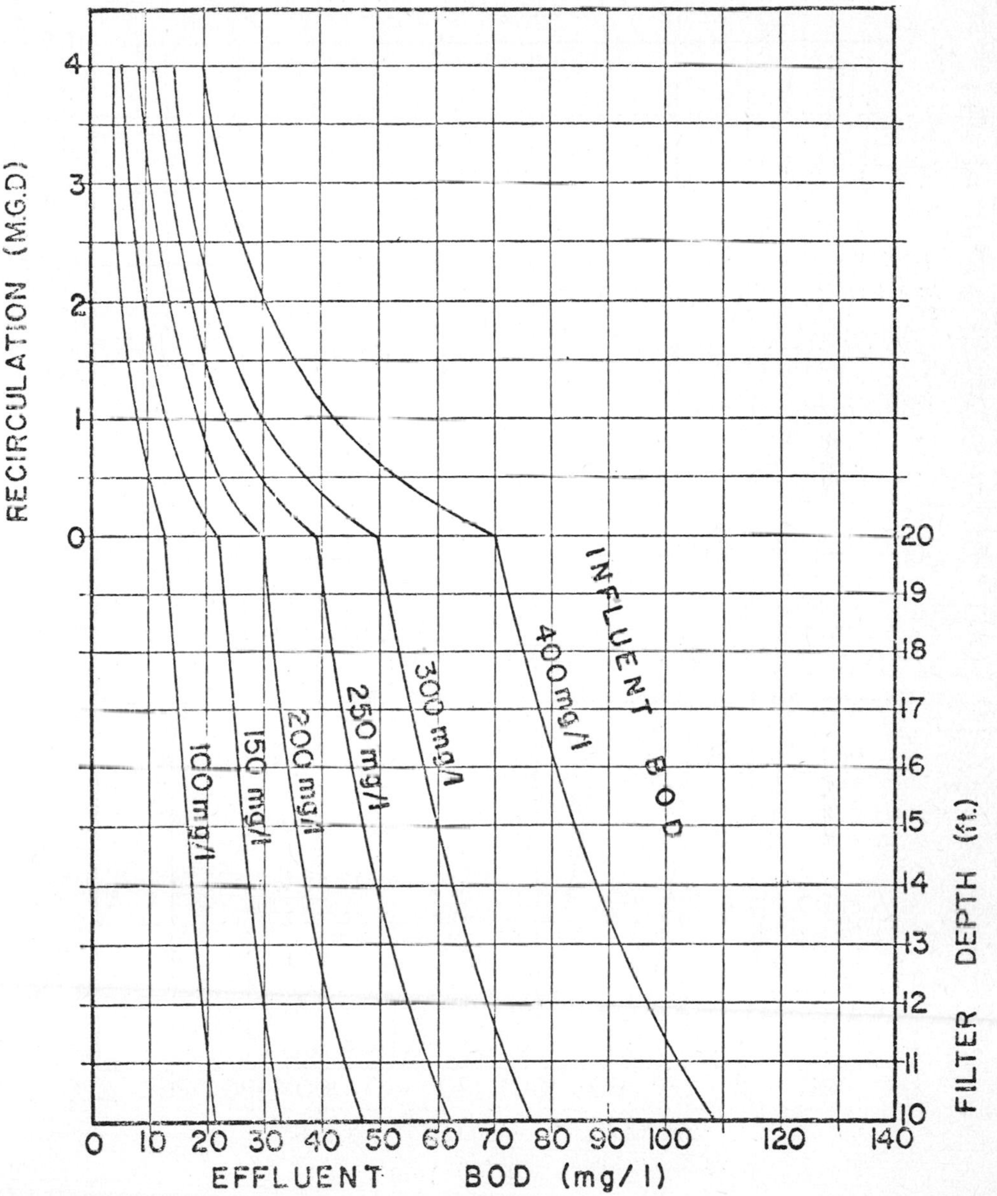

RECIRCULATION (M.G.D)

4
3
2
1
0

INFLUENT BOD
400 mg/l
300 mg/l
250 mg/l
200 mg/l
150 mg/l
100 mg/l

20
19
18
17
16
15
14
13
12
11
10

FILTER DEPTH (ft.)

0 10 20 30 40 50 60 70 80 90 100 120 140

EFFLUENT BOD (mg/l)

FIG. 6. TRICKLING FILTER DESIGN CHARACTERISTIC AT 30 M.G.A.D. AND 18° C. MAX. DEPTH 10 FT. - PLANT INFLUENT 1M.G.D. NRC MODEL

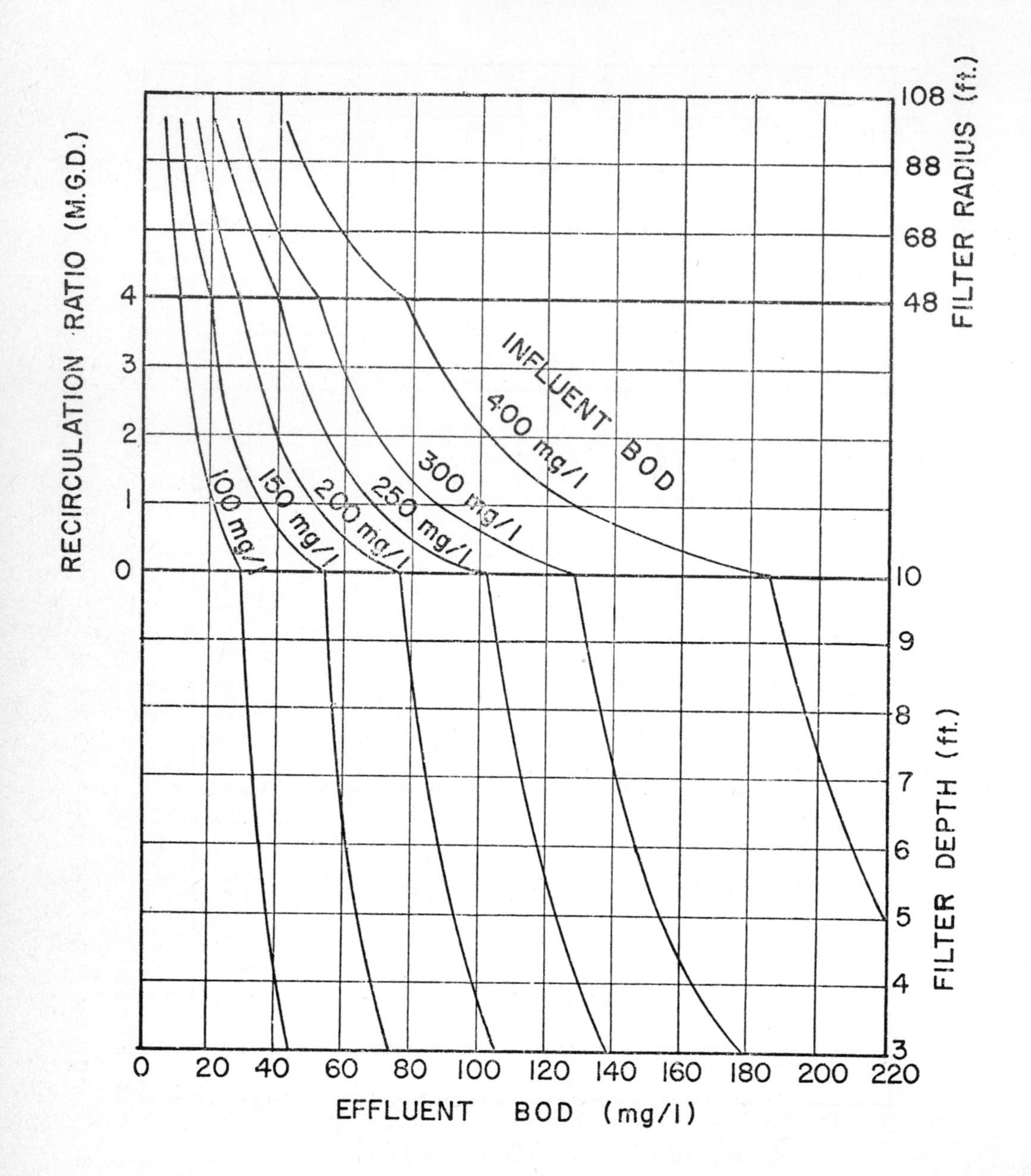

of the organic loading raised to a power. W is equivalent to L_i and

$$E = \frac{(L_i - L_e)(i + r)}{iL_i + r\ (L_i - L_e)} \tag{27}$$

Thus the effluent BOD is

$$L_e = \frac{(i + r)\ K\ L_i^{3/2}}{i + (i + r)\ K\ L_i^{1/2}} \tag{28}$$

Figure 7 presents the Eckenfelder Model. It is seen that only one curve is developed because in this model the effluent BOD is a function of the influent BOD raised to unit power. This means that for any given BOD removal efficiency, the filter design is related only to filter depth and recirculation. If it can be proven that BOD removal is related only to Q and the loading has no effect on BOD removal, this will provide one design for all BOD loadings for a constant Q. In the above models the filter radius or volume increases after the maximum depth and recirculation constraints are reached.

Figure 8 presents the Mississippi River - Great Lakes Board model. It is again seen that only one curve is developed because the effluent BOD is a function of the influent BOD raised to unit power. The filter depth can be at the minimum allowed since there is no interaction between BOD and depth in this model. The volume of the filter is established by the maximum loading criteria of 110 lbs. BOD per 1000 cu. ft. of filter, and the area by the hydraulic loading of 30 mgad for a filter depth from 5 ft. to 7 ft. The curve is valid only for a recirculation ratio of 0 to 4, or the maximum allowed, since the recirculation rate must be such as to provide an influent BOD not greater than three times the effluent BOD. Inspection of

FIG. 7. TRICKLING FILTER DESIGN CHARACTERISTIC AT 30 M.G.A.D. AND 18° C. MAX. DEPTH 10 FT. - PLANT INFLUENT 1M.G.D. ECKENFELDER MODEL

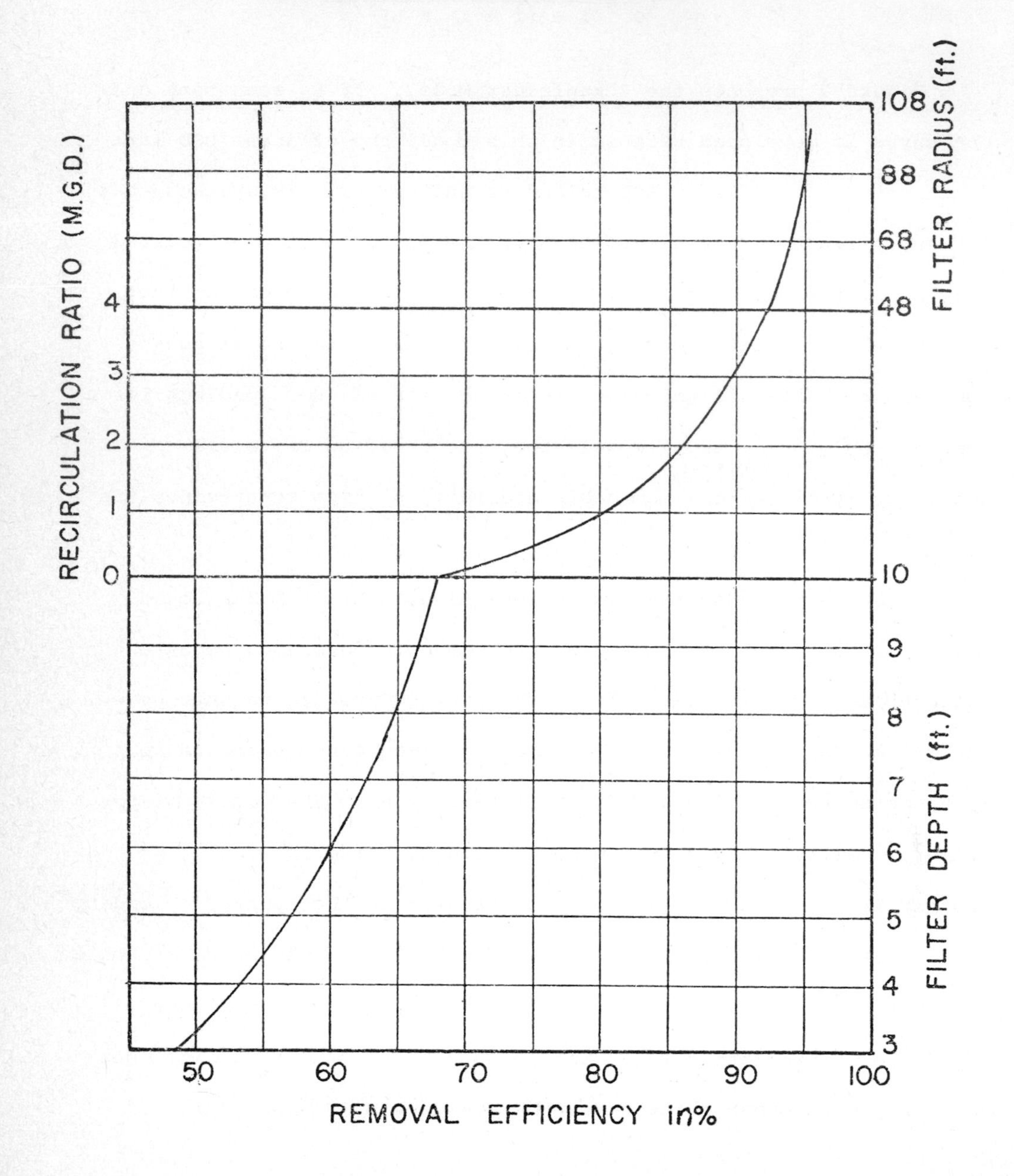

FIG. 8. TRICKLING FILTER DESIGN CHARACTERISTIC AT 30 M.G.A.D. AND 18° C. MAX. DEPTH 10 FT. – PLANT INFLUENT 1M.G.D. UPPER MISSISSIPPI RIVER-GREAT LAKES MODEL

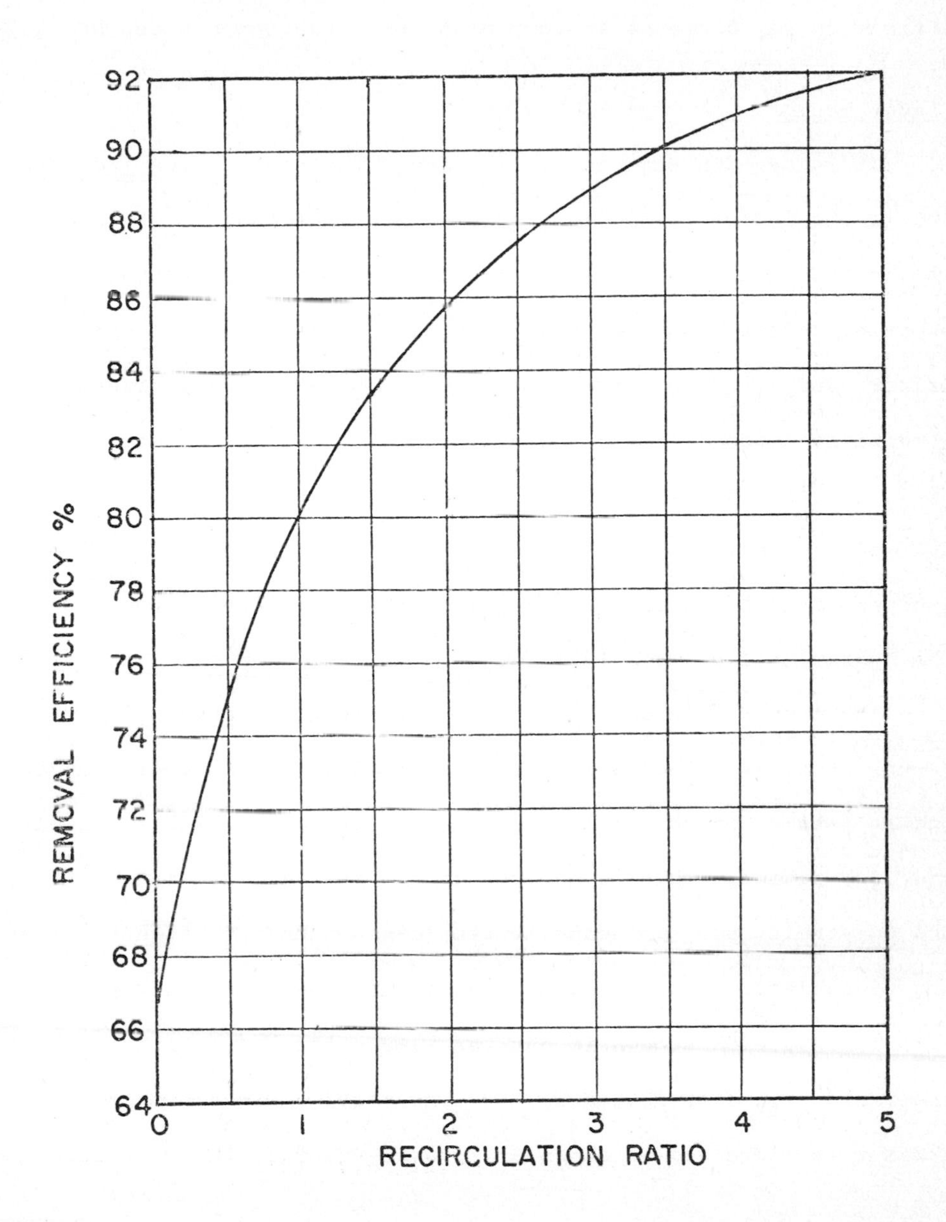

the equation for the model shows that the minimum removal efficiency will be 66%, and as the recirculation rate approaches infinity, the efficiency approaches 100%. Thus, using this formulation it is possible to obtain 100 percent removal in a filter of infinite diameter, which is not the case in practice, since it has been found that the efficiency may decrease at very high recirculation ratios (NRC [17]).

Filter Design

It is seen in Figures 2, 3, 6 and 7 that the curves indicate the BOD in the influent and the abscissa shows the concentration of BOD in the effluent. The ordinate is divided into three parts. Progressing up the ordinate, the lower part represents the depth of the filter, varying from 3 feet to 10 feet. Once the maximum depth constraint is reached, the depth of the filter remains constant at maximum and recirculation now commences and increases to a maximum of four volumes per volume of influent. The recirculation then remains constant at four until the maximum radius is reached, thus reducing the hydraulic flow rate in mgad.

It is seen that as the fraction of the BOD remaining in the effluent decreases, the variables are forced to their upper boundaries, one at a time, in the following order: (a) depth; (b) recirculation; and (c) radius. This is because depth has the least interactions in the constraint set, and contributes less to the cost than the remaining variables of recirculation and radius. Recirculation enters second because, although it has the same number of interactions as the radius, it contributes less to the cost. When the depth constraint is increased as shown in Figures 4 and 5, the most economical filter depth reaches the maximum depth when the hydraulic gradient of the plant is such that pumping the influent to a deep filter is not required, and sufficient ventilation for the necessary oxygen supply is available. Ten feet has been generally considered to be the maximum allowable depth. The primary reason for shallow filters is to assure adequate oxygen supply. There is considerable evidence that it is possible to have adequate ventilation at filter depths above ten feet. If it is necessary to cover the filters and provide forced ventilation with odor and bacterial control of the exiting air, the deeper filters are more economical. If pumping is required, and covering the filter is unnecessary, the shallower filters may be more economical.

Analysis of the use of forced air ventilation with blowers at a rate of two cubic feet per gallon of influent, showed that the filter variables follow the curves plotted in Figures 4 and 5. The studies indicated that if the recirculation required for a shallow filter is greater than 0.5 times the influent rate, a deeper filter with air blowers is more economical than the shallow filter. No analysis has been made of the cost effect of covering the filter on the optimum depth. It is obvious that deeper filters would be more economical when enclosing the filter and treating the exit air with ozone to reduce odors is necessary.

In determining the raduis or size of the filter, optimum cost analyses show that a single large filter is more economical than several small filters. Limitations imposed by available equipment and operating flexibility made the provision of multiple filter installations desirable. To determine the radius of a filter at

$Q = 30$ mgad, in which $Q = Q_{max}$ and $r \leq 4$, the following equation is solved,

$$a = \sqrt{\frac{43,560\ (i + r)}{\pi\ Q_{max}}} \qquad (29)$$

or the results can be obtained from Figure 9 after determining the recirculation ratio. If more than one filter is required, the influent volume is divided appropriately between the number of filters for design purposes.

When Q is less than Q_{max} (represented by the part of the ordinate above $r = 4$), the radius of the number of filters required when i is greater than one can be found by

$$a = a_p\sqrt{i} \qquad (30)$$

in which a_p is read from the appropriate Figure 2, 3, 4, 5, 6 or 7, depending upon the model and conditions stipulated.

FIGURE 9

MAXIMUM RECIRCULATION RATIOS FOR VARIOUS PLANT INFLUENT RATES FOR VARYING FILTER RADII

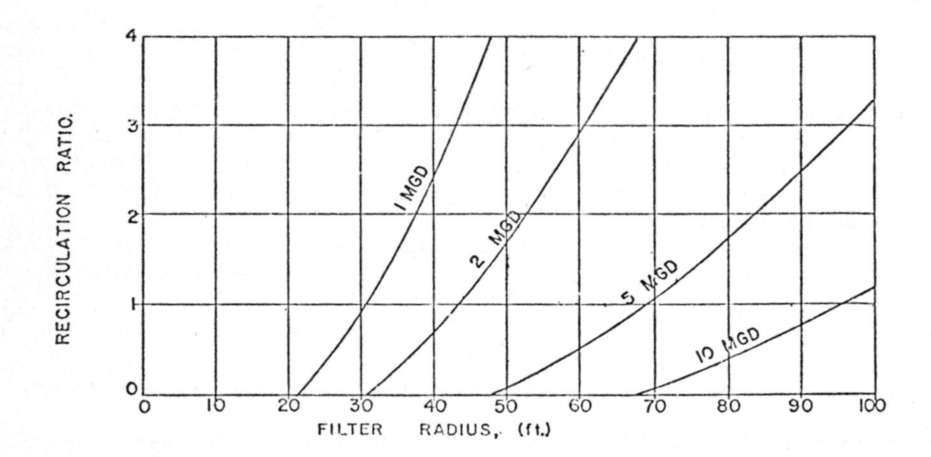

If pumping of the influent is necessary due to the gradients of the plant site, depth of influent conduit and the filter depth, the following method can be used to determine whether a deep or shallow biological filter is more economical:

1. Determine the depth radius and recirculation rate for the shallow filters using the Figure for the desired model which included depth;
2. determine the annual cost of the shallow filter, C_s, and that of the deep filter, C_d, based on local conditions;
3. determine the cost to pump the influent to each, P_s for the shallow and P_d for the deep filter;
4. if $(C_s + P_s) > (C_d + P_d)$, the deep filter is more economical and
5. if $(C_s + P_s) < (C_d + P_d)$, the shallow filter is more economical.

Figure 8, which presents the Upper Mississippi River-Great Lakes Board Model, gives the recirculation ratio required for a given efficiency of BOD removal based on providing a filter influent BOD concentration not greater than three times that of the effluent. The area of filter is then determined to permit a hydraulic loading not less than 10 mgad, nor more than 30 mgad. A filter depth of not less than five feet, nor more than seven feet is specified, as well as a maximum total BOD loading of 110 lbs. per 1000 cubic feet of filter.

Trickling filters with Settling Basins

If the recirculated liquid is to pass through the secondary

settling tank before recirculation, the cost of the secondary tank will depend on the volume of recirculation. Also, if the recirculated liquid is directed to the inlet of the primary settling tank as shown in Figure 10, the primary tank is a function of recirculation. Hence, the trickling filter and the two settling basins become the unit to be investigated for optimum cost. Only the effects on rectangular basins were studied, but in the opinion of the authors, the results would be the same for circular basins as for rectangular.

FIGURE 10

TRICKLING FILTER WITH RECTANGULAR SETTLING TANKS

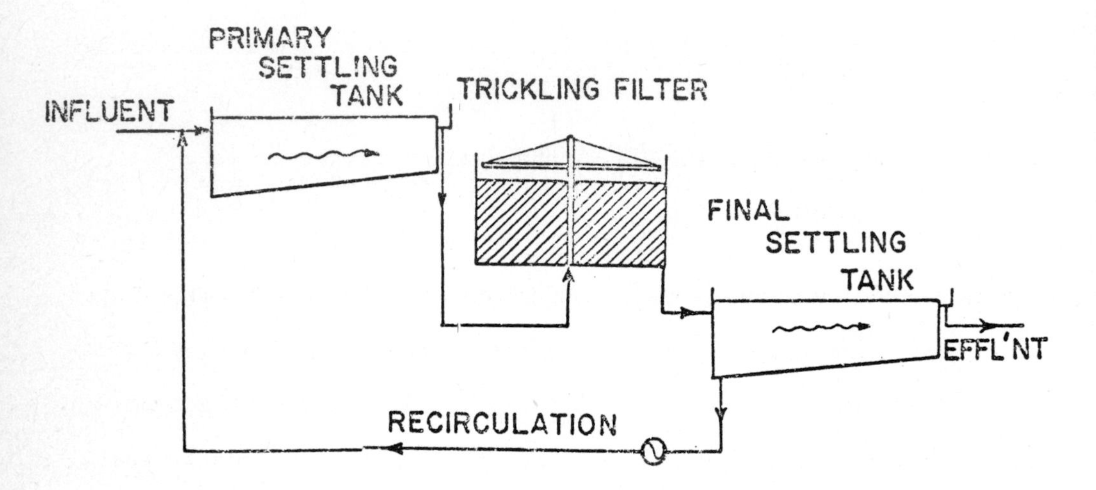

The studies show that the variables investigated for the trickling filters and settling basins show the same relationships as shown in the figures for the models of the trickling filter. Thus, the consideration of settling basins did not alter the design characteristics of the trickling filters and the most economical design of the basins to handle the acquired liquid quantities. With recirculation of sludge underflow of the secondary basin, the basin can be designed for the influent rate instead of the influent rate plus recirculation. Although the settling basin and recirculation arrangement did not change the optimum filter design to achieve minimum plant cost, the settling basins should be designed as small as permitted to meet the design requirements for the recirculation arrangement used. There have been some cases where recirculation of underflow containing suspended solids having good settling characteristics from the secondary settling tanks seemed to help obtain greater BOD removal from the primary effluent in the primary settling tanks. However, this effect has not been generally established.

EXAMPLE OF FILTER DESIGN USING THE FOUR MODELS

It is desired to design the biological filters for a waste treatment plant having an influent of 10 mgd with a BOD to the filters of 200 mg per liter at a temperature of 18°C. An effluent having a BOD of not more than 30 mg per liter is desired. Shallow filters will be designed.

Caller-Gotaas Model

1. Using Figure 2, find the point on the 200 mg per liter influent loading curve where the effluent BOD is 30 mg per liter. At this point the depth is 10 feet and the recirculation ratio is 1.1.
2. Since the recirculation is less than 4.0, by dividing the total flow (21.0 mgd) by 30 mgad to determine the filter area or by finding the 10 mgd influent curve and the point

where the recirculation ratio is 1.1 on Figure 9, a radius of 98.0 ft. is determined. Had the recirculation ratio been over 100 ft. and the two filters would be required. Since two filters are desirable for flexibility in plant operations, let the influent rate be 5 mgd for each filter, and as shown above the radius of each filter is 70.0 feet.

Hence, the design has two filters having a depth of 10 feet, a radius of 70 feet and a recirculation ratio of 1.1 with a hydraulic load of 30 mgad.

NRC Model

1. On Figure 6 the 200 mg per liter influent loading curves reaches an effluent BOD of 30 mg per liter when the depth is 10 feet, and the recirculation ratio is 3.7.
2. On Figure 9 it is seen that the filter radius is over 100 feet when the recirculation ratio is 3.7 if two filters were used, hence three filters are required.
3. Using three filters, each treating 3.33 mgd, as shown the radius of each filter is 72 feet.

Hence, the design using the NRC model has three filters having a depth of 10 feet, a radius of 72 feet and a recirculation ratio of 3.7 with a hydraulic loading of 30 mgad.

Eckenfelder Model

1. An effluent BOD of 30 mg per liter requires a removal efficiency of 85.0%. Using Figure 7, the curve for efficiency of BOD removal shows that a filter depth of 10 ft. and a

recirculation ratio of 1.7 is required for 85% BOD removal by the filter.

2. By computing the required area for a 30 mgad hydraulic loading or using Figure 9, each of two filters treating 5 mgd with a recirculation ratio of 1.7 must have a radius of 79.0 feet.

Hence, the design using the Eckenfelder model has two filters having a depth of 10 feet, a radius of 79 feet and a recirculation ratio of 1.7 with a hydraulic loading of 30 mgad.

Upper Mississippi River-Great Lakes Board Model

1. From Figure 8 it is seen that a BOD removal efficiency of 85% requires a recirculation ratio of 1.8 to meet the requirement that the BOD concentration in the filter influent should not be more than three times that of the effluent.
2. The requirements of the model stipulate that the depth of the filter must be not less than 5 ft. or more than 7 ft., the hydraulic loading rate between 10 mgad and 30 mgad, and the BOD loading not more than 100 lbs. per 1000 cubic feet per day.
3. Analysis shows that the model requirements of a maximum hydraulic loading of 30 mgad and minimum depth of 5 feet determine the size of filter. For a maximum hydraulic loading of 30 mgad, and a recirculation ratio of 1.8, the filter area required for 10 mgd is $\frac{18 \text{ mgd} + 10 \text{ mgd}}{30 \text{ mgad}}$ =

0.942 acres = 40,600 square feet. Considering two filters the radius of each will be $\sqrt{\frac{20,300}{3.1416}}$ = 80 feet.

4. Since the area and radius are determined by the recirculation ratio and 30 mgad hydraulic loading required, the minimum allowable filter depth of 5 feet to minimize the volume can be used.

5. The BOD applied to the filters per day is 10 x 8.34 x 200 = 16,667 lbs. The BOD loading in lbs. per 1000 cubic feet is $\frac{16,667 \times 1000}{2 \times \pi \times 80^2 \times 5}$ = 83.0. This is considerably lower than the maximum permitted for the model of 110 lbs. per 1000 cubic feet.

Hence, the design using the Upper Mississippi-Great Lakes Board model has two filters each having a depth of 5 feet, a radius of 80 feet, and a recirculation ratio of 1.8 with a hydraulic loading of 30 mgad.

The filter volumes for the optimum design example for each of the four models are: Galler-Gotaas Model--10,400 cu. yds.; NRC Model--181,000 cu. yds.; Eckenfelder Model--14,500 cu. yds.; and Upper Mississippi River-Great Lakes Board Model--14,900 cu. yds. The filter volume required for the different models varies considerably.

DISCUSSION

A detailed evaluation of the merits of the different biological filter models will not be attempted. Evaluation of seven models was made by Robertson, _et. al._[15] in the treatment of Kraft wastes and

Boxboard wastes. Non linear regression methods were used to fit the models to 290 observations of Kraft waste treatment on biological filters, and 225 observations of Boxboard waste treatment to light different models. They found that the Galler-Gotaas model was much superior to any of the others in formulating their observations. This is probably because this model was developed from 322 observations of the treatment of domestic sewage on trickling filters under conditions were more of the variables could be isolated than are included in the other models. They found that the constant and exponents relating to the different variables varied with the characteristics being treated. Table I shows the variations of the constants and exponents relating to the different variables varied with the characteristics being treated. Table I shows the variations of the constants in the Galler-Gotaas Model for the three different wastes.

TABLE I

VARIATION OF THE CONSTANTS IN THE GALLER-GOTAAS MODEL WITH WASTE CHARACTERISTICS

Waste (1)	Constants					
	Constants (2)	L_o Exponents (3)	Recirculation Exponent (4)	Q/A Exponent (5)	Depth Exponent (6)	Temperature Exponent (7)
Domestic Sewage	0.464	1.19	0.28	0.13	-0.67	-0.15
Kraft Waste	0.218	1.18	0.07	0.21	-0.19	-0.08
Boxboard Waste	0.663	1.12	0.18	0.14	+0.02	-0.31

The best fit equations of the other models to the observations showed variations in the constants between sewage, the Kraft waste and the Boxboard waste treatment. However, it was not possible to obtain as good fit of the data to the other models after changing the constants. Again, this is probably due to the presence of fewer variables in the models.

Hanumanula [7] studied the application of the four models listed above and the Velz model to covered filters 12 feet and 18 feet deep without recirculation, with forced ventilation and 5-day 37° BOD influent concentrations of 225 to 380 mg per liter and loadings of about 50 pounds to 80 pounds per 1000 cu. ft. Removal efficiencies and effluent BOD concentrations considerably larger than those predicted by any of the five models were observed. The computed efficiencies for the Eckenfelder and Galler-Gotaas models were much closer to the observed than for the other models when recirculation was used, and the mean ratio of the observed efficiency to the computed efficiency was 1.06 for the Eckenfelder model and 1.056 for the Galler-Gotaas model, which is not significant.

The observed efficiencies for BOD removal on the filters without recirculation deviated more from the Eckenfelder and the Galler-Gotaas models than from the other models. Apparently without recirculation the top layers of the filter are actively contributing to BOD removal. The lower layers do not come into full play and hence the depth of the filter is not as important a factor in the design and performance of a filter without recirculation. Hanumanula concluded that the significance of the depth term increases with the volume of recirculation, and that the exponent of the depth terms in the filter model equations should itself be a function of recirculation. Hence, he modified the Galler-Gotaas model to the form

$$L_e = \frac{K\ (iL_i + rL_e)^{1.19}}{(i+r)^{0.78}\ (D+1)^{0.48\ +\ 0.35R}} \tag{30}$$

where R is the recirculation ratio, concluding that this modified model includes the primary variables and is equally applicable to deep filters with and without recirculation.

Baker and Graves [1] studied filter volume relationships for different loadings and removal efficiencies for the three models, NRC, Eckenfelder and Galler-Gotaas. They found considerable variation between efficiency and volumes at filter depths of 6 feet and 10 feet for the different models, and show curves expressing these variations. A decrease in volume does not occur with increased depth in the NRC model as in the Eckenfelder and Galler-Gotaas models. The Galler-Gotaas model differed more from the NRC model than did the Eckenfelder model, because depth is a stronger variable in the Galler-Gotaas model than in the Eckenfelder. Examination of Figures 2 and 6 indicates the difference in the effect of depth on BOD removal for the two models. They conclude:

> The trickling filter volume computed by the National Research Council and Eckenfelder formulas will vary directly with the flow rate and will vary with the efficiency functions $E(1 - E)^{-2}$. Volume will decrease with increases in the recirculation in both formulas, reaching practical limiting recirculation ratios of 4 to 5. Volume varies directly with influent BOD in the NRC formula, but influent BOD is not a parameter in the Eckenfelder formula. Volume varies with the depth function $D^{-0.33}$ in the Eckenfelder formula, which causes the required volume to decrease with depth increase.

> In the Galler-Gotaas formula, the required volume will increase with increased flow rate, influent BOD and required efficiency. It will decrease with increases in temperature and depth. A volume decrease is shown only with higher efficiencies, where a practical limiting ratio of 4 or 5 is reached.

Much additional reliable performance data from existing plants are needed to more completely establish the validity of the filter models. The choice of the model to be used is a judgment consideration for the designer based on his analysis as to which model best fits the requirements of his design problem, except when regulatory agency standards for filter design specify a design criteria which do not conform to the factual information concerning the biological filter variables. Whichever model is used for the design of a biological filter system, the optimization design will provide the most economical filter for that model.

In Galler's studies [6], twenty-one cost vectors and ten interest vectors were used to test the sensitivity of the optimization model to variations in these parameters. It was found that the solutions for optimum design figures shown also provide optimum design characteristics for the range of costs and interest rates described in the cost conditions.

SUMMARY AND CONCLUSIONS

An optimization method to achieve the most economical design of biological sewage filters has been applied to the Galler-Gotaas, NRC, Eckenfelder and Upper Mississippi-Great Lakes Board Models for filter design. The cost optimization procedure provides a more effective method for treating the different variables associated with biological filters to achieve the most efficient design.

It is seen that there is considerable variation between the models. However, for optimum design conditions, the hydraulic rate through the filter is at the maximum level permitted until the recirculation ratio is approximately **4** to 1. After this ratio is achieved the hydraulic rate will decrease.

Filter depth interacts differently with the other variables in the different models. The depth factor variation has no influence in the NRC and Upper Mississippi River-Great Lakes models, some influence in the Eckenfelder model, and of greatest influence on removal in the Galler-Gotaas model. When pumping of influent is not a factor, the most economical filter should be designed as deep as possible while maintaining satisfactory oxygen ventilation in the filter. Deep filters with forced ventilation are more economical than shallow filters only for high reductions of BOD and lose their advantage if pumping of influent is required in order to use the deep filter. Deep filters, and forced ventilation are necessary for covered filters.

For a recirculation ratio lower than the maximum effective limit of about four, the cost of increasing the diameter and the volume of the filter is greater than increasing the recirculation to obtain increased BOD removal. Apparently the type of waste influences the effect of some of the different variables and recirculation, and the effectiveness of depth. The Eckenfelder and Galler-Gotaas models can be modified to these conditions.

Theoretically, optimum cost conditions are met by one filter having a diameter up to the maximum permitted by the availability of distributor equipment; however, operating flexibility requires multiple units except for very small plants. The most economical design will have a minimum number of filters consistent with the necessary operating flexibility.

Variation of the cost parameters within reasonable limits does not affect the design variables when the cost is being minimized.

The inclusion of settling basins in the trickling filter problem does not alter the optimum design characteristics of the filter. The settling basins should be designed at minimum size consistent with the constraints related to them. The design curves for the different models presented provide optimum biological filter design for the constraints stated.

APPENDIX I

DEVELOPMENT OF THE COST FUNCTION

Cost of Wall = (Area) (Depth) $(c_1/27)$ = $(2\ Wa\ D + 2\ F\ Wa + F\ W^2) \cdot \pi\ c_1/27$.

Cost of Floor = Area (c_2) = $\pi\ a^2 c_2$.

Cost of Packing = Volume $(c_3/27)$ = $\pi\ a^2\ D\ c_3/27$.

Cost of Distribution System = $2\ c_4\ a$.

Cost of Power = $c_5(8.34)(365)(r)(D+1)/(2.65p)$; Here (8.34) (365)/ 2.54 converts dollars per killowatt hr. to dollars per million gallons per year.

Pump Cost = $r/c_6 + c_7 r$.

Line Total Annual Cost = Total annual cost = $(2c_1\ \pi\ WaD/27 + 2c_1 \pi W\ Fa/27 + c_1\ \pi\ W^2 D/27 + c_2\ \pi\ a^2 + c_3\ \pi\ a^2\ D/27 + 2\ c_4\ a + r/c_6 + c_7\ r) + c_5\ (8.34)(365)(r)(D+1)/(2.65p)$. The terms containing no variables are neglected because they will not affect the optimization.

After Transformation: $C_1 = c_5\ i(8.34)(365)(d-1)/(2.65p)$; $C_2 = 4\ c_1\ \pi\ W^2\ 27 - 4\ i\ c_5(8.34)(365)/(2.65p)$; $C_3 = 20\ c_4 \lambda + 20\ c_1 \pi\ W\ \lambda\ (F-1)/27$; $C_4 = 4\ ic_7(8.34)(365)/(2.65)$; $C_5 = 80\ c_1\ \pi\ W\ \lambda/27$; $C_6 = 400\ c_3\ \pi\ \lambda/27$; $C_7 = 100\ \pi(c_2 - c_3/27)\lambda$; $C_8 = c_1\ \pi\ W^2\ \lambda(f-1) - C_1$ = constant. Therefore, C_8 need not be considered in the optimization.

λ is capital recovery factor.

APPENDIX II

SAMPLE PROBLEM

It is desired to design a trickling filter for a sewage treatment plant having an influent rate of 1 mgd and BOD of 200 mg per liter. The influent temperature is 18°C. The plant effluent must contain an organic loading not exceeding 30 mg per liter. The maximum filter depth is 10 ft. The wall thickness and freeboard are 1 ft. The following costs or cost factors are to be used:

$c_1 = 80$; $c_2 = 4$; $c_3 = 10$; $c_4 = 53$; $c_5 = 3.2 \times 10^{-4}$;
$c_6 = 5.55 \times 10^{-4}$; $c_7 = 0.01$.

If recirculation is to be used, the pump would supply a head of 4 ft. in excess of the depth and have an over-all efficiency of 70%. The plant is to be paid for in 20 yr. at an interest rate of 4%.

1. Calculate the functional coefficients:

$$C_1 = \frac{(1)(0.01)(8.34)(365)(4-1)}{(2.65)(0.7)} = 49.3$$

$$C_2 = \frac{(4)(80)(\pi)(1)^2(0.07358)}{27} - \frac{(4)(1)(0.01)(8.34)(365)}{(2.65)(0.7)}$$

$$C_2 = -62.96$$

$$C_3 = (20)(53) + \frac{(20)(80)(\pi)(1-1)}{27} 0.07358$$

$$C_3 = 39.00$$

$$C_4 = \frac{(4)(1)(0.01)(8.34)(365)}{(2.65)(0.70)} = 65.70$$

$$C_5 = \frac{(80)(80)(\pi)(1)(0.07358)}{27} = 54.8$$

$$C_6 = \frac{(400)(10)(\pi)(0.07358)}{27} = 34.20$$

$$C_7 = (11)(\pi)(0.07358)\ (4 - \frac{10}{27}) = 84.0$$

2. Evaluate constraint constants:

   (a) Let Q_{max} = 30 mgd; $x_2 \le \min \{\ln(30)(100)^2(\pi)/43{,}560, \ln 5\}$;

   $x_2 \le \ln 21.8$. Therefore, $x_2 \le \ln 5$; and $x_2 \le 1.60944$.

   (b) $x_3 \le \ln 11/4$; therefore, $x_3 \le 1.01160$.

   (c) $x_4 \le \ln 10$; $x_4 \le 2.30259$.

   (d) $x_4 - (\frac{1}{2})x_2 \ge \frac{1}{2} \ln(4{,}356/30\ \pi)$; $x_4 - \frac{1}{2}(x_2) \ge 0.765$.

   (e) $\ln L_e = \ln K_1' + 1.8896\ x_1 - 0.78408\ x_2 - 0.66916\ x_3 -$

   $0.25452\ x_4$; $\ln L_e = \ln 30 = 3.40120$.

$$K_1' = \frac{(0.463)\left(\frac{43{,}560}{\pi}\right)^{0.12226}}{(1)^{0.27762}(18)^{0.15440}(10)^{0.25452}(1)^{0.78408}(4)^{0.66916}}$$

$$K_1' = -.219,\ \ln K_1' = 0.51868$$

3. Linearization of the functional for imbedding the nonlinear functional into the constraint matrix:

   Let x_5 be the new functional.

Let

$$g_f(X) = C_1 e^{x_2} + C_2 e^{x_3} + C_3 e^{x_4} + C_4 e^{x_2+x_3} + C_5 e^{x_3+x_4}$$

$$+ C_6 e^{x_3+2x_4} + C_7 e^{2x_4} + \frac{i\left(e^{x_2}-1\right)\lambda}{c_5 + i\, c_6\left(e^{x_2}-1\right)}$$

$$-x_5 \le 0 \quad \ldots\ldots\ldots\ldots \quad (15)$$

To linearize this, let

$$g_f(X) + g_{m+1}\cdot(x - x^u) \le 0 \quad \ldots\ldots\ldots\ldots \quad (16)$$

and

$$g_f \cdot (x-x^u) = \left(C_1 e^{x_2^u} + C_4 e^{(x_2^u + x_3^u)} + \frac{\lambda\ i\ C_5 e^{x_2^u}}{\left(C_5 + iC_6 (e^{x_2^u} - 1)\right)^2} , \right.$$

$$C_2 e^{x_3^u} + C_4 e^{(x_2^u + x_3^u)} + C_5 e^{(x_3^u + x_4^u)} +$$

$$C_6 e^{(x_3^u + 2x_4^u)} , C_3 e^{x_4^u} + C_5 e^{(x_3^u + x_4^u)} +$$

$$\left. 2C_6 e^{(x_3^u + 2x_4^u)} + 2C_7 e^{2x_4^u} , -1 \right)$$

$$\cdot \begin{bmatrix} x_2 - x_2^u \\ x_3 - x_3^u \\ x_4 - x_4^u \\ x_5 - x_5^u \end{bmatrix} \quad \ldots\ldots\ldots\ldots\ (17)$$

4. Linearization of nonlinear constraint:

$$x_1 = \ln \left[i\ L_i + i\ L_e \left(e^{x_e} - 1 \right) \right] \quad \ldots\ldots\ldots\ldots\ (18)$$

$$g_7(X) = -x_1 + \ln \left[i\ L_{i_3} + i\ L_{e_3} \left(e^{x_2} - 1 \right) \right] = 0 \quad \ldots\ldots\ (19)$$

The new constraint is

$$g_7(X^u) + \nabla\ g_{4.13a}(X-X^u) \leqq 0 \quad \ldots\ldots\ldots\ldots\ (20)$$

$$\text{which } \nabla g_7(X-X^u) = \left(-1, \frac{L_e\ e^{x_2}}{L_i + L_e \left(e^{x_2} - 1 \right)} \right) \cdot \begin{bmatrix} x_1 - x_1^u \\ x_2 - x_2^u \end{bmatrix} \quad \ldots\ (21)$$

5. Solution algorithm of the linear programming problem: Because of the nature of linear programming problems, constraints can be added

any point, provided that they are properly updated. Redundant constraints can be removed at the same time without affecting the problem or solution. Also, it is not necessary to show the basic matrix (the identity part of matrix), if the proper steps are used in exchanging rows and columns, thus shrinking the size of the matrix required for calculation. The following algorithm is used to solve the problem.

(a) Determine the nonlinear constraint most disobeyed by the present solution. Linearize this constraint.

(b) Update the new constraint. If the variable to which a coefficient is attached is not in the basis, place the coefficient in the m + 1 row, in the column occupied by the variable. If the variable is in the basis, place the coefficient into the n + 1 column in the row occupied by the basic variable. Multiply the n + 1 column by -1. The updated coefficients in the m + 1 row: $\bar{A}_{m+1,v}$ are found by

$$\bar{A}_{m+1,v} = A_{m+1,v} + \sum_{u=1}^{m} A_{u,v} \cdot A_{(u,n+1)},$$

$$v = 1, 2, \ldots, m$$

in which $A_{u,v}$ are the positions in the matrix. Let m = m + 1.

(c) Apply the dual method on the problem. Find all negative elements in the first column for u = 1, ..., m. Find the quotient $A_{1,v}/A_{u,v}$, v = 2, ..., n for all negative elements in the rows having negative values for $A_{u,1}$. Find the minimum value quotient and let the element in the denominator be the pivot element. This element is now in the jth row and k^{th} column; thus, $A_{j,k}$ is the pivot element. To determine $\bar{A}_{j,u}$, multiply each element in the jth row by $1/A_{j,k}$. Now multiply each element in the kth column by $-1/A_{j,k}$ to determine $\bar{A}_{u,k}$. Now

$$\bar{A}_{j,k} = \frac{1}{A_{j,k}} \qquad \text{.........} \quad (23)$$

and

$$\bar{A}_{u,v} = A_{j,v}\bar{A}_{u,k} + A_{u,v} \qquad u \neq j,\ v \neq k \qquad \text{..........} \quad (24)$$

Repeat until all elements of the first column are positive, then solution is optimum.

(d) Test the solution.

6. The problem solution: Following is the initial coefficient matrix with basic and non-basic variable indicated. Let σ_{100} be an artificial variable and let all other σ be slack variables. The first tabulation is shown in Table 2. Here $g_f(X^o) = 787.94$; $g_7(X^o) = 5.30$; and

$$X^o = \begin{bmatrix} 0 \\ 0 \\ 0 \\ 0 \\ 0 \end{bmatrix}$$

TABLE 2

Basis		x_1	x_2	x_3	x_4	x_5
σ_{100}	4.920	1.189	-0.784	-0.669	-0.255	-1
σ_1	1.609		1			
σ_2	1.012			1		
σ_3	2.303				1	
σ_4	-0.765		0.5		-1	

The final tabulation is shown in Table 3. Here, $g_f(x^4) = 0.0$; $g_7(x^4) = 0.0$; and

$$\begin{bmatrix} 5.456 \\ 0.762 \\ 1.012 \\ 1.146 \\ 2157. \end{bmatrix}$$

The radius, a, = 31.46 ft.; the recirculation, r, = 1.142 mgd; and the depth, D, = 10.0 ft.

TABLE 3

Basis		σ_8	σ_4	σ_1	σ_2	σ_7
	2157.000	-3876.000	-1085.000	-3260.000	-2082.000	-1.
x_3	1.012		1.000			
x_4	1.146	-0.988	-0.556	-0.831	-0.788	
σ_3	1.156	0.988	0.556	0.831	0.788	
x_2	0.762	-1.976	-1.112	-1.662	0.423	
σ_5	0.848	1.976	1.112	1.662	-0.423	
x_5	2157.000	-3876.000	-1085.000	-3260.000	-2082.000	-1.
x_1	5.456	-1.515	-0.290	-0.433	0.110	

REFERENCES

1. Baker, J. M. and Q. B. Graves "Recent Approaches for Trickling Filter Design," Journal of the Sanitary Engineering Division, ASCE, Vol. 94, No. SA1, 65-82 (Febraury 1968).

2. Bulletin, Upper Mississippi River Board of Public Health Engineers and Great Lakes Board of Public Health Engineers (1952).

3. Eckenfelder, W. W. Jr. "Trickling Filter Design and Performance," Journal of the Sanitary Engineering Division, ASCE, Vol 87, No. SA4, Proc. Paper 2860, 33-45 (July 1961).

4. Galler, W. S. and H. B. Gotaas "Analysis of Biological Filter Variables," Journal of the Sanitary Engineering Division, ASCE, Vol. 90, No. SA6, Proc. Paper 4194 (December 1964).

5. Galler, W. S. and H. B. Gotaas "Optimization Analysis for Biological Filter Design," Journal of the Sanitary Engineering Division, ASCE, Vol. 92, SA1, Proc. Paper 4684, 163-182 (February 1966).

6. Galler, W. S. "Optimization Analysis for Biological Filters," Doctoral Thesis, Northwestern University (June 1965).

7. Hanumanula, V. "Performance of Deep Trickling Filters by Five Methods," Journal, Water Pollution Control Federation, Vol. 42, 1446-1457 (1970).

8. Howland, W. E. "Flow over Porous Media as in a Trickling Filter," Engineering Bulletin Extension Series No. 94, Proceedings of the 12th Industrial Waste Conference 1957, Purdue University, Lafayette, Ind., Vol. 42, 435-465 (1958).

9. Kelley, J. E. Jr. "The Cutting Plane Method for Solving Convex Programs," Journal of the Society for Industrial and Applied Mathematics, Vol. 8, 703-712 (1960).

10. Lemke, C. E. "The Dual Method of Solving the Linear Program Problem," Naval Research Logistics Quarterly, Vol. 1, 36-47 (1954).

11. Lumb, C. and P. K. Eastwood "The Recirculation Principle in Filtration of Settled Sewage - Some Notes and Comments on its Application," Journal and Proceedings of the Institute of Sewage Purification, Part 4, 380-398 (1958).

12. Lynn, W. A., J. A. Logan and A. Charnes "Systems Analysis for Planning Wastewater Treatment Plants," Journal, Water Pollution Control Federation, Washington, D.C., Vol. 34, 565-581 (1962).

13. Moore, W. A., R. S. Smith and C. C. Ruchoft "Efficiency Study of a Recirculating Sewage Filter at Centralia, Mo.," Sewage and Industrial Wastes, Vol. 22, 184-189 (1950).

14. Rankin, R. S. "Performance of Biofiltration Plants by Three Methods," Proc. Sep. No. 336, ASCE, Vol. 79 (1953).

15. Robertson, P. G., P. D. Wilson and J. C. Liebman "Statistical Comparison of Trickling Filter Equations using Paper Mill Effluent Data," Journal of the Water Pollution Federation, Vol. 40, No. 10, 1764-1768 (1968).

16. Schulze, K. L. "Load and Efficiency of Trickling Filters," Water Pollution Control Federation, Vol. 32, 245-261 (1960).

17. "Sewage Treatment at Military Installations," NRC Subcommittee Report, Sewage Works Journal, Vol. 18, 791-1028 (1946).

18. Velz, C. J. "A Basic Law for the Performance of Biological Filters," Sewage Works Journal, Vol. 20, 607-617 (1948).

C.

CONTROL OF INFECTIOUS DISEASE

SAMPLING & MAPPING RESERVOIRS & VECTORS OF HUMAN DISEASE

A. A. Arata
Vector Biology and Control
World Health Organisation

The problem of sampling reservoir and vector species have existed since it was first recognized that animals played a role in the transmission of human disease. Biological and mathematical descriptions of infectious cycles have followed, but are both highly dependent upon field sampling and the biological variables involved. This paper surveys the development and present status of sampling techniques used in ecological studies using those animals involved in zoonotic cycles as examples. It also traces the development of biographic mapping techniques of animal-borne diseases. The implications of these techniques for surveillance of epizootic and/or epidemic patterns are discussed.

INTRODUCTION

One of the fundamental problems in animal ecology is the sampling of populations that vary in time and space. Animal populations may vary in density and structure during brief periods (e.g. mosquitos), or by regular seasonal or annual cycles (e.g. insects in general and most mammalian species), and the long term population cycles seen in some rodents, other mammals and some forest insects. Spatial movement causes sampling problems whether it is the long range movements observed in birds or locusts, or the shorter distances traversed by field rodents. As many species of insects and mammals serve as vectors and reservoirs of human diseases, the dynamics of these populations are not only of academic concern.

Diseases such as malaria, onchocerciasis, yellow fever,

filiariasis, plague, and the rickettsial and viral zoonoses are still major public health problems in many developing countries. Considering that the most reliable and cheapest measures of control are often reduction of the numbers of animal vectors or reservoirs, the sampling and mapping of such populations is of major concern to public health authorities. Developing countries often spend a considerable portion of their per capita health budgets on vector and reservoir control, and the planning application and evaluation of the success of control programmes (whether by chemical, biological, genetic or environmental measures) is heavily dependent upon the accuracy of measures of density and spatial distribution involved. Thus the manner and accuracy of measurement and mapping of vector and reservoir can be of serious economic concern and may ultimately affect the national health budgets of many countries in a major way. Also as potential disease epidemics do not respect national borders, the distribution and densities of certain species, such as the stegomyia vectors of yellow fever, can be of serious international importance. Similar problems are encountered in the field of agriculture where pest control expenses, or the cost of losses due to the inability or failure to control such pests, is great. The resultant reduced agricultural productivity and/or subsequent destruction of food stores generally occurs in areas where such losses cannot be afforded, and may have secondary health effects where animal-borne and other diseases are intensified because of malnutrition and its concurrent disabilities.

Thus the problems of sampling and mapping populations of animals of public health and economic importance should be of major concern to national and international economic and policy planners. If it is not of concern, it is in part due to a lack of contact between ecologists concerned with quantitative management of animal populations and the appropriate governmental planners. Further, there is a lack of co-ordination between those concerned with manipulation of populations (reduction, maintenance or increase), public health workers, agriculturists and conservationists. The same ecological principles are in operation, and many of the same techniques for sampling, analyzing and mapping populations are applicable in all three fields.

The material presented in this report is that resulting from various projects conducted for public health purposes by WHO.

It should be understood that much of this work is preliminary, unpublished, and the results of the efforts of numerous individuals.

RABIES IN EUROPE*

The occurrence of wildlife rabies in Europe since the 1940's has been of concern to many countries as the disease has slowly spread southwest presently invading France and Switzerland. The concern for human and domestic animal health promoted massive

---

*This work was done as part of a joint WHO/FAO Wildlife Rabies Study, and the results (unpublished) have been compiled by the various national groups and WHO staff members.

fox (the principle reservoir and vector) control programs in certain areas that have been controversial between conservation and hunting groups, and public health workers. Using a simple arithmetic model (Figure 1) based on known reproductive rate and population density to estimate the natural recovery rate of fox populations controlled by gassing or intensified hunting. As there is good evidence that: 1) fox populations average about $1/km^2$, 2) fox populations reduced (either by control measures or disease) to approximately $0.25/km^2$ will not support transmission of rabies, and 3) reproductive potential allows 30-40% annual increase, it can be shown that control measures are not permanently detrimental to fox populations, and that "natural" population levels can be restored in a brief time. Analysis of the spread of over 3000 cases of wildlife rabies in southern Germany showed the spread to be about 35-40 km/annum, irrespective of land use patterns and topography.

Considering that rabies is a fatal disease of foxes, the epizootic lasts until transmission is broken by reduced populations. The data studies should help determine for public health authorities what degree of control is necessary, and for the hunter or conservationist, the length of time that control measures or disease will restrain the population concerned.

GLOBAL MAPPING OF DISEASE VECTORS*

The increased rates of international travel pose a problem for those concerned with spread of disease. The transport of the vector, the pathogen, or an infected host could result in epidemic spread of diseases such as yellow fever, dengue hemorrhagic fever, filiariasis, or plague. In certain cases, the vectors and the host (man), exist in areas where the pathogen has not been introduced, or become established. Computerized mapping of potential high risk areas, based on available data, is essential. Yellow fever is a good example. Man is highly susceptible, as is the primary mosquito vector, _A_. _aegypti_. Both occur in areas where yellow fever is not known (S. E. Asia and W. Pacific). Then, the mapping of vector distribution in areas at high risk is of considerable importance (Figs. 2 & 3). Every known record for _A_. _aegypti_ has been mapped either by presence or absence (P or X) including early records (E) of presence before 1945. Three commonly used indices of larval infestation were countered empirically to a scale from 1-9, utilizing 171 records from which all three were obtained at the same time and place.

Such mapping, supported by local collecting or monitoring system in endemic areas also provide the bases for surveillance

---

*This mapping program was developed by Dr. A.W.A. Brown, VBC/WHO. The results have been duplicated and distributed (VBC/72.6), and will be formally published by Dr. Brown in the future.

or "early-warning" system to predict outbreaks of disease, as seasonal (winter-summer or wet-dry) data become available. The utility is obvious, but the difficulty is not technical, but convincing national health planners that it is economically feasible.

The corollary to the above-mentioned work is that of pesticide resistance. The longterm use of pesticides for agricultural and public health purposes has stimulated the genetic resistance of many vector species to various classes of pesticides. Figures 4 & 5 detail the levels of resistance that A. aegypti have developed against DDT and malathion throughout the world. This is of considerable importance as the decision, often on national levels, to use one pesticide or the other is often based on economic, not biological grounds. It is the purpose of this program to provide the most recent global data on the potential utility of various pesticides so that national authorities can be well advised.

CALCULATION OF RODENT POPULATIONS*

As rodents are reservoirs of many human diseases, a program was initiated in 1970 to compare the various methods used to determine population densities and the time/space relationships between individuals. Estimates based upon five different

---

*This work was conducted in Switzerland to develop techniques that would subsequently be used in developing countries in disease studies. It was conducted jointly with the Swiss Federal Agricultural Station, Nyon. This computer mapping program was done in collaboration with the Institute of Public Health Research, University of Teheran, Iran.

variations of the mark-release technique provide similar relative estimates between seasons, although they differ by as much as 100% within trapping periods (Figs. 6-9). Population turnovers at a rate of 4-5 months, with maximum life expectancy of 9-10 months were demonstrated.

The observations do not fit the classical concept of "home-range" based on a central place theory. Lacking in this theory is the temporal aspect, as "home-ranges" shift over time. This is of considerable interest to students of epidemiology concerned with the localization of foci of disease. In addition it demonstrates the problem faced by the field biologist. The area studied was a grid of 200 x 200 m (4 ha. or 10 acres) with 314 fixed points and 628 traps operated for a three day period each three weeks (25 sampling periods in 18 months). Although rodents are often considered to have low mobility, over 25% (of 1000 observed) showed movement patterns within the grid of at least 100 m, indicating that our sample frame was spatially too small, although logistically we could not have increased it.

In a different study (1969-1971), a survey of wild mammal-borne diseases was conducted in Iran. Over 20 zoonoses were diagnosed, some recognized for the first time in the area. Distribution of all wild mammals in the country were mapped (Figs. 13 & 14) including data on relative abundance of each species. The manner in which the distributions are presented could easily serve as the basis for a national monitoring system for changes in rodent populations and potential outbreaks of epizootics and/or epidemics.

It is interesting to note that although 5000 animals were sampled for 20 diseases at 50 localities the data were insufficient to allow statistical analysis of time-space-habitat relationships.

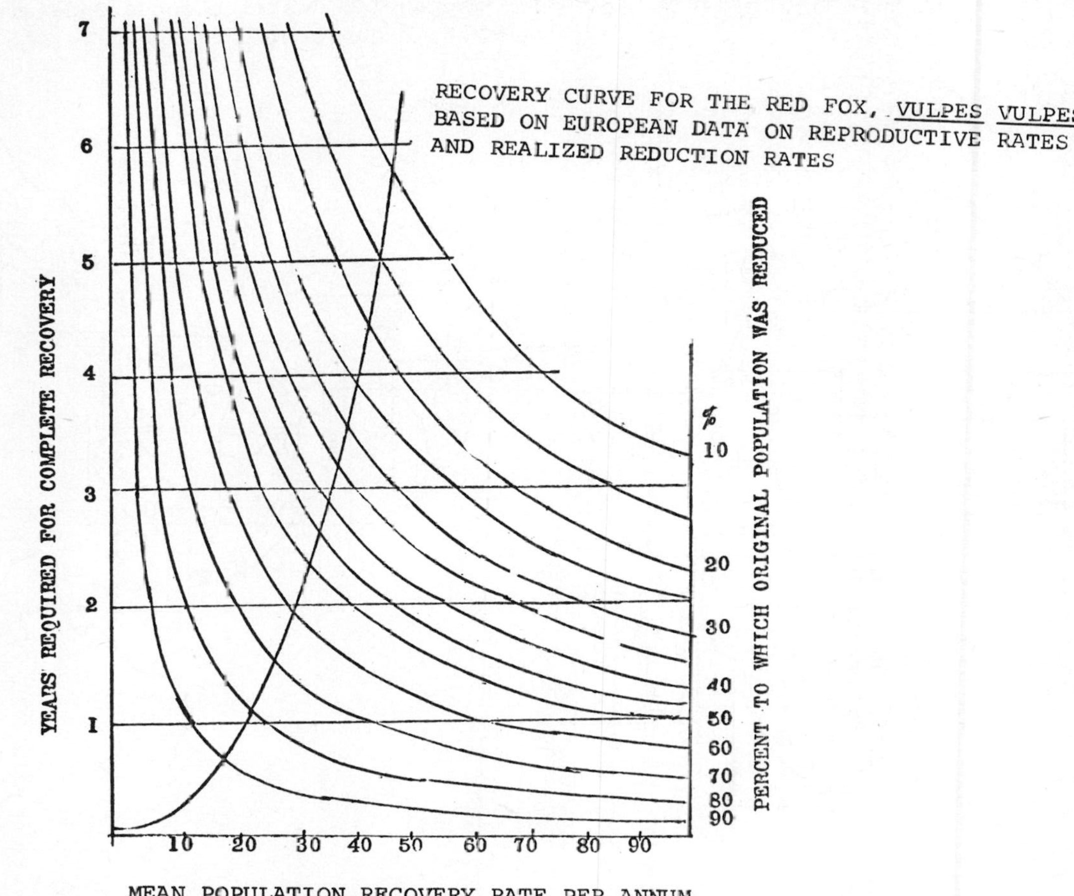

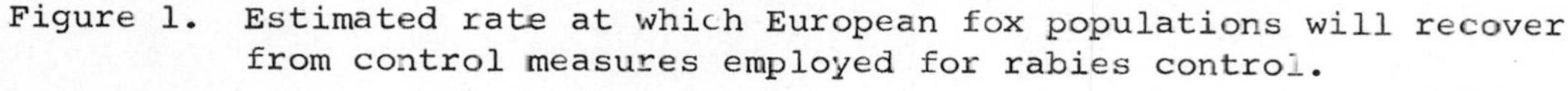

Figure 1. Estimated rate at which European fox populations will recover from control measures employed for rabies control.

Figure 2 Areas of the world mapped for distribution and density of Stegomyia mosquitoes.

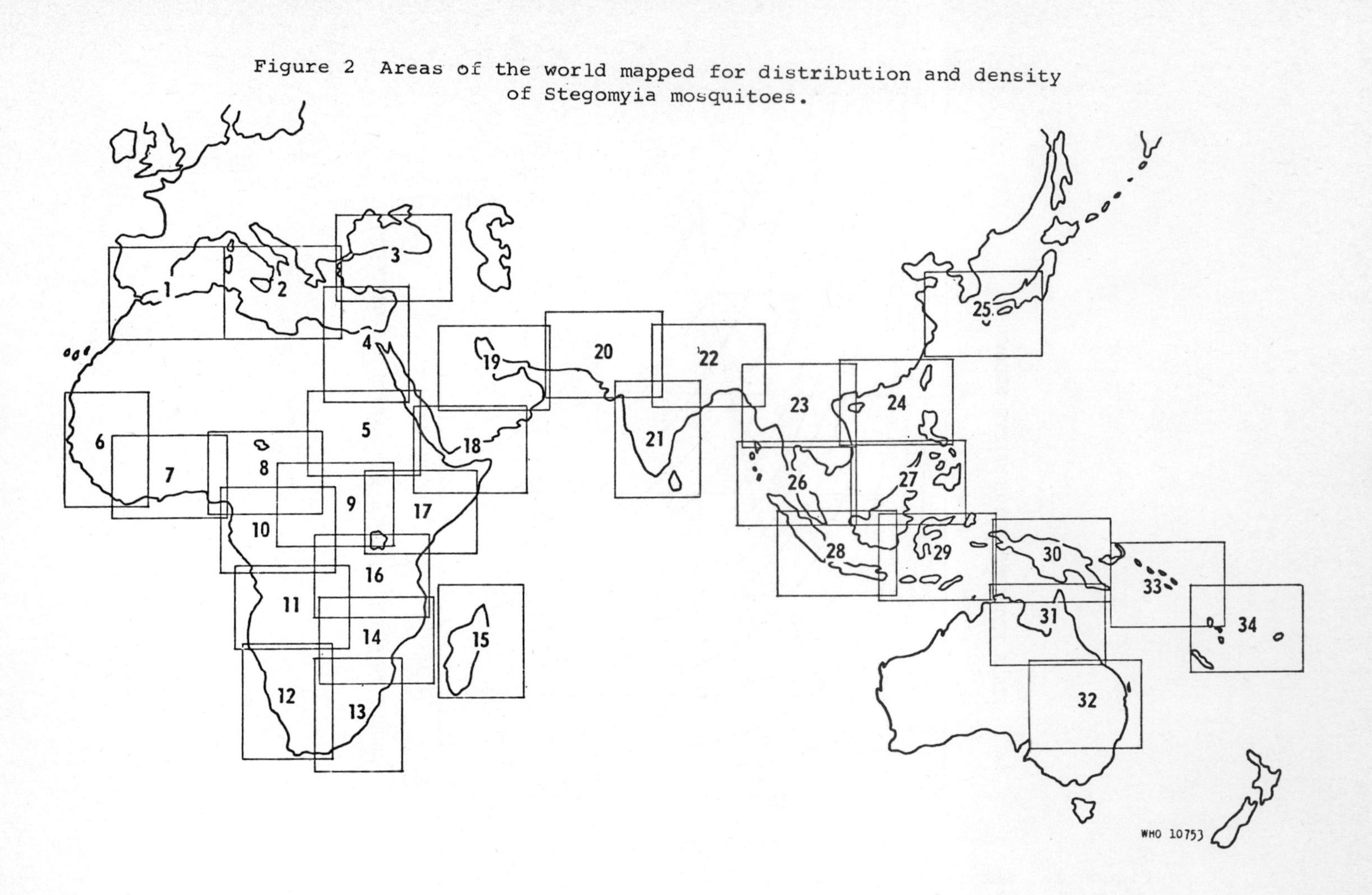

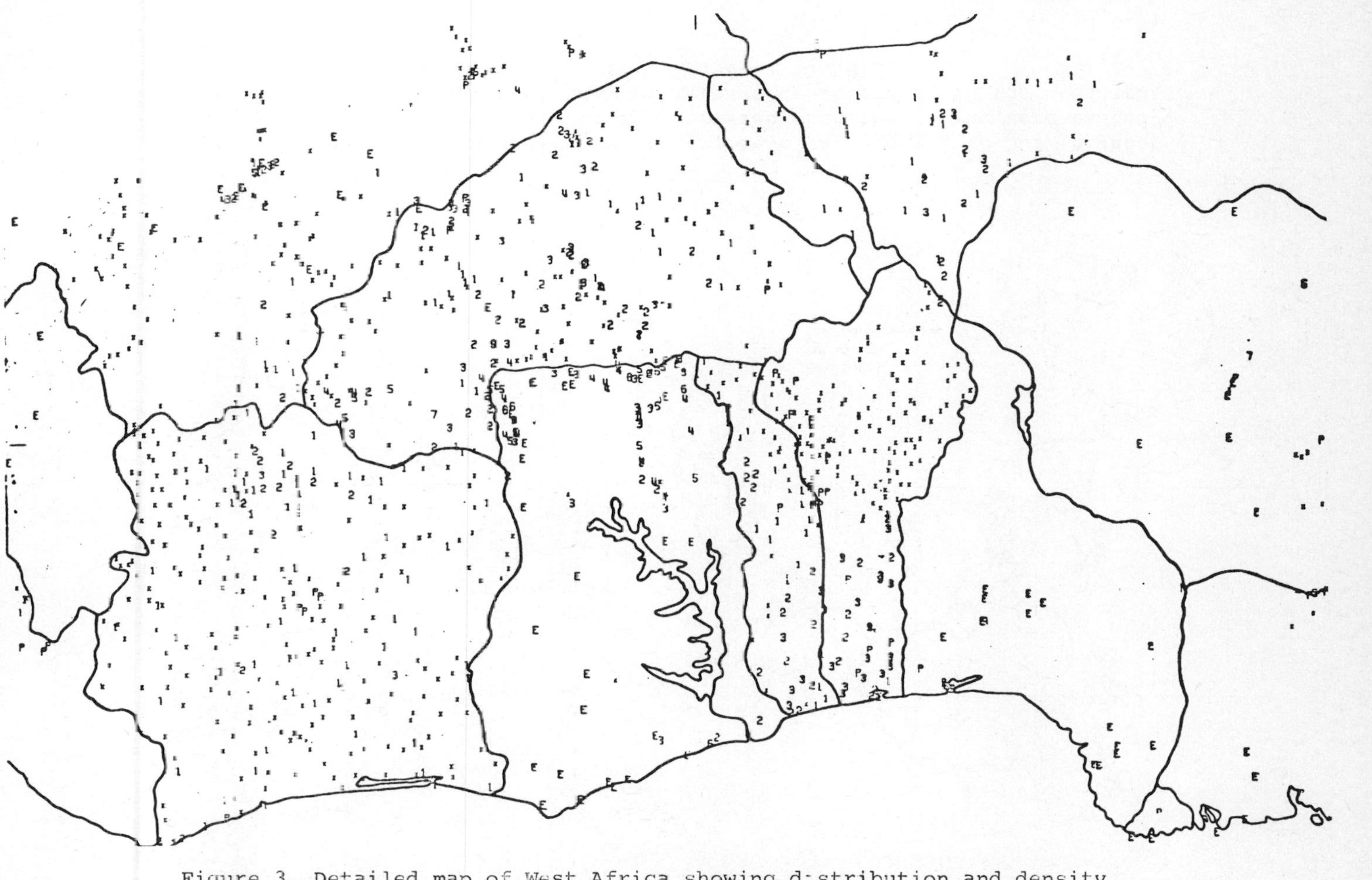

Figure 3 Detailed map of West Africa showing distribution and density of <u>Aedes aegypti</u>, the main vector of Yellow Fever.

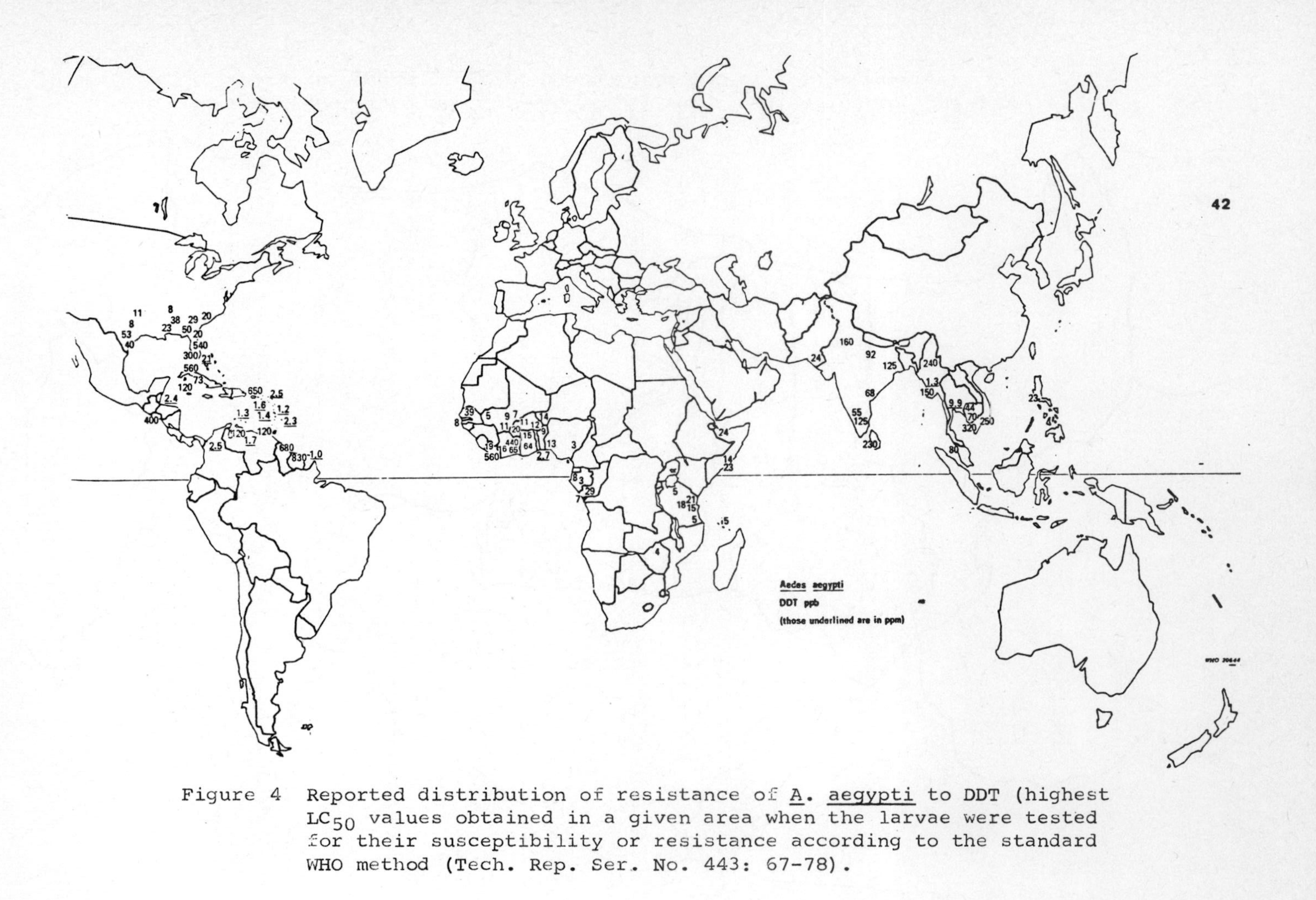

Figure 4 Reported distribution of resistance of A. aegypti to DDT (highest LC_{50} values obtained in a given area when the larvae were tested for their susceptibility or resistance according to the standard WHO method (Tech. Rep. Ser. No. 443: 67-78).

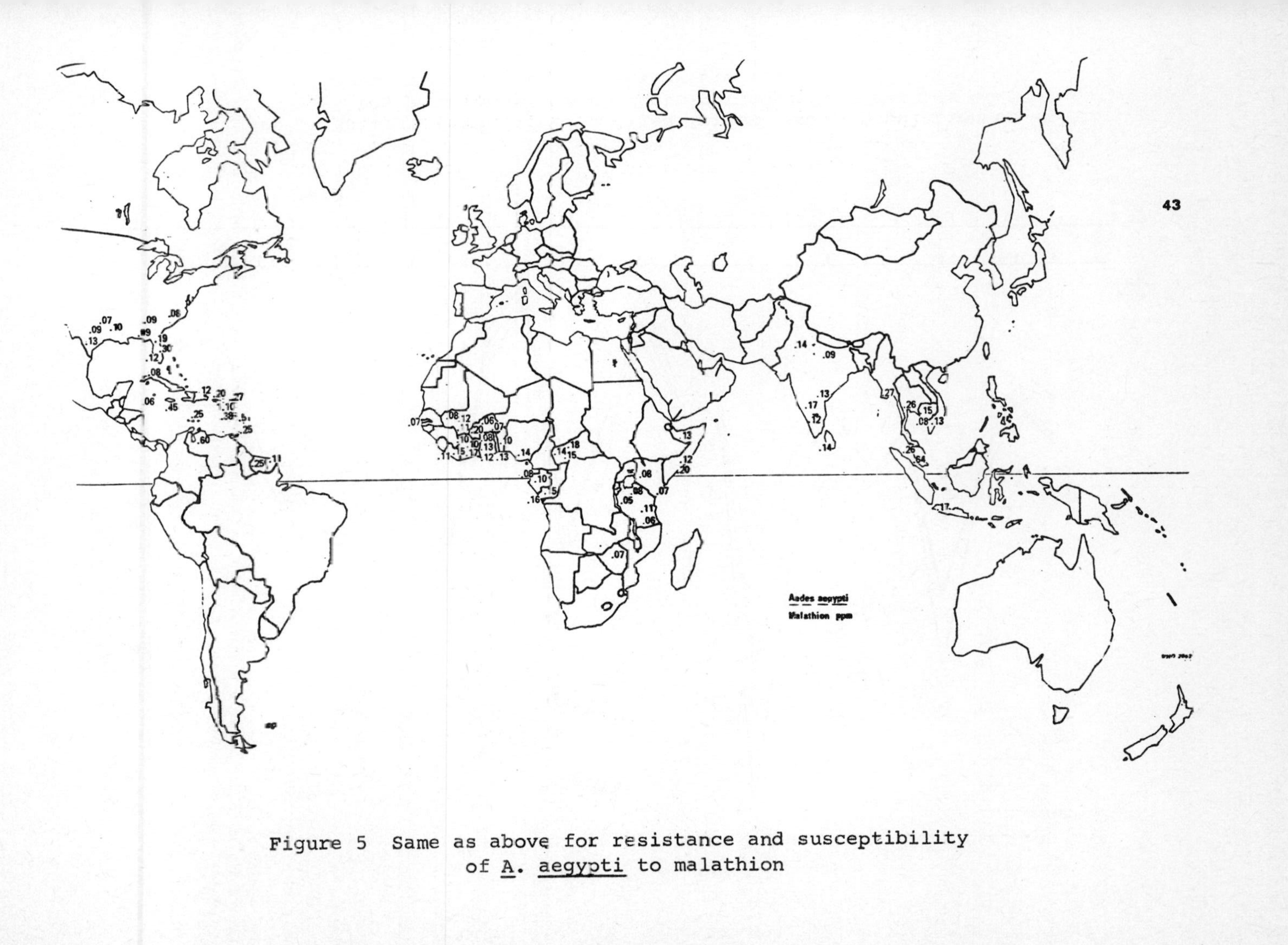

Figure 5 Same as above for resistance and susceptibility of A. aegypti to malathion

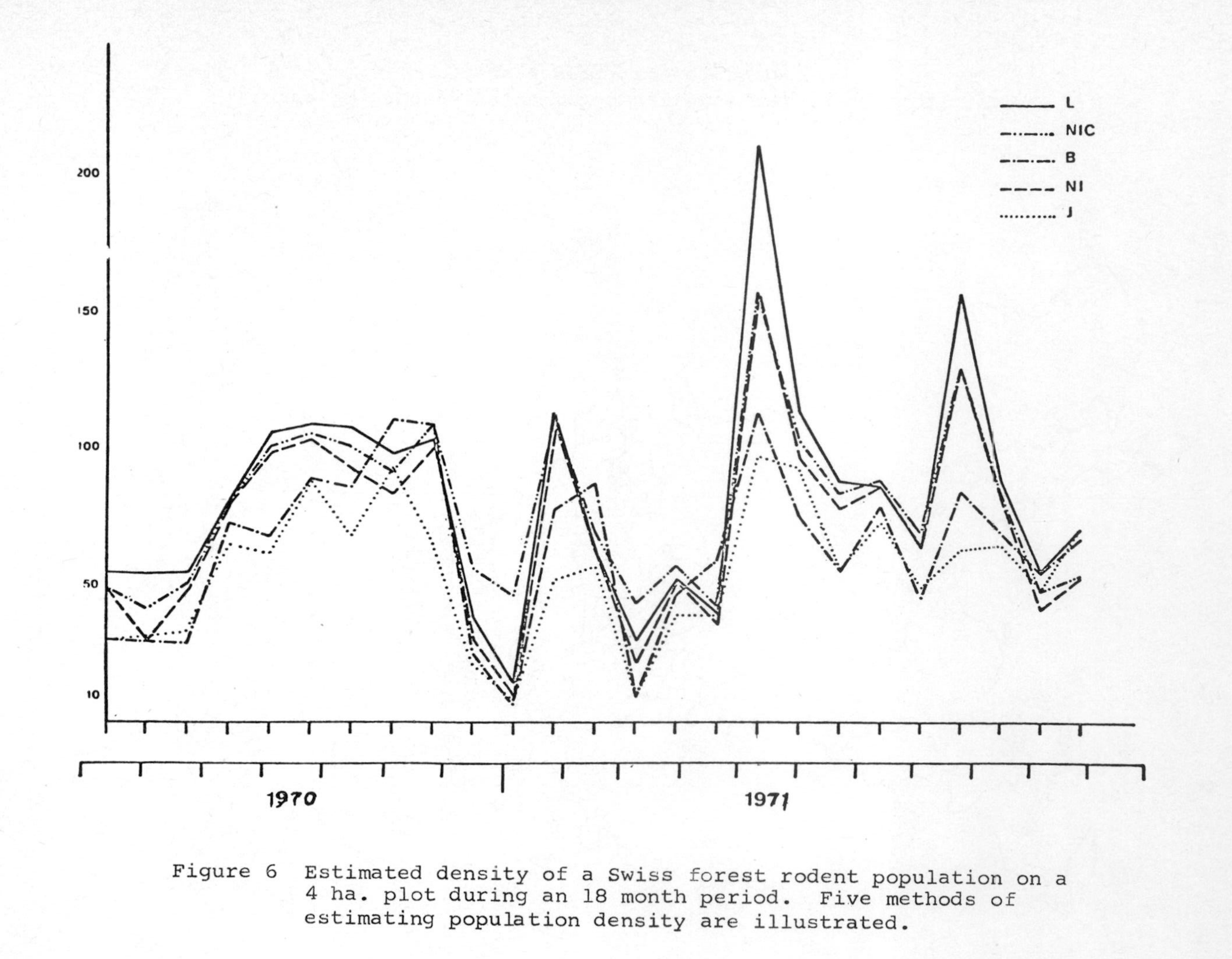

Figure 6 Estimated density of a Swiss forest rodent population on a 4 ha. plot during an 18 month period. Five methods of estimating population density are illustrated.

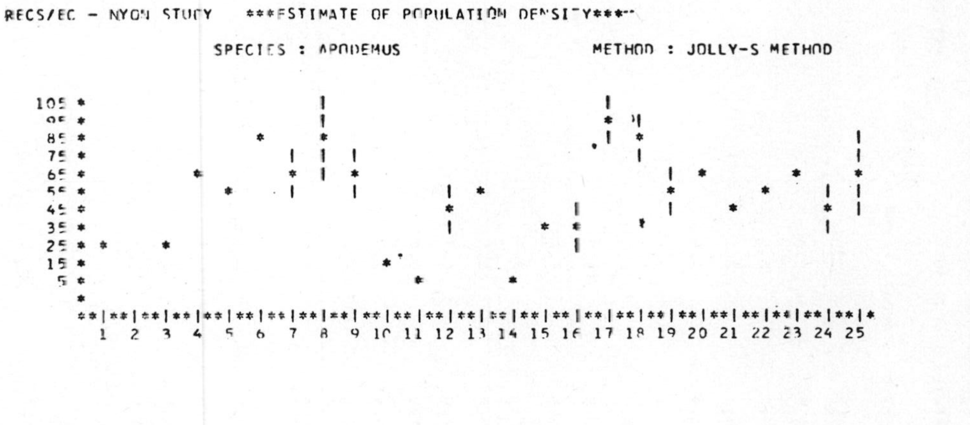

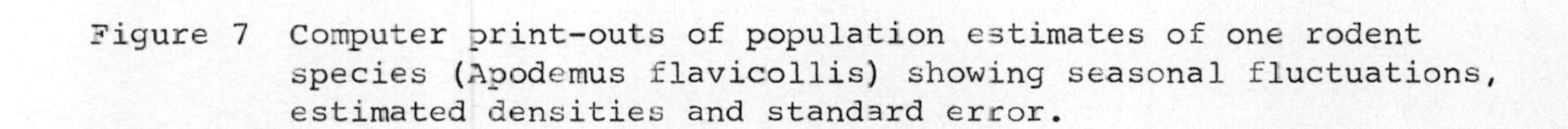

Figure 7 Computer print-outs of population estimates of one rodent species (Apodemus flavicollis) showing seasonal fluctuations, estimated densities and standard error.

RECS/EC - NYON STUDY ***ESTIMATE OF POPULATION DENSITY***

SPECIES : APODEMUS METHOD : JOLLY-S METHOD

PERIOD	ESTIMATED N	STANDARD ERROR
1	29.0	0.0
2	0.0	0.0
3	33.0	2.5
4	65.1	2.1
5	61.4	2.6
6	85.4	3.7
7	68.9	8.3
8	92.9	15.2
9	65.6	7.8
10	21.6	3.4
11	8.0	0.0
12	52.0	6.7
13	57.0	0.0
14	10.5	2.0
15	39.5	4.4
16	39.3	8.6
17	98.3	10.7
18	92.7	9.3
19	55.8	5.9
20	73.5	2.5
21	49.8	1.7
22	63.2	4.0
23	65.1	3.8
24	48.7	11.1
25	70.8	21.8

Figure 8 Computer print-outs of population estimates of one rodent species (*Apodemus flavicollis*) showing seasonal fluctuations, estimated densities and standard error.

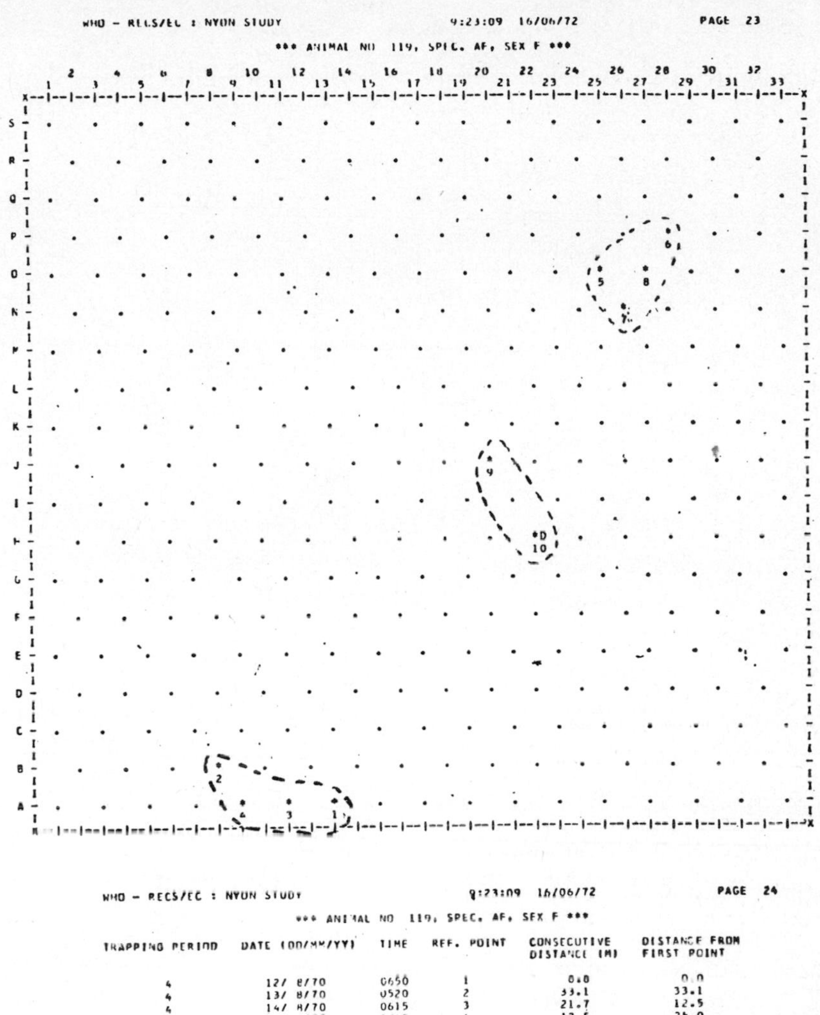

WHO - RECS/EC : NYON STUDY 9:23:09 16/06/72 PAGE 24

*** ANIMAL NO 119, SPEC. AF, SEX F ***

TRAPPING PERIOD	DATE (DD/MM/YY)	TIME	REF. POINT	CONSECUTIVE DISTANCE (M)	DISTANCE FROM FIRST POINT
4	12/ 8/70	0650	1	0.0	0.0
4	13/ 8/70	0520	2	33.1	33.1
4	14/ 8/70	0615	3	21.7	12.5
4	14/ 8/70	1245	4	12.5	25.0
5	2/ 9/70	1035	5	181.6	169.2
5	2/ 9/70	1625	5	0.0	169.2
5	3/ 9/70	1630	6	21.7	187.6
5	4/ 9/70	1055	6	0.0	187.6
6	23/ 9/70	1140	7	25.0	162.6
6	23/ 9/70	1815	7	0.0	162.6
6	24/ 9/70	0740	7	0.0	162.6
6	25/ 9/70	0655	7	0.0	162.6
6	25/ 9/70	1410	8	12.5	175.1
7	14/10/70	0930	9	69.6	106.8
7	15/10/70	0925	10	25.0	94.4

MEAN = 78.8

Figure 9 Time-space pattern observed for one individual (Apodemus)

WHO - RECS/EC : NYON STUDY 9:23:12 16/06/72 PAGE 31

*** ANIMAL NO 168, SPEC. CG, SEX F ***

WHO - RECS/EC : NYON STUDY 9:23:13 16/06/72 PAGE 32

*** ANIMAL NO 168, SPEC. CG, SEX F ***

TRAPPING PERIOD	DATE (DD/MM/YY)	TIME	REF. POINT	CONSECUTIVE DISTANCE (M)	DISTANCE FROM FIRST POINT
4	14/ 8/70	0655	1	0.0	0.0
5	2/ 9/70	0835	2	25.0	25.0
5	3/ 9/70	0800	1	25.0	0.0
5	4/ 9/70	0755	1	0.0	0.0
6	23/ 9/70	0945	3	65.0	65.0
6	24/ 9/70	0955	3	0.0	65.0
6	24/ 9/70	1140	4	12.5	54.5
6	25/ 9/70	0935	5	21.7	50.0
7	14/10/70	1120	6	21.7	69.6
8	4/11/70	0225	7	12.5	57.3
12	27/ 1/71	1705	8	33.1	66.2
15	31/ 3/71	0125	9	57.3	33.1
17	12/ 5/71	0530	10	25.0	45.1
17	12/ 5/71	1725	10	0.0	45.1
17	13/ 5/71	0855	8	33.1	66.2
17	13/ 5/71	1545	8	0.0	66.2
				MEAN = 22.1	

Figure 10 Time-space pattern observed for one individual (<u>Clethrionomys</u>)

Figure 11

FIGURE 11

Distribution of Captures of Mice in Switzerland on a Grid of 4 Ha.
During a Trapping Period in September 1970 when Densities were High.

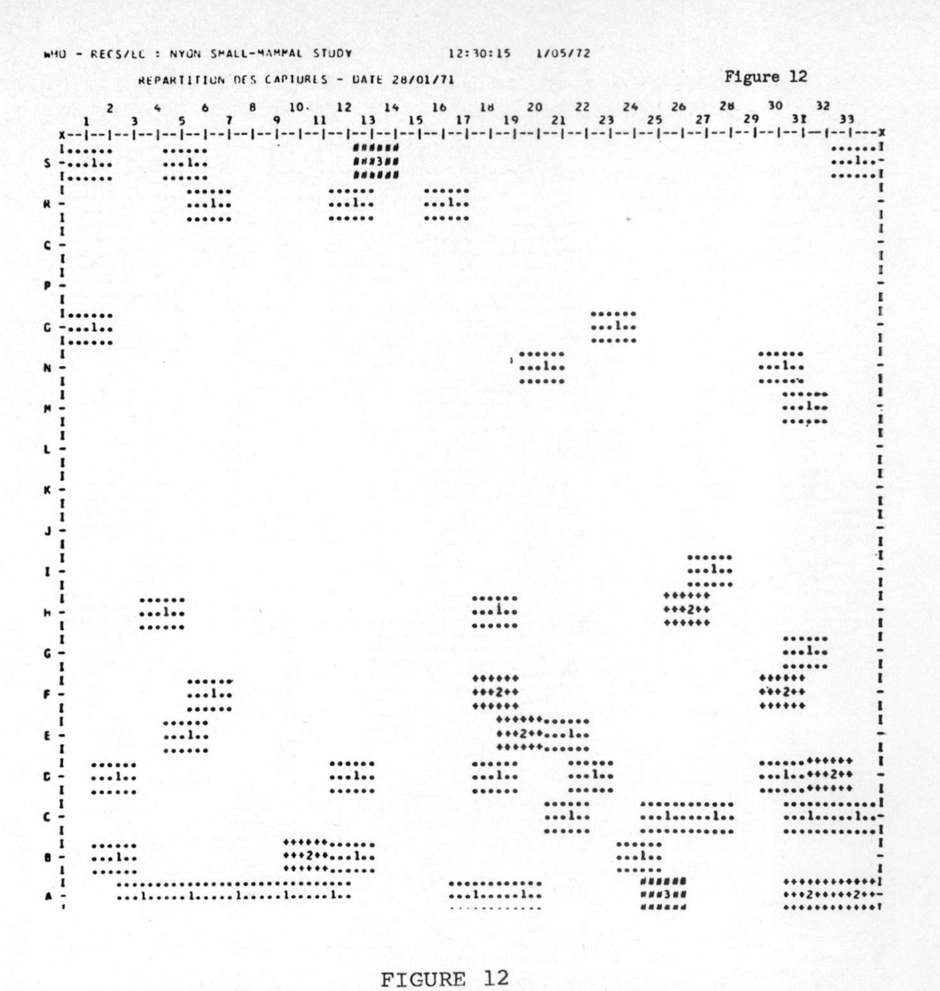

FIGURE 12
As above, but in January 1971, when densities were low.

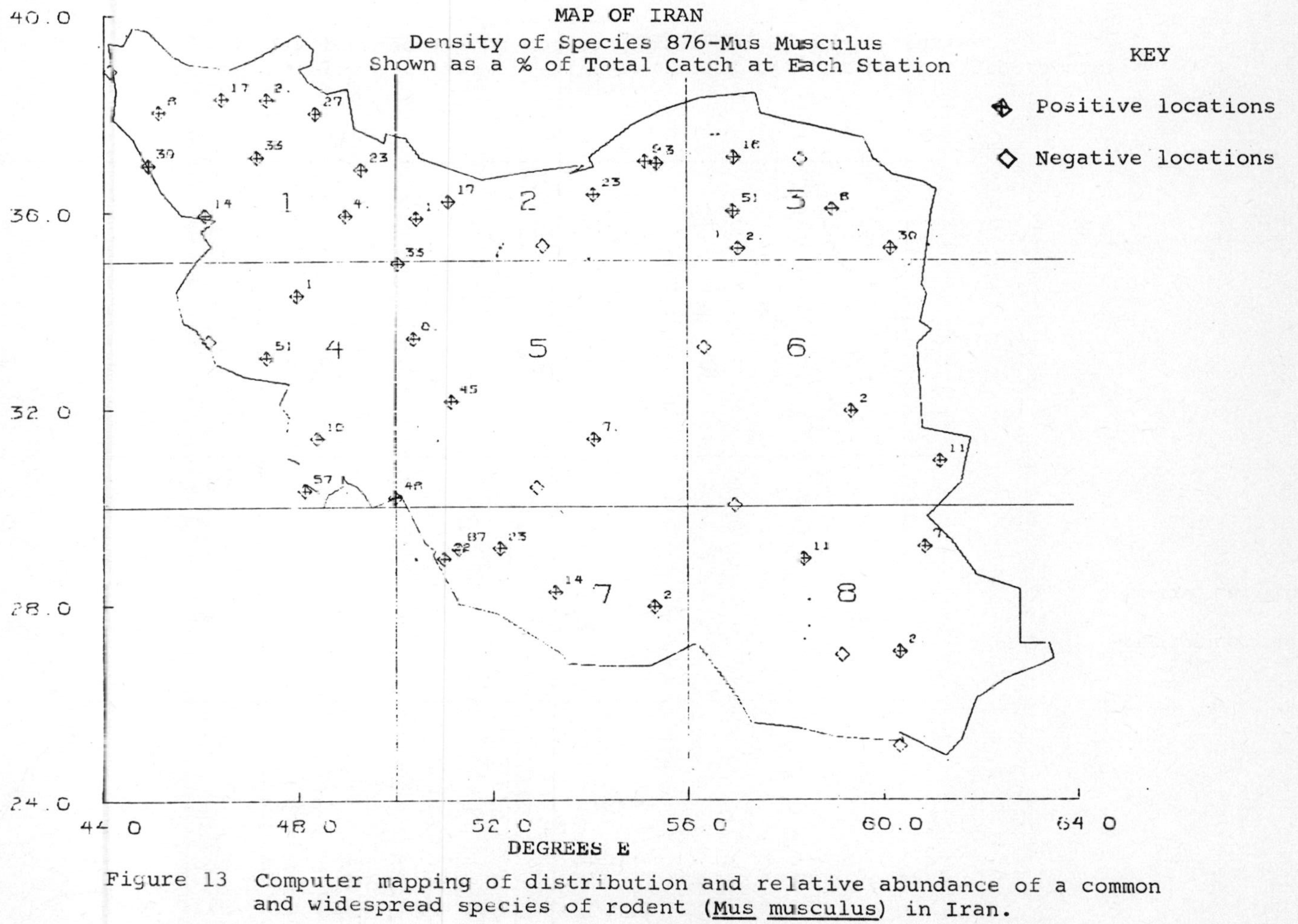

Figure 13 Computer mapping of distribution and relative abundance of a common and widespread species of rodent (Mus musculus) in Iran.

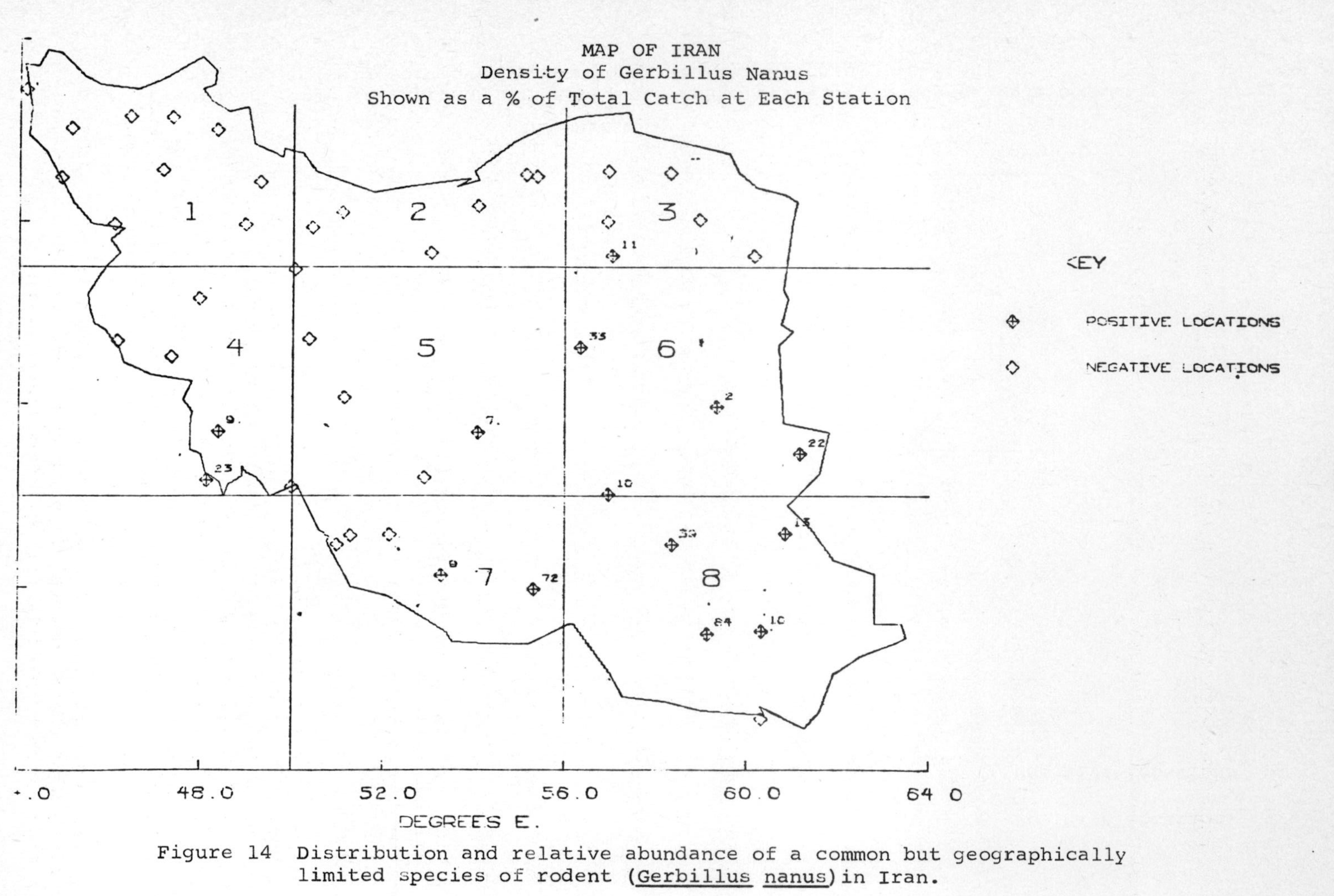

Figure 14 Distribution and relative abundance of a common but geographically limited species of rodent (Gerbillus nanus) in Iran.

THE TRANSMISSION AND CONTROL OF SCHISTOSOME INFECTIONS*

Warren M. Hirsch
Courant Institute of Mathematical Sciences, New York University

and

Ingemar Nåsell
Bell Laboratories, Holmdel, New Jersey

The following is a summary of the biomedical content of the paper presented (by the first author) at the NATO Conference, Istanbul. For precise mathematical formulations and details the reader is referred to [6].

1. Introduction and Biological Background

Schistosomiasis (known also as bilharzia or bilharziasis) is one of the great endemic diseases of the tropical and subtropical regions. In 1965 it was estimated by the World Health Organization [8] to affect 180-200 million persons throughout the world. Moreover, unlike most other infections, schistosomiasis is on the increase (Farooq [1], Jordan [2]) particularly in the developing countries, where man-made changes in the

---

*The work of the first author was supported by the Office of Naval Research, NR042-206, as principal investigator under Contract N00014-67-A-0467-0004. Much of his work was performed while he was Visiting Fellow, Yale University, where he enjoyed the kind encouragement and support of Professor Francis J. Anscombe, Department of Statistics, Dr. Gene Higashi, Department of Pathology, and Professors Curtis L. Patton and David Weinman, Department of Microbiology, to each of whom he acknowledges his appreciation.

The work of the second author was supported by Bell Laboratories under their Doctoral Support Plan. His present address is Royal Institute of Technology, Stockholm, Sweden.

environment are creating ideal breeding sites for various species of aquatic snails that serve as intermediate hosts for the parasite responsible for the infection. Since this parasite has a complex life-cycle, many different opportunities for the disturbance of this cycle present themselves. A central problem, therefore, is to develop methods that make it possible to compare the relative efficacies of various proposed strategies for control or eradication. The present paper is intended as a contribution in this direction, its main goal being to initiate the development of a mathematical methodology for studying the course of the infection as a function of the various biological and environmental factors that determine prevalence, intensity, and incidence.

Schistosomiasis is caused by a flatworm, a digenetic trematode, whose life-cycle involves one stage in a mammalian host (man or other animal) and one stage in a mollusc. Among the various species of schistosomes that can parasitize man, there are three of major medical significance: *Schistosoma haematobium*, *Schistosoma mansoni*, and *Schistosoma japonicum*. Although these species have similar life-cycles, they differ in host specificity and may therefore present different problems from the point of view of control or eradication. *S. japonicum*, which is prevalent in China, Japan, Taiwan, and the Phillipine Islands, infects not only man but a wide range of animals as well. The non-human mammalian hosts provide a reservoir for the parasite, which may play a significant role in transmission of the infection to man.

S. mansoni, which is widespread in Africa, the Middle East, South America, and the Caribbean Islands, is more restricted in its mammalian host distribution. In many foci of infection man is the principal host, but we cannot exclude the possibility that in some foci various other verterbrates play a role in transmitting the infection to man. S. haematobium is markedly but not exclusively host-specific to man, and while some natural animal infections have been discovered, they are not believed to play an important part in the transmission of the infection to humans. We restrict our discussion to S. mansoni and S. haematobium in endemic foci of infection where man is the only significant mammalian host.

The schistosomes are dioecious worms, that is to say, there are two adult forms, male and female, which mate bisexually. Although unisexual infections are possible, typically the infected person contains worms of both sexes, which seem to have little difficulty in finding each other in the liver of the definitive host, where pairing occurs. The paired female is carried by her male mate to a fine blood vessel, where they set up housekeeping, the male attached to the interior wall of the vessel, the female lying in a ventral groove (the gynecophoric canal) of the male, both freely suspended in the flowing blood and copulating regularly. Some of the fertilized eggs (there may be as many as 300-900 per day) succeed in passing through the wall of the blood vessel in which they are laid, thence through the wall of the intestine or bladder (depending on the species of schistosome) until they fall into the lumen of the organ and are voided with the excreta (feces or urine). Because of low hygienic standards, both personal and public, some of the eggs ultimately may be deposited in fresh water streams or lakes, where small ciliated larvae (miracidia) emerge. These swim actively, and if one comes into contact with an appropriate molluscan host (a snail of the genus Biomphalaria for S. mansoni and of the genus Bulinus for S. haematobium), it rapidly penetrates the snail tissue. By a peculiar process of repeated asexual multiplication within the snail, thousands of a second larval form (cercariae) are produced, all cercariae resulting from a single miracidium being of the same sex. After a prepatent period (4-6 weeks) the cercariae are shed by the snail, at first only a few being shed daily, but the number increasing until a reasonably constant daily level is reached (McClelland [4]). This level is usually maintained until a short time before the death of the snail, or as occasionally happens, the snail is spontaneously cured of the infection (Jordan and Webbe [3]). After being shed, the cercariae enter a free-swimming stage designed for invasion of the definitive host. On coming into contact with a human a cercaria attaches itself to the skin and quite rapidly penetrates it, while at the same time sloughing larval structures to become a juvenile schistosome (schistosomulum). This is followed by migration to the liver, maturation, pairing (if a worm of the opposite sex is available), migration to the permanent vascular abode, copulation, and oviposition, which begins the insidious cycle all over again.

As has been observed by the late Dr. George Macdonald [5], the interplay among the various biological and environmental characteristics that govern the transmission of schistosomiasis is so subtle and so complicated that without mathematical analysis it is virtually impossible to assess the relative effects of each of these characteristics on the prevalence, intensity, and incidence of the infection.

> . . . Whereas bacterial, virus and protozoal infections may multiply with relative ease from small origins, owing to the absence of immunity against them, and find growth restricted when they become common, the [bisexual] helminth inevitably to some extent experiences the reverse, the probability of the two sexes meeting being less when they are rare than when they are common, with consequent restriction from a small start. . . . However, this is not the only point of principle on which helminth infections differ from others. The reaction of immunity to many of them is low; superinfection is not only possible, but the general rule; multiple infections occur in considerable numbers and the infectivity of the host to others is related to the order of multiple infections carried, and consequently varies greatly from host to host and from time to time. The unit of infectivity is, in fact, the number of parasites harbored, whereas in most other infections it is the number of infected hosts. . . . Other characteristics to be taken into account are that most helminths do not complete multiplication within the vertebrate host, and require an alternative environment. When this is a living alternative host such as a snail or mosquito, superinfection may be possible but is necessarily limited if only by the relative size of the two organisms, with the result that infectivity of the alternative host cannot for long remain proportional to the number of infections which it has received. Its status as regards infectivity resembles that of the vertebrate host of bacterial and similar infections, an infective unit rather than the carrier of an almost infinitely variable number of discrete infections. The implications of these characteristics and their working are too complicated to follow by direct reasoning or to infer from field observation

> without a preliminary hypothesis as a background on which to work. The most effective way of studying them is through creation of a model by strict definition of these characteristics, their interpretation into a form which can be manipulated mathematically, analysis of the manner in which the different characteristics influence dynamic happenings, and study of the effect of changes in them on the total volume of transmission. Such a model does not primarily depend on a quantitative knowledge of these characteristics, but rather on their nature, and, by indicating which are the most significant, it may be a sound guide to the general policies of both research and operational prevention. [pp. 489, 490]

Viewed schematically (see Fig. 1) the transmission dynamics of schistosomiasis can be seen to depend on two interrelated stochastic flows, one a drift of eggs from human to snail, the other a stream of cercariae from snail to human. The intensities of these flows, while determining the level of endemicity of the infection, are themselves determined by a number of identifiable biological and environmental parameters. These include, for example, the sizes of the human and snail populations, the frequency and duration of human contacts with polluted waters, the magnitude of cercarial outputs of infected snails, the longevity and fecundity of coupled schistosomes, and so forth. As happens so often in applied mathematics, this functional dependence can be described implicitly by a system of differential equations whose solution describes the entire history of the infection and makes possible comparison of the relative efficacy of various procedures aimed at control or eradication. Since schistosomiasis is primarily an endemic infection, we have based our epide-

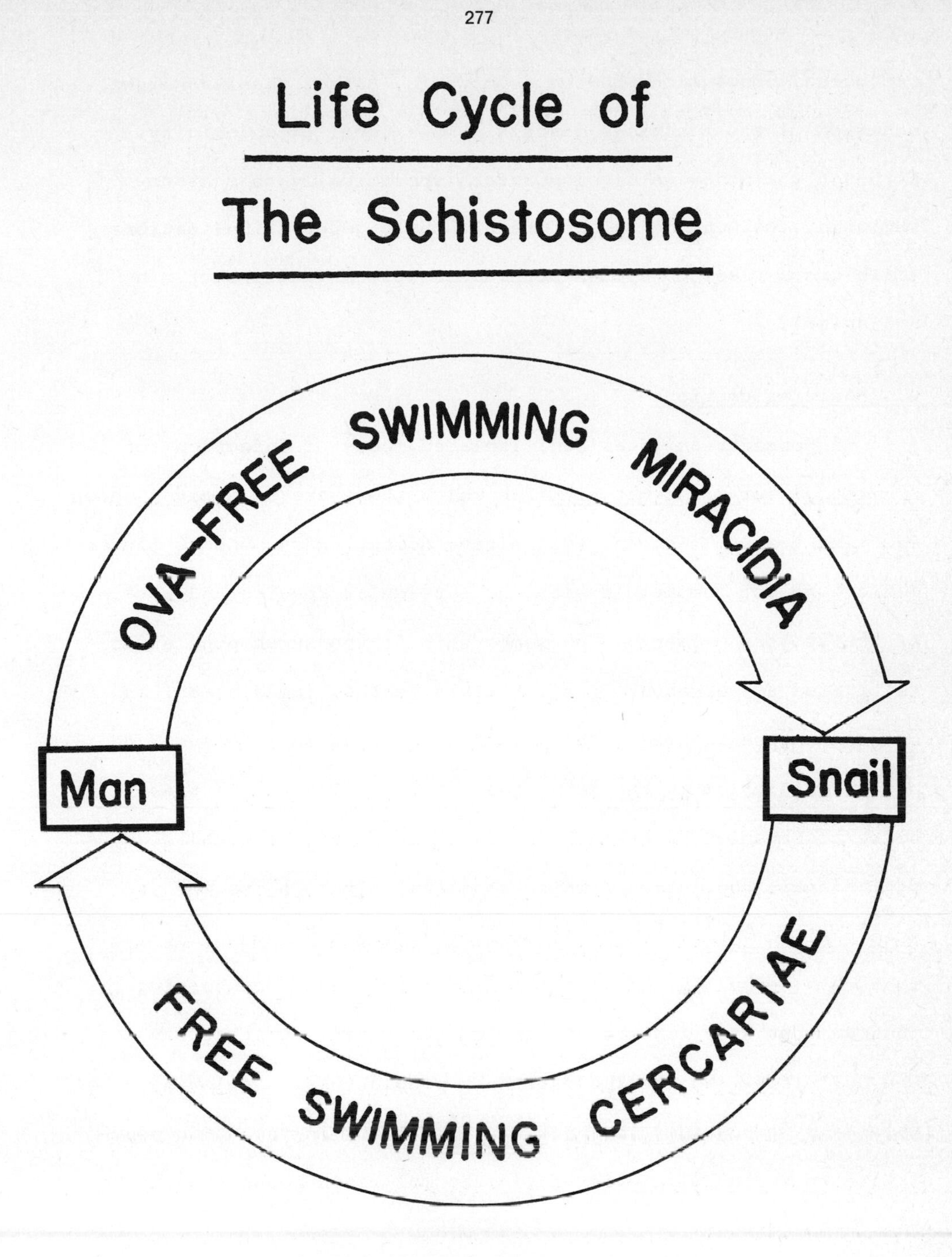

Life Cycle of
The Schistosome
OVA-FREE SWIMMING MIRACIDIA
Man
Snail
FREE SWIMMING CERCARIAE

FIGURE I

miological conclusions on the asymptotic, rather than transient, behavior of the functions describing the level of endemicity. Although the model we develop refers specifically to schistosomiasis, the methodology and results have general implications for a broad class of vector-borne parasitic infections of man and animal.

2. Basic Assumptions

We consider an idealized, isolated focus of infection for *S. mansoni* (or *S. haematobium*) in which there are at a given epoch labelled $t = 0$, H_2 humans (definitive hosts) and H_1 snails (intermediate hosts) compatible with the particular species and strain of schistosome present. The membership of the human population is treated as permanent, i.e., births, deaths, immigration, and emigration are ignored. The molluscan population is assumed to be in equilibrium in the sense that each snail death is presumed to be accompanied by a snail birth; in this way the total size of the snail population remains constant, although its membership varies over time. To describe the course of schistosomiasis in such a community, it is necessary to introduce bookkeeping devices that keep track over time of the status of infection in the definitive and intermediate host populations. Following Macdonald we consider the unit of infectivity in the human population to be the worm pair and in the molluscan population to be the infected snail.

We imagine that each human has attached to him a device

(schistosome counter) sensitive to penetrations of the human by schistosomes and to their deaths. This counter contains two long rolls of graph paper, on each of which a record of events is traced by a pen. The pens attached to the two rolls operate independently of one another, one being sensitive to events pertaining to male schistosomes and the other to female schistosomes. A rectangular coordinate system is printed on each roll of graph paper, the horizontal axis, X, referring to time and the vertical axis, Y, referring to schistosomes. At epoch $t = 0$ the pens are affixed, respectively, to the points on the Y-axis that represent the number of schistosomes of each sex then alive in the given definitive host. As time passes the pens trace straight lines in the positive direction of the X-axis. When a schistosome penetrates the host, at the instant of penetration the appropriate pen jumps one unit in the positive direction of the Y-axis and continues to trace a straight line parallel to the X-axis at this new higher level. If a schistosome dies, the appropriate pen instantaneously jumps one unit in the negative direction of the Y-axis and continues to trace a straight line parallel to the X-axis at the lower level. The resulting graphs, which are step functions, provide a history of theinfection in the given host. It is assumed that all penetrating cercariae develop into mature schistosomes instantaneously, that the number of pairs at time t is the minimum of the number of males and the number of females present, and that pairing is monogamous and permanent unto death. Accordingly we can infer from the graphs the number of infective units in each definitive host as a function of time and by summing obtain the number of infective units in the human population at each instant.

Plainly the processes that determine immigrations and deaths of schistosomes are stochastic, so that the graphs printed by the counter are random functions. We assume that these functions have the Markov property, namely, that knowing the male (female) worm burden at a time s the conditional probability of a death or an immigration of a male (female) worm at a future time t is the same as if we also knew the male (female) worm burdens at times prior to s.

It seems reasonable to postulate that during a sufficiently short time interval [s, t] the probability that the worm burden of a given human will change by more than 1 is negligible. Hence, if this burden consists of m males and f females at time s, the only transitions with significant probabilities are

(a) $m \to m-1$ (male schistosome death)

(b) $m \to m+1$ (penetration by a male schistosome)

(c) $m \to m$ (no change in male worm burden)

(d) $f \to f-1$ (female schistosome death)

(e) $f \to f+1$ (penetration by a female schistosome)

(f) $f \to f$ (no change in female worm burden).

Clearly the probability of (a), denoted by $P_{m,m-1}(s,t)$, depends on the number of male worms alive at time s and on the length of the time interval. We assume that it is approximately proportional to each of these, more precisely, that there is a constant $\mu_1 > 0$ and a small number ε_1 depending on s, t, and m such that

$$P_{m,m-1}(s,t) = (\mu_1 m + \varepsilon_1)\ (t-s), \tag{1}$$

where $\varepsilon_1 \to 0$ as $s \to t$ or $t \to s$. It is easily seen that $1/\mu_1$ represents the mean life of a male schistosome.

The probability of (b), denoted by $P_{m,m+1}(s,t)$, depends on the extent of infection in the snail population at time s and on t-s. Measuring the former by the expected proportion of infected snails at time s, denoted by y(s), we assume that there is a constant $\nu_1 > 0$ and a small quantity ε_2 depending on s, t, and m such that

$$P_{m,m+1}(s,t) = \{\tfrac{1}{2}\ \nu_1 H_1 y(s) + \varepsilon_2\}\ (t-s) \tag{2}$$

where $\varepsilon_2 \to 0$ as $s \to t$ or $t \to s$. This equation expresses the assumption that the probability of an additional male worm penetrating a given human during [s,t] is approximately proportional to the expected number of infected snails at time s and to the duration of the interval. The constant ν_1 is a measure of intensity of the cercarial flow per infected snail to a given human. If c_1 is the shed rate per infected snail and e_1(exposure factor) is the probability that a given cercaria infects a given human, then

$$\nu_1 = c_1 e_1 .^{1}$$

The transition (c) carries essentially all of the remaining probability, i.e., there is a number ϵ_3 depending on s, t, and m such that

$$P_{m,m}(s,t) = 1 - \{\mu_1 m + \frac{1}{2} \nu_1 H_1 y(s) + \epsilon_3\}(t-s) \quad (3)$$

where $P_{m,m}(s,t)$ denotes the probability of (c) and $\epsilon_3 \to 0$ as $t \to s$ or $s \to t$. This equation makes precise the assumption that a change of more than one in the male worm burden during a sufficiently short time interval has negligible probability.

It is assumed that equations (1), (2), and (3) hold for all human hosts. Thus, although the humans may differ markedly in infectious status at epoch t = 0, we have idealized the situation by postulating that thereafter they are equally exposed to risk.

A similar set of equations governs the transitions (d), (e), (f) in each human, where we assume that the average length of life of a female schistosome is the same as that of a male, and that the shed rate and exposure factor do not depend on the sex of the worm. We complete our description of worm transitions in the human by postulating that changes in the male and female worm burdens of a given host are independent of each other.

---

[1] Under the assumption that about ½ of the infected snails shed male cercariae, it is necessary to use the proportionality factor $(\frac{1}{2})\nu_1$ in equation (2), since the equation refers to male schistosomes.

Turning to the molluscan population, we picture a device (infected snail counter) that is sensitive to the penetration of uninfected snails by miracidia. It operates like the schistosome counters previously described, except that there is only one snail counter for the entire snail population. The graph it prints out shows the number of infected snails as a function of time. The counter is set initially at the number of infected snails at epoch $t = 0$. Consistent with our assumption that the infected snail is the infectious unit in the intermediate host population, the infected snail counter ignores superinfection of already infected snails. It is assumed that the infected snail graph is a random function that has the Markov property, that is, knowing the number of infected snails at a time s the conditionalprobability of a death of an infected snail or of an additional infected snail at a future time t is the same as if we also knew the number of infected snails prior to time s. Parallel to our treatment of worm burden in the human, we postulate that in a sufficiently small time interval, $[s,t]$, the number of infected snails has only a negligible probability of changing by more than 1. Thus, if there are i infected snails at time s, the transitions with significant probabilities are

(a') $i \to i-1$ (infected snail death)

(b') $i \to i+1$ (additional snail infection)

(c') $i \to i$ (no change in number of infected snails).

Plainly the probability of (a'), denoted by $\Pi_{i,i-1}(s,t)$, depends on the number of infected snails at time s and on the length of the time interval. We assume that it is approximately proportional to each of these quantities, i.e., that there is a constant $\mu_2 > 0$ and a small number ε_4 depending on s, t, and i such that

$$\Pi_{i,i-1}(s,t) = \{\mu_2 i + \varepsilon_4\}(t-s), \tag{4}$$

where $\varepsilon_4 \to 0$ as $t \to s$ or $s \to t$. The quantity $1/\mu_2$ is the mean life of an infected snail.

The probability of (b') depends on the extent of infection in the human population and on the number of uninfected snails (targets) at time s, as well as on the interval length, t-s. Measuring the infectious status of the humans by the expected worm pair burden in the human population, denoted by x(s), we assume that there is a constant $\nu_2 > 0$ and a small number ε_5 depending on s, t, and i such that

$$\Pi_{i,i+1}(s,t) = \{\nu_2 x(s)(H_1-i) + \varepsilon_5\}(t-s) \tag{5}$$

where $\varepsilon_5 \to 0$ as $t \to s$ or $s \to t$. This equation asserts that $\Pi_{i,i+1}(s,t)$ is approximately proportional to each of the quantities on which it depends. The proportionality factor ν_2 is a measure of the intensity of the flow of miracidia per worm pair to a given uninfected snail. If e_2 is the oviposition rate of a paired female schistosome, and c_2 is the probability that an oviposited egg infects a given uninfected snail (contamination factor), then

$$\nu_2 = c_2 e_2.$$

Since the probability of a change greater than 1 in the number of infected snails is negligible, we must assign essentially all the remaining probability to (c'). Thus, $\Pi_{i,i}(s,t)$, the probability of (c'), satisfies the equation

$$\Pi_{i,i}(s,t) = 1 - \{\mu_2 + \nu_2 x(s)(H_1-i) + \epsilon_6\}(t-s) \qquad (6)$$

where ϵ_6 is a small number depending on s, t, and i, and $\epsilon_6 \to 0$ as $t \to s$ or $s \to t$.

Plainly the preceding postulates simplify or ignore many aspects of transmission; for example, prepatent periods in the snail and the human, possible immunity or development of resistance to infection, dependence of the exposure and contamination factors on age and sex, density dependence of the worm death rate, and so forth. Nevertheless, we trust that the resulting caricature sufficiently resembles the real world to portray, though perhaps in distorted manner, some of its features. It is also an aid in polarizing thinking and posing sharp questions. The task for the future is to modify the assumptions in the direction of reality and to identify the conclusions that are permanent under increasing refinements of the model. It is our hope that among these will be the threshold effect and control efficiency results described in Section 6.

3. Determinants of Endemicity

By deriving a system of differential equations for the transition probabilities, it is possible to prove that our assumptions completely and uniquely determine the course of the infection. The nature of this course depends on the initial status of the infection in the definitive and intermediate host populations and on the various biological and environmental parameters that affect transmission. We summarize these factors below:

(1) initial worm burdens in each definitive hose, specified as H_2 pairs of numbers, $(m_1,f_1),(m_2,f_2), \ldots, (m_{H_2},f_{H_2})$, where m_k, f_k denote, respectively, the initial male and female worm burdens of host k

(2) initial prevalence of infection in the snail population, specified as a number y_0, $0 \leqq y_0 \leqq 1$, representing the proportion of infected snails at epoch $t = 0$

(3) $\frac{1}{\mu_1}$, mean life of a schistosome

(4) c_1, shed rate per infected snail

(5) e_1, exposure factor

(6) H_1, size of snail population

(7) $\frac{1}{\mu_2}$, mean life of an infected snail

(8) e_2, oviposition rate per paired female schistosome

(9) c_2, contamination factor

(10) H_2, size of human population.

The central problem is to elucidate the role of each of these factors in determining the level of endemicity of the infection and to study the change resulting from alteration of the various biological and environmental parameters.

4. Epidemiological Relations

The intensity of infection in the human population is measured by mean worm burden and mean worm pair burden; we denote the values of these quantities at time t by, respectively, $\beta(t)$ and $\gamma(t)$. Prevalence in the human population is interpreted to mean the expected proportion of humans having at least one worm pair; its value at time t is denoted by $p(t)$. Incidence means the rate at which new cases are developing. As a measure of "spread of infection" it is certainly of epidemiological interest, but as a measure of transmission or of public health conditions it is less satisfactory, since it refers only to changes occurring in the population of uninfected individuals. For schistosomiasis a better measure of transmission or of the state of public health at time t is the mean risk per human (infected or uninfected) at time t of an increase of 1 in worm pair burden. We denote this latter quantity by $r(t)$ and call it the "public health factor". While both incidence and the public health factor are relevant to the study of schistosomiasis, in the present work we consider only the latter.

In the molluscan population the prevalence of infection is measured by the expected proportion of infected snails; this quantity at time t is denoted by $y(t)$.

It can be proved that under the assumptions described in the preceding section the level of endemicity of schistosomiasis in a focus of infection tends to a steady state in the sense that as $t \to \infty$ all of the quantities $\beta(t)$, $\gamma(t)$, $p(t)$, $r(t)$, and $y(t)$

tend to limiting values; we denote these, respectively, by β_∞, γ_∞, p_∞, r_∞, and y_∞. Although these equilibrium values depend on the ten factors listed in the preceding section, it turns out that only certain algebraic combinations of these factors are important. More precisely, let us put

$$T_1 = \frac{1}{\mu_1} c_1 e_1 H_1$$

and

$$T_2 = \frac{1}{\mu_2} c_2 e_2 H_2 .$$

The level of endemicity in the steady state depends only on T_1, T_2, $1/\mu_1$, $1/\mu_2$ and the initial conditions in the sense that these quantities uniquely determine β_∞, γ_∞, p_∞, r_∞, and y_∞. T_1 and T_2 have suggestive interpretations. The quantity $\frac{1}{\mu_1} c_1 e_1$ measures the ability of an infected snail to deliver viable schistosomes, while H_1 may be thought of as the capacity of the reservoir containing infected snails. Therefore, T_1 represents the potential of the intermediate host population to deliver live schistosomes to a given definitive host. Similarly, $\frac{1}{\mu_2} c_2 e_2$ measures the ability of a paired female schistosome to deliver a viable miracidium to a given uninfected snail, while H_2 may be interpreted as the size of the reservoir that harbors paired female schistosomes. Therefore, T_2 represents the potential of the definitive host population to deliver a live miracidium to a given uninfected snail. We refer to T_1 and T_2, respectively, as the snail and human transmission factors.

The interpretation of T_1 is supported by a relation between β_∞ and T_1 that can be derived from the differential equations referred to in Section 3, namely,

$$0 \leqq \beta_\infty < T_1 .$$

This asserts that the mean steady state worm burden of a human host cannot exceed the potential of the snail population to deliver worms to that host. As might have been anticipated, the quantities β_∞, γ_∞, p_∞, r_∞ and y_∞ satisfy various algebraic relations. For example,

$$y_\infty = \frac{1}{T_1} \beta_\infty .$$

Thus, from a knowledge of the prevalence of infection in the snail population it is possible to estimate mean worm burden in the human population (assuming of course that the snail population size, oviposition rate, contamination factor, and mean life of a schistosome can be estimated). Since $\beta_\infty < T_1$, it follows that

$$y_\infty < 1.$$

(No matter how intense the infection is in the human population, not all snails are infected.) From mean worm burden it is possible to compute mean worm pair burden, namely there is a universal function f such that

$$\gamma_\infty = f(\beta_\infty) .$$

The function f is increasing. For large values of its argument it behaves like x/2, which means that in heavy infections almost all worms are paired. For small values of its argument it behaves like $\frac{1}{4}x^2$, which means that in light infections only a small fraction of worms are paired. For example, if mean worm burden in a given focus is 50, mean worm pair burden is approximately 22, while if mean worm burden is 0.5, mean worm pair burden is only 0.05.

From the mean worm burden, β_∞, we can compute also the prevalence, p_∞, of the infection in the human population through the relation

$$p_\infty = \left(1 - e^{-\frac{1}{2}\beta_\infty}\right)^2 < 1.$$

According to this relation, if β_∞ is very large, then $p_\infty \cong 1$ (the infection is holoendemic), while if β_∞ is very small, $p_\infty \cong \frac{1}{4}\beta_\infty{}^2$. To give numerical examples, for $\beta_\infty = 5$ we have $p_\infty = 0.84$; for $\beta_\infty = 1$, we have $p_\infty = 0.15$; and for $\beta_\infty = .2$, we have $p_\infty = .009$.

From the preceding it follows that the proportion of infected humans is related to the proportion of infected snails through the equation

$$p_\infty = \left(1 - e^{-\frac{1}{2}T_1 y_\infty}\right)^2 .$$

The public health factor is related to mean worm burden through the equation

$$r_\infty = \tfrac{1}{2}\,\mu_1\beta_\infty f'(\beta_\infty),$$

where the prime denotes differentiation. Like f, f' is also an increasing function.

The precise algebraic forms of the preceding relations should not be taken too seriously, since they may change as the model is modified. What is of great interest, however, is the qualitative fact if one of the quantities β_∞, γ_∞, p_∞, r_∞, y_∞ can be estimated, then estimates of all the others can be obtained (assuming, of course, that the parameters T_1, μ_1 can also be estimated). This could lead to practical indirect methods for estimating quantities that are difficult to get at directly; for example, it might be possible to estimate mean worm pair burden from the prevalence of the infection in the snail or the human population, without making fecal egg counts.

5. Nature of the Steady State

In this section we describe the nature of the dependence of the level of endemicity in the steady state on the transmission factors, T_1, T_2 the longevities, $\frac{1}{\mu_1}$, $\frac{1}{\mu_2}$, and the initial conditions, y_0, (m_k, f_k) $k = 1, \ldots, H_2$.

Let $T_1^* = \frac{1}{\mu_1^*} c_1^* p_1^* H_1^*$, $T_2^* = \frac{1}{\mu_2^*} c_2^* p_2^* H_2^*$ be the values of the transmission factors in a given focus of infection. We may plot the point (T_1^*, T_2^*) in a cartesian plane whose horizontal and vertical axes indicate, respectively, the snail and human transmission factors. The precise location of this point in the first quadrant determines the possible steady state values of the quantities that describe the prevalence and intensity of infection. It can be proved that there exists a curve (threshold curve), $T_2 = g(T_1)$ (see Fig. 2), such that if (T_1^*, T_2^*) falls below this curve the infection will ultimately vanish ($\beta_\infty = 0$, $\gamma_\infty = 0$, $p_\infty = 0$, $r_\infty = 0$, $y_\infty = 0$), regardless of the values of $1/\mu_1$, $1/\mu_2$, and no matter how prevalent and intense the infection is at epoch $t = 0$.

If (T_1^*, T_2^*) falls on the threshold curve, there are two

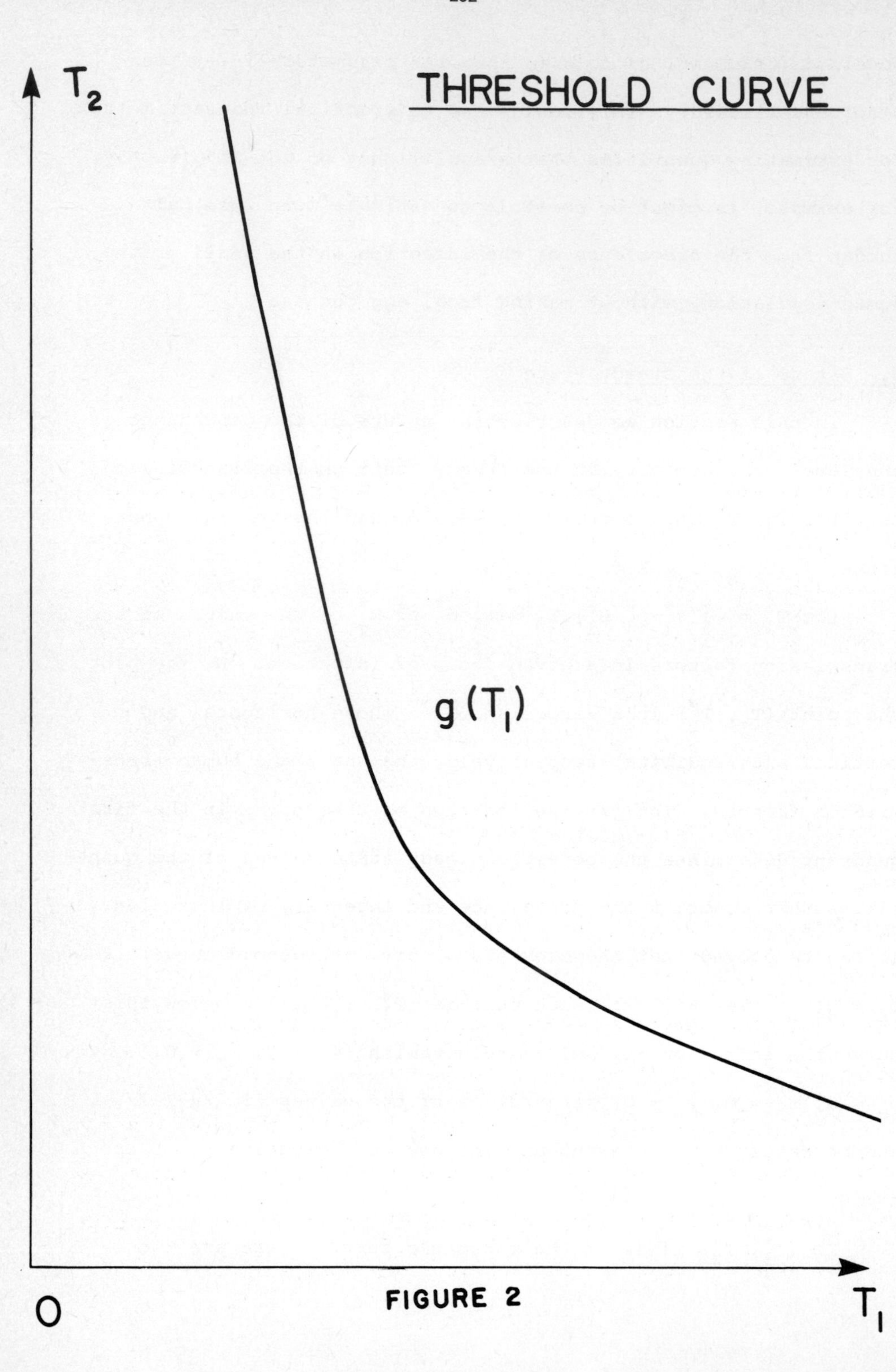

FIGURE 2

possible steady states. Either the infection vanishes, or it is driven to a specific positive level. Which of these two possibilities actually occurs depends on the initial conditions and the longevities, as well as on the point (T_1^*, T_2^*). Thus, if $T_2^* = g(T_1^*)$, either $\beta_\infty = 0$, $\gamma_\infty = 0$, $p_\infty = 0$, $r_\infty = 0$, $y_\infty = 0$, or $\beta_\infty = \beta^* > 0$, $\gamma_\infty = \gamma^* > 0$, $p_\infty = p^* > 0$, $r_\infty = r^* > 0$, $y_\infty = y^* > 0$, where β^*, γ^*, p^*, y^*, are computable from T_1^*, T_2^*, and r^* is computable from T_1^*, $1/\mu_1^*$, and T_2^*.

If (T_1^*, T_2^*) lies above the curve, there are three possible steady states. Either the infection ultimately vanishes, or it is driven to one of two positive levels. Which of these possible eventualities is realized depends on the initial conditions and longevities as well as on T_1^*, T_2^*. Thus, either $\beta_\infty = 0$, $\gamma_\infty = 0$, $p_\infty = 0$, $r_\infty = 0$, $y_\infty = 0$, or $\beta_\infty = \beta_1 > 0$, $\gamma_\infty = \gamma_1 > 0$, $p_\infty = p_1 > 0$, $r_\infty = r_1 > 0$, $y_\infty = y_1 > 0$, or $\beta_\infty = \beta_2 > 0$, $\gamma_\infty = \gamma_2 > 0$, $p_\infty = p_2 > 0$, $r_\infty = r_2 > 0$, $y_\infty = y_2 > 0$, where β_i, γ_i, p_i, y_i are computable from T_1^*, T_2^*, r_i is computable from $1/\mu_1^*$, T_1^*, and T_2^*, and $\beta_2 > \beta_1$, $\gamma_2 > \gamma_1$, $p_2 > p_1$, $r_2 > r_1$, $y_2 > y_1$. Mathematical analysis suggests (although it has not yet been rigorously established) that when the steady state level of endemicity is positive, the higher of the two possible positive levels is, in general, the one that will be established.

The threshold curve is monotonically decreasing. It behaves like the curve c/T_1^2 for small values of T_1 and like the rectangular hyperbola d/T_1 for large values of T_1, where c and d are constants. The existence of this curve is the mathematical explanation of George Macdonald's discovery that in computer simulations of the course of infection there is a "break point", a point below which the infection is unable to maintain itself [5]. The threshold curve shows that Macdonald's break point is not (as has been erroneously suggested) an artifact of the particular parameter values used in his programs, but an inherent property of the transmission dynamics of schistosomiasis (as modelled).

6. Control and Eradication

Although control may be a more realistic goal than eradication, the threshold curve shows that the latter is possible even without total elimination of the snail population or the worm population. Suppose, for example, that (T_1^*, T_2^*) lies above the threshold curve. By suitable interventions (mollusciciding, chemotherapy, sanitary engineering, etc.) it may be possible to alter the values of some of the biological or environmental parameters that affect transmission, thereby bringing about a new set of transmission factors (T_1^{**}, T_2^{**}). If this new point lies below the threshold curve (and can be maintained there) the infection will ultimately disappear. We may compare the relative efficacies of two different interventions, one of which reduces T_1^* and the other of which reduces T_2^*, in the following way. Consider an action like mollusciciding, which diminishes H_1^* (number of snails) and thereby reduces the value of T_1^*. Suppose that this is carried out until the point (T_1^*, T_2^*) is transformed into a new point

(T_1^{**}, T_2^*) which lies just below the threshold curve; the infection will then ultimately vanish. Alternatively, we might improve methods for disposal of human excreta, which diminishes p_2^* (contamination factor) and thereby reduces the value of T_2^*. Suppose that this is done to the extent that the point (T_1^*, T_2^*) is transformed into a new point (T_1^*, T_2^{**}) which lies just below the threshold curve; once again we may conclude that the infection will ultimately disappear. It can be proved that no matter where the point (T_1^*, T_2^*) is located above the threshold curve, it is an intrinsic property of the curve that

$$\frac{T_1^* - T_1^{**}}{T_1^*} < \frac{T_2^* - T_2^{**}}{T_2^*} ,$$

which asserts that the relative change in T_1^* required to being about eradication is smaller than the relative change in T_2^* required for the same purpose. This suggests that interventions that decrease T_1, the snail transmission factor, are more effective than those that decrease T_2, the human transmission factor. It supports Macdonald's conclusion, derived from computer simulation, that "safe water supplies are more important than latrines" [5, p. 502].

We regard the objectives of a control program to be reductions in prevalence, intensity, and transmission. Plainly such an undertaking is not a short term affair; control activities must be consistently maintained for many years to produce successful results. The analytical relations between the epidemiological indices and the transmission factors provide a useful theoretical tool for comparing the relative worths of various control strategies. Suppose that (T_1^*, T_2^*) lies above the threshold curve, and that the conditions are such that the level of endemicity stabilizes at the higher of the two possible positive steady state levels. It can be proved that the proportionate change in T_1^* required to bring about a small reduction in any one of the five epidemiological indices is always less than the proportionate change in T_2^* required to achieve the same reduction. Thus, once again it is suggested that interventions like the use of molluscicides, provision of fresh water supplies, or use of cercariacides, all actions which diminish T_1^*, are superior to interventions like improvement of sanitary engineering or use of miriacidiacides, which reduce T_2^*. Similarly, chemotherapy aimed at reducing the mean life of the schistosome, which reduces T_1^*, is better than chemotherapy aimed at reducing the oviposition rate, which diminishes T_2^*.

Finer discriminations can be made from a closer study of the public health factor. It can be proved that the proportionate change in $c_1^* p_1^* H_1^*$ required to produce a small reduction in r_∞ is less than the proportionate change in $1/\mu_1^*$ required to bring about the same reduction. Thus, for example, although schistosomicides that diminish mean life and molluscicides are both aimed at reducing T_1^*, the latter has a theoretical advantage over the former, since it acts by driving down H_1^* rather than $1/\mu_1^*$.

It is of interest to note that similar but not identical results hold for helminthic infections in which the parasite's life cycle resembles that of the schistosome, but where each parasite contains both testes and ovaries, and any pair can mate. For such infections it was shown by the present authors in [7] that T_1 is more important than T_2 from the point of view of control, but eradication through a decrease in T_2 is just as efficient as it would be through T_1. It appears, therefore, that the method of mating has serious implications for the approach to the problem of eradication.

The preceding results provide a theoretical frameowkr within which various proposed strategies for control and eradication can be thought about systematically. Although the results are tentative and must be viewed only as intimations of weak points in the life-cycle of the parasite, they do hint at the kind of potentially important breakthrough that may be attained by refining and carrying further this type of study. The factors in T_1 all relate to the parasite's ability to establish and maintain a viable infection in the snail. It is therefore tempting to speculate that in schistosomiasis the snail → human arc is more sensitive to intervention than the human → snail arc.

REFERENCES

1. Farooq, M. "Recent developments and trends in epidemiology and control of schistosomiasis," J. Trop. Med. Hyg. 72, 210-211 (1969).

2. Jordan, P. "Epidemiology and control of schistosomiasis," Br. Med. Bull. 28, 55-59 (1972).

3. Jordan, P. and Webbe, G. Human Schistosomiasis, William Heinemann Medical Books, Ltd., London (1969).

4. McClelland, W. J. F. "The production of cercariae by schistosoma mansoni and S. haematobium and methods for estimating the numbers of cercariae in suspension," Bull. Wld. Hlth. Org. 33, 2, 270-276 (1965).

5. McDonald, G. "The dynamics of helminth infections, with special reference to schistosomes," Trans. R. Soc. Trop. Med. Hyg. 59, 489-506 (1965).

6. Nåsell, I. and Hirsch, W. M. "The transmission dynamics of schistosomiasis," Comm. Pure Appl. Math. XXVI, No. 4, 395-453 (July 1973).

7. Nåsell, I. and Hirsch, W. M. "A mathematical model of some helminthic infections," Comm. Pure Appl. Math. XXV, No. 4 459-477 (July 1972).

8. WHO Expert Committee on Bilharziasis, Third Report WHO Techn. Rep. Ser., No. 229 (1965).

SIMULATION MODELS FOR GENETIC CONTROL ALTERNATIVES*

K. Dietz
World Health Organization

The problems in relation to the use of pesticides (e.g. environmental pollution, selection of resistant strains) have stimulated research in genetical methods of control and/or eradication of pest and vector populations. These methods aim at reducing the fertility by releasing either sterile males or strains with translocations. The paper surveys the mathematical models describing the various genetical control methods and predicting the outcomes of specified release strategies. It is shown that success and failure depend crucially on the density-dependent mechanism that regulates the natural population densities. Computer simulations allow us to optimize the timing of the releases in relation to the seasonal pattern of the carrying capacity of the environment. Particular reference is made to the genetic control of mosquito vector populations. In practice, the release of partially fertile males and even females together with the sterile males cannot be avoided. Under certain assumptions it can be shown that the release ratios have to be between certain boundaries in order to achieve population reduction. Transmission models help to assess the effects of a certain reduction in vector density on the transmission of the disease.

1. INTRODUCTION

Genetic control methods for insect populations have recently been reviewed extensively (Smith and Borstel [5]). The present paper is concerned with a few problems which arose in the context of an ICMR/WHO research project on the genetic control of mosquitos in New Delhi. We shall study the effects of immigration, density-dependent regulation and contamination of sterile male releases by partially fertile females on the outcome of a sterile male release programme and the effects of replacing a wild strain by an integrated strain

---

*The author is grateful to Dr. C. B. Cuellar and Dr. C. F. Curtis for stimulating discussions, and to Mr. A. Thomas for the programming in relation to the Tables.

which has a bidirectional cytoplasmic incompatibility and reduced fertility due to translocations. The two genetic methods, release of sterile males (RSM) and the replacement technique (RT) may be described by the same model which is presented in Section 2. In Section 3 we shall assume that the genetic method used will be perfect, i.e. there will be no fertile matings in the target area. We shall determine the equilibrium density which will be maintained by a certain rate of immigrating inseminated females. This equilibrium density represents a lower boundary to which the population may be reduced by any genetic control method. For many situations this value will be so high as to render the application of genetic methods useless. The following sections therefore make the assumption that the target population is isolated, so that time-limited release programmes could lead to eradication or replacement. Previous calculations about RSM assumed full sterility of the released males and perfect sezing such that no females were released. In practice it can often not be avoided that some partially fertile females are released and that the males are not fully sterile. We shall derive optimal release strategies for the situation in Section 4. In Section 5 we shall study the RT proposed by Laven and Aslamakhan [2] and compare it to the RSM.

2. A GENERAL MODEL FOR RSM AND RT

Before defining the variables of the model we shall briefly describe the principles of the two methods.

RSM: The sterility in the released material is induced either by exposure to radiation or a chemical mutagenic agent causing dominant lethal mutations. Sterile males are released into a target population in numbers far exceeding the wild males so that the probability that a female will be mated by a sterile male is large. If the sterility achieved is high enough to overcome the reproduction potential of the population then eradication is possible.

RT: Releasing both sexes of a strain which is incompatible in both crossing directions with the target population in sufficient numbers causes the target population to be replaced by the released strain, since inter-breeding between the released males and females is possible. The released strain carries certain translocations causing lower fertility which is retained in all its offspring. The variables of the model are given without the time index:

A_1 number of native females mated by a native male

A_2 number of native females mated by a released male

B_1 number of released females mated by a native male

B_2 number of released females mated by a released male

C_1 number of native males

C_2 number of released males

We make use of the fact that females mate only once and that males may mate several times. Thus for the females the fertility of the eggs laid throughout their life is determined by the single mating. We assume no bound for the number of matings a male can perform so that the chances of mating with a native or a released male simply depend on their relative numbers and on the competitiveness.

Let c_1 and c_2 be the competitiveness of released males against native males to mate with a native or a released female, respectively. Then the probabilities p_1 and p_2 that a native or released female, respectively, mates with a released male are given by

$$p_j = C_2 c_j / (C_1 + C_2 c_j) \qquad j = 1, 2$$

We assume that all females are mated.

In order to take into account the regulation of the larval and adult population by density dependent factors, we assume that the population is dependent on the availability of suitable breeding surface b. (Letting b vary with time allows simulation of seasonal changes.)

- $Z_1(i)$ number of aquatic forms developing into native forms of age i, i = 1, ..., n, where n is the duration of the aquatic stage (in days).
- $Z_2(i)$ number of aquatic forms developing into partially sterile forms of age i, i = 1, ..., n. ($Z_2(i)$ is non-zero only for RT.)

For the adults A_j, B_j, C_j, j = 1, 2 we suppose a constant age-independent death rate v. The adult female density F is defined as

$$F = (A_1 + A_2 + B_1 + B_2)/b$$

and the larval density D as

$$D = \sum_{j=1}^{2} \sum_{i=1}^{n} Z_j(i) \quad /b.$$

Ecological observations suggest that the probability of survival of the aquatic forms from egg to adult decreases with increasing density. In the absence of more precise information we assume that the daily probability of survival q of aquatic forms decreases exponentially as a function of D, i.e.

$$q = q_o \exp(-kD),$$

where q_o is the maximum daily survival probability (for D tending to zero) and k is the rate of decline of q. This implies the following relation between the numbers of aquatic forms of successive ages:

$$Z_j(i+1) = q\, Z_j(i), \quad i = 1, \ldots, n-1, \quad j = 1, 2.$$

Similarly to Fujita [1], we assume that the rate of oviposition, a, approaches a saturation level as a function of the adult female density F:

$$a = a_o\,(1 - \exp(-wF/a_o)),$$

where a_o is the maximum number of eggs oviposited per day per unit area and w is the rate of laying eggs per female per day.

The two genetic methods differ only in the definition of the fertility of the four different types of females. For the sterile male technique let m be the fertility of native females when mated by treated males and let g be the fertility of treated and released females when mated by native males. The fertility of treated females mated by treated males is assumed to be the product of m and g. Hence, according to the definitions of A_1, A_2, B_1 and B_2 we get the daily addition to the aquatic forms for RSM:

$$Z_1(1) = a(A_1 + A_2\, m + B_1\, g + B_2\, mg)/F. \qquad (1)$$

In the case of RT m and g are zero, i.e.

$$Z_1(1) = a\, A_1/F.$$

All the offspring of the released females will retain their low fertility due to the translocation. Therefore it is necessary to count the aquatic forms produced by released females separately. Let h denote the fertility of the integrated strain. Then

$$Z_2(1) = a\ B_2 h/F.$$

Let E_1 be the emergence rate of native males or females and let E_2 be the emergence rate of incompatible males or females (for RT) assuming a 1:1 sex ratio. Then

$$E_j = 0.5\ qZ_j(n), \quad j = 1, 2.$$

The model is discrete in time and assumes a one day iteration interval. The difference equations for subsequent values of the variables are as follows $(\Delta F = F(t + 1) - F(t))$:

$$\Delta C_1 = E_1 - vC_1$$

$$\Delta C_2 = E_2 + r - vC_2,$$

where r is the rate of release of males into the target population.

$$\Delta A_1 = E_1(1 - p_1) + I - vA_1,$$

where I is the rate of immigration of inseminated females.

$$\Delta A_2 = E_1 p_1 - vA_2$$

$$\Delta B_1 = (E_2 + rf)(1 - p_2) - vB_1,$$

where f is the proportion of virgin females included in the releases.

$$\Delta B_2 = (E_2 + rf)p_2 - vB_2.$$

3. IMMIGRATION OF INSEMINATED FEMALES

The problem of immigration of already inseminated females into the target area has so far been completely ignored in theoretical studies of genetic control methods. Field experiments are preferably performed on islands where immigration is no problem. Otherwise one has to establish a barrier zone around the target area to minimize immigration. In this section we would like to determine the density of females assuming a perfect genetic control method, i.e. all matings in the target area are fully sterile, such that all the females present in the target area at any point in time consist of the immigrated females plus the female offspring of the previous generation of immigrants. We shall calculate this density for a range of immigration rates and density-dependent regulation (DDR) mechanisms. The strength of DDR depends both on the maximum reproduction rate for densities tending to zero and on the slope of the probability of survival at equilibrium density. The maximum reproduction rate is given by the following formula

$$R = 0.5w\ q_o^n\ /\ v,$$

since w/v is the average number of eggs a female is laying throughout her life, half of which are female and q_o^n is the maximum proportion surviving from egg to adult age.

Assuming a daily mortality v of 0.2 and a rate of laying eggs w of 20, a female would produce an average of 50 female eggs throughout her life. If the aquatic cycle is given the value of 15 days, then the maximum daily survival probabilities 0.81, 0.9 and 1 give approximately reproduction rate of 2, 10 and 50, respectively. For k we choose the three values 10^{-8}, 10^{-6} and 10^{-4}. The maximum number of eggs laid per day per unit surface of breeding area has been set to 10^5. For these nine DDR mechanisms equilibrium adult and larval densities have been determined. In order to get comparable results we normalize the breeding surfaces such that the adult female density is equal to 1 million for all nine mechanisms in the absence of immigration. Table 1 gives the adult female density maintained by immigration in the target area assuming a constant immigration rate as given in the first row (in thousands). The density is expressed both in absolute numbers and as a percentage of the equilibrium density which would be maintained by immigration without genetic control. In the absence of DDR one would expect a density of $(R + 1)(I/V)$ in the case of perfect genetic control, i.e. there would be a linear relationship between density and immigration rate. We observe however either an approach of a saturation level or even a decline after a peak has been reached (DDR 9). If it were the aim of the genetic control programme to reduce density by 99%, we could only succeed in a situation with 200 females immigrating per day and DDR's 1-3 or with 20 females immigrating. In order to maintain this reduction in density releases would have to continue indefinitely on a massive scale. What is even more discouraging is the fact that for some DDR mechanisms one might end up with more females with control than without control (all those cases where the percentage is more than 100). This paradoxical phenomenon is due to higher survival of aquatic forms at lower larval densities due to a reduced input into the breeding places.

4. CONTAMINATED RELEASES

In genetic control programmes against disease vectors, especially mosquitos, it is necessary to separate males from females before releases in order to minimize the release of females. For *C. fatigans*

TABLE 1. Number (in thousands) of females maintained by immigration in the target area. In brackets: percentage of equilibrium without control

			Immigration rate of inseminated females				
DDR	q	k	0.02	.2	2	20	200
1	0.81	10^{-8}	.31 (0.03)	3.11 (0.3)	30.66 (3.0)	270.6 (24)	1702 (77)
2	0.81	10^{-6}	.31 (0.03)	3.12 (0.3)	30.98 (3.1)	291.9 (25)	1978 (92)
3	0.81	10^{-4}	.31 (0.03)	3.11 (0.3)	31.01 (3.1)	294.3 (26)	2000 (95)
4	0.9	10^{-8}	1.12 (0.1)	10.79 (1.1)	81.33 (8.1)	378.6 (34)	1638 (82)
5	0.9	10^{-6}	1.13 (0.1)	11.22 (1.1)	106.4 (11)	726.7 (68)	2078 (119)
6	0.9	10^{-4}	1.13 (0.1)	11.25 (1.1)	108.6 (11)	801.3 (75)	2001 (121)
7	1.0	10^{-8}	4.65 (0.5)	29.80 (3.0)	114.3 (11)	394.4 (36)	1653 (83)
8	1.0	10^{-6}	5.08 (0.5)	49.33 (5.0)	390.3 (39)	1557 (147)	2391 (141)
9	1.0	10^{-4}	5.09 (0.5)	50.31 (5.0)	448.7 (47)	2077 (198)	2004 (130)

a sex separator has been developed (Sharma et. al.[4]) which is based on the different sizes of male and female pupae. The females are bigger in the average than the males, so that mainly the males may pass through a grid of a certain width when they try to get to the water surface for breathing. If the grid is narrow, then very few females may be wrongly released, but then there is also a considerable wastage of males which are too big to pass through. On the other hand, a wider opening increases the output of males with the disadvantage of releasing more females. In the following we shall study the implications of this problem for the RSM, which is particularly important for the chemosterilized material, where there is a differential dose/response relationship for the two sexes. At doses sufficient to achieve 99% sterility in males, females may still be more than 50% fertile.

In order to get some analytical expression for the optimal release strategy we ignore for the sake of simplifying the argument that the reproduction rate is density-dependent.

This means that we try to minimize the input into the breeding surface given by formula (1).

If we denote the ratio of released males to emerging males (or females) r/E_1 by S, and assume that the released females are virgin and have the same chance of mating with a native male as with a released male (i.e. $c_1 = c_2 = c$) then we have to minimize the following expression

$$(1 + cSm)(1 + Sfg)/(1 + cS) \qquad (2)$$

for the fertility of the eggs laid.

Let us first assume that m, g, f, and c are fixed. Then (2) takes its minimum value for

$$S_{opt} = (\sqrt{1 + (c-cm-fg)/mfg} - 1)/c.$$

Thus the optimal strategy is to maintain this release ratio throughout the release programme, which means that the numbers to be released daily have to be reduced as the target population declines.

The optimal release ratio is only leading to a reduction of the target population, if the fertility expressed by (2) multiplied by the maximum reproduction rate is less than 1, i.e. if in the average one female is replaced by less than one female in the next generation. This condition yields an upper bound for the reproduction rates which are compatible with a successful release programme. This upper bound is

$$(1 + cS_{opt})/(1 + (cm + fg)S_{opt} + cmfgS^2_{opt}).$$

It also allows us to calculate lower and upper bounds for the release ratios which will give a population reduction. These bounds are given by the solution of the quadratic equation

$$Rcmfg\,S^2 + (Rcm + Rfg - c)S + R - 1 = 0.$$

So far we have assumed the proportion of females to be fixed. In practice one may switch from coarse sexing (high release rate of males and females) to fine sexing (low release rate of males and females) by choosing a finer grid size of the sex separator. The problem consists of finding the critical emergence rate below which coarse sexing should be replaced by fine sexing. Let K_1 and K_2 be the maximum daily output of males with coarse and fine sexing, respectively, and let $r = K_2/K_1$. Since the output for coarse sexing is larger than for fine sexing, r is less than 1. Let f_1 and f_2 denote the proportion of females with coarse and fine sexing respectively. The critical release rate for coarse sexing S^* satisfies the equation

$$(1 + cmrS^*)\ (1 + f_2grS^*)/(1 + crS^*) \quad =$$

$$(1 + cmS^*)(1 + f_1gS^*)/(1 + cS^*)$$

This leads to the quadratic equation

$$AS^{*2} + BS^* + C = 0$$

with

$$A = c^2rmg\ (f_1 - rf_2)$$

$$B = cg[m(f_1 - r^2f_2) + r(f_1 - f_2)]$$

$$C = g(f_1 - rf_2) - c(1 - r)(1 - m).$$

The relevant solution is given by

$$S^* = (-B + \sqrt{B^2 - 4AC}\ )/\ 2A$$

The critical emergence rate below which coarse sexing should be replaced by fine sexing is given by K_1/S^*. As soon as the optimal release ratio for fine sexing is reached, the output with fine sexing should be reduced at the same rate as the emergence rate of the target population declines. Table 2 illustrates these calculations with a numerical example.

So far we have neglected the DDR in order to get some simple analytical results for optimal release strategies. We can test by simulation, how much the true optimum release ratio for each DDR mechanism differs from the one which is independent of DDR. Let us

TABLE 2. Optimal release strategies depending on the number of males emerging per day in the target population and the competitiveness _c_ of the released males.

c	Number of males emerging per day in target population				
	20000	10000	5000	500	200
0.5	100%C	100%C 6762*	100%F 970**	52%F	21%F
0.2	100%C	100%C 6324*	100%F 622**	80%F	32%F
0.1	100%C	100%C 6891*	100%F	100%F 446**	45%F
0.035	100%C 16961*	100%F	100%F	100%F 278**	72%F

Daily output with fine (F) sexing = 225,000, proportion of females released : = 0.006
" " " coarse (C) sexing = 375,000, proportion of females released : = 0.02

* critical emergence rate below which output with coarse sexing should be replaced by fine sexing.

** critical emergence rate below which the output with fine sexing should be reduced.

Fertility of females = 60%, fertility of males = 1%.

take the example $c = 0.5$, $m = 0.01$, $f = 0.006$ and $g = 0.5$. The formula for the optimum release ratio yields approximately 254. The maximum reproduction rate for which this release programme could be successful is about 32. Thus DDR's 7-9 with a reproduction rate of about 50 cannot be overcome by this RSM. For DDR's 1-6 the optimum release ratios are somewhat lower than the value predicted under the assumption of no DDR:

DDR	1	2	3	4	5	6
optimal release ratio	227	207	203	242	229	208

The daily rate of reduction for DDR's 1-3 is about 12% whereas for DDR's 4-6 it is only about 5.6% which means that the duration of the release programme until virtual eradication depends crucially on the strength of the DDR mechanism.

5. REPLACEMENT BY AN INTEGRATED STRAIN

The results of Section 2 on immigration of inseminated females indicate already that reduction of fertility may lead to a high adult density due to DDR. This phenomenon also occurs if a fully fertile native strain is replaced by another strain with lower fertility. Table 3 lists the equilibrium densities after successful replacement expressed as a percentage of the natural equilibrium density of 1 million females for the same 9 DDR mechanisms as in Table 1. We only get eradication in those cases where the fertility times the reproduction rate is less than one, e.g. for a maximum reproduction rate of 10 the fertility of the released strain would have to be less than 10%. Such a low fertility raises some problems for the mass rearing of such a strain, since one needs at least 10 times more adult females to produce the same output as with a fully fertile strain. On the other hand, simulations show that one requires fewer mosquitos to be released in order to achieve replacement as compared to the RSM.

TABLE 3. Equilibrium densities after successful replacement expressed as a percentage of the normal equilibrium for 9 DDR mechanisms and 3 levels of fertility of the replacing strain

	Fertility of replacing strain		
DDR	5%	10%	25%
1	0	0	0
2	0	0	0
3	0	0	0
4	0	2%	24%
5	0	15%	100%
6	0	22%	117%
7	5%	12%	29%
8	118%	173%	180%
9	197%	223%	186%

We compare the results of two release strategies against 9 DDR mechanisms:

1. Release of a fully competitive integrated strain with 25% fertility at a constant rate for 90 days;
2. Release of chemosterilized males with 99% sterility and 50% competitiveness at a constant rate for 90 days assuming perfect sexing.

For all nine DDR mechanisms the target population has an equilibrium size of 1 million females. There is no immigration. At the start of releases both the adult and larval populations are reduced to 5% of the equilibrium by nonresidual insecticides.

Since the release of females should be kept minimal in the case of mosquito vectors, we also explored the effects of using a biased sex ratio in the releases on the total numbers required for replacement. As Table 4 shows, for some DDR's a 10:1 male to female sex ratio in the releases would require the smallest releases but from the public health point of view a 100:1 sex ratio is preferable. The last column of Table 4 shows the numbers of sterile males to be released daily for 90 days in order to reduce the number of females in the target population to less than one. According to the deterministic model used, all populations will eventually regenerate and return to the equilibrium size of releases are not resumed. The numbers required increase fast with increasing strength of DDR and are larger than those required for replacement even if one takes into account the lower competitiveness.

6. CONCLUSIONS

The simulations have demonstrated a number of crucial factors influencing the outcome of genetic control programmes:

1. Immigration of inseminated females into the target area has to be zero if eradication by time limited releases is the objective. If immigration cannot be prevented, then releases have to be continued indefinitely, but for some levels of immigration, genetic control may lead to an increase in density. High degrees of density reduction can only be expected for very low immigration rates.
2. Sterile male releases contaminated by partially fertile males and females may lead to an increase in the target population for high reproduction rates. Lower and upper boundaries for the release ratios as well as optimal release ratios may be determined which lead to eradication.
3. Eradication by an integrated strain is only possible if the new reproduction rate is less than one. The numbers required for release are less than those for the sterile male technique, but for some density-dependent regulations the new density is higher than the normal one.
4. The simulations have shown again the importance of the need to know the strength of the density-dependent regulation for the prediction of the effect of a particular release programme.

TABLE 4. Comparison of genetic control techniques for different density-dependent regulations.

	Sex ratio in the release								
DDR	100:1	50:1	20:1	10:1	5:1	3:1	2:1	1:1	(1)
1	$2.48 \cdot 10^4$	$2.39 \cdot 10^4$	$2.30 \cdot 10^4$	* $2.26 \cdot 10^4$	$2.26 \cdot 10^4$	$2.33 \cdot 10^4$	$2.43 \cdot 10^4$	$2.75 \cdot 10^4$	$6.66 \cdot 10^4$
2	$3.51 \cdot 10^4$	$3.40 \cdot 10^4$	$3.28 \cdot 10^4$	* $3.23 \cdot 10^4$	$3.27 \cdot 10^4$	$3.39 \cdot 10^4$	$3.56 \cdot 10^4$	$4.11 \cdot 10^4$	$9.53 \cdot 10^4$
3	$3.64 \cdot 10^4$	$3.52 \cdot 10^4$	$3.40 \cdot 10^4$	* $3.35 \cdot 10^4$	$3.39 \cdot 10^4$	$3.52 \cdot 10^4$	$3.70 \cdot 10^4$	$4.28 \cdot 10^4$	$9.92 \cdot 10^4$
4	$1.38 \cdot 10^5$	$1.26 \cdot 10^5$	$1.08 \cdot 10^5$	$9.38 \cdot 10^4$	$8.16 \cdot 10^4$	$7.54 \cdot 10^4$	$7.05 \cdot 10^4$	* $6.75 \cdot 10^4$	$1.03 \cdot 10^6$
5	$3.76 \cdot 10^5$	$3.44 \cdot 10^5$	$2.98 \cdot 10^5$	$2.64 \cdot 10^5$	$2.34 \cdot 10^5$	$2.18 \cdot 10^5$	$2.10 \cdot 10^5$	* $2.07 \cdot 10^5$	$3.59 \cdot 10^6$
6	$4.30 \cdot 10^5$	$3.94 \cdot 10^5$	$3.40 \cdot 10^5$	$3.01 \cdot 10^5$	$2.67 \cdot 10^5$	$2.47 \cdot 10^5$	$2.38 \cdot 10^5$	* $2.33 \cdot 10^5$	$4.38 \cdot 10^6$
7	$2.20 \cdot 10^5$	$1.90 \cdot 10^5$	$1.52 \cdot 10^5$	$1.26 \cdot 10^5$	$1.04 \cdot 10^5$	$9.22 \cdot 10^4$	$8.52 \cdot 10^4$	* $7.89 \cdot 10^4$	$6.05 \cdot 10^6$
8	$1.20 \cdot 10^6$	$1.00 \cdot 10^6$	$7.74 \cdot 10^5$	$6.32 \cdot 10^5$	$5.22 \cdot 10^5$	$4.64 \cdot 10^5$	$4.32 \cdot 10^5$	* $4.09 \cdot 10^5$	$3.22 \cdot 10^8$
9	$1.44 \cdot 10^6$	$1.19 \cdot 10^6$	$9.07 \cdot 10^5$	$7.32 \cdot 10^5$	$5.95 \cdot 10^5$	$5.22 \cdot 10^5$	$4.80 \cdot 10^5$	* $4.42 \cdot 10^5$	$3.34 \cdot 10^{10}$
	Minimum number of mosquitos of both sexes to be released daily for 90 days in order to achieve replacement								

* smallest number in the row (the corresponding sex ratio is optimal for that DDR).

(1) Minimum number of sterile males to be released daily for 90 days in order to reduce the number of females to less than one assuming that no females are released.

REFERENCES

1. Fujita, H. An interpretation of the changes in the type of population density effect upon the oviposition rate, Ecology, 35, 253-257 (1954).

2. Laven, H., Aslamakhan, M. Control of *Culex pipiens pipiens* and *Culex pipiens fatigans* with integrated genetical systems, Pakistan J. Sci. 22, 303-312 (1970).

3. Pal, R. LaChance, L.E. The operational feasibility of genetic methods for control of insect vectors of medical and veterinary importance, Ann. Rev. Entomology 19, 768-791 (1974).

4. Sharma, V.P., Patterson, R.S., Ford, H.R. A device for the rapid separation of male and female pupae, Bull. Wld. Hlth. Org. 47, 429-432 (1973).

5. Smith, R.H., von Borstel, R.C. Genetic control of insect populations, Science 173, 1164-1174 (1972).

MATHEMATICAL MODELS OF THE STERILE MALE TECHNIQUE OF INSECT CONTROL

William G. Costello
Howard M. Taylor

Cornell University
Center for Environmental Quality Management

The sterile male technique is a biological pest control method in which sterilized members of a target insect species are released into an infested region in order to dilute the native population's reproductive capacity. Under certain conditions, this can lead to extinction of the target population. Experience with the method, including the critical 1954 Curacao experiments and the recent outbreak of screwworm flies in the Southwestern U.S. after years of successful control, suggest a possible threshhold between success and failure. We present a stochastic model, based on the birth-and-death process, generalizing a deterministic model due to E.F. Knipling. Computations with our model show a sharp threshhold in the number of sterile males needed for success, depending on parameters such as net reproduction rate and environmental capacity, and where success is measured by the criterion mean lifespans to population extinction. The model predicts that adding steriles beyond this threshhold will yield little additional return in reducing the mean time to extinction, while supplying less than the threshhold number will have virtually no effect in terms of this criterion. From a preliminary statistical analysis of data from the Curacao experiments we obtain estimates of population size and net reproduction rate. Using these estimates in our stochastic model yields predictions that are consistent with the Curacao results. Suggestions for future work are offered.

INTRODUCTION

The sterile male technique for the suppression of insect pest populations has become an effective, and sometimes spectacular, method of biological control. It offers a precise means for dealing specifically with a single species, without the harmful direct effects which can occur with large-scale pesticide usage. The first successful field test, conducted on the island of Curacao, Netherlands Antilles against the screwworm fly (Baumhover *et. al.* [2]), involved the release of sterilized adult screwworm flies from aircraft. Pens of goats were maintained on the island to provide controlled oviposition sites, where the experimental team periodically counted the number of egg masses and determined how many of these were sterile. For the first several weeks, the sterile flies were released at the rate of approximately 100 per square mile per week. Later, an increased rate of approximately 400 per square mile per week was used for several weeks. Figure 1 shows the total number of egg masses counted in each week for the periods corresponding to the two release rates. In the first case, with the lower release rate, there appears to be an increasing population trend, despite the presence of sterilized flies. In the second case, accompanied by the higher rate of release of sterile flies, no egg masses were found after 9 weeks, nor for several weeks thereafter. In other words, the local screwworm fly population had been driven to extinction within a time period which is about 3 times the average life span of that species! These remark- experiments proved the usefulness of the sterile male technique in the field, paving the way for its later large-scale application.

This sharp contrast between success and non-success was of great interest to us. The sterile males seemed to have virtually no effect on the natural population at the low release rate, but the higher release rate led to rapid extinction. Might there be a critical number, or threshold, such that releasing at least that many sterile flies would lead to rapid extinction, while a lesser amount would not appreciably affect the population size? Was the extinction fortuitous, brought about by a combination of factors such as the unfavorable dry weather and declining population levels intervening between the two experiments? If there was a threshold level of

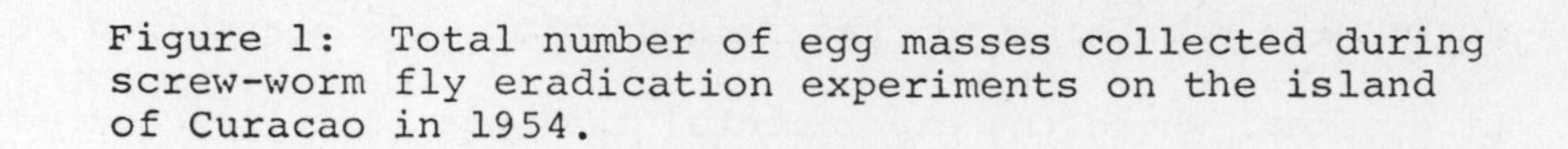
Figure 1: Total number of egg masses collected during screw-worm fly eradication experiments on the island of Curacao in 1954.

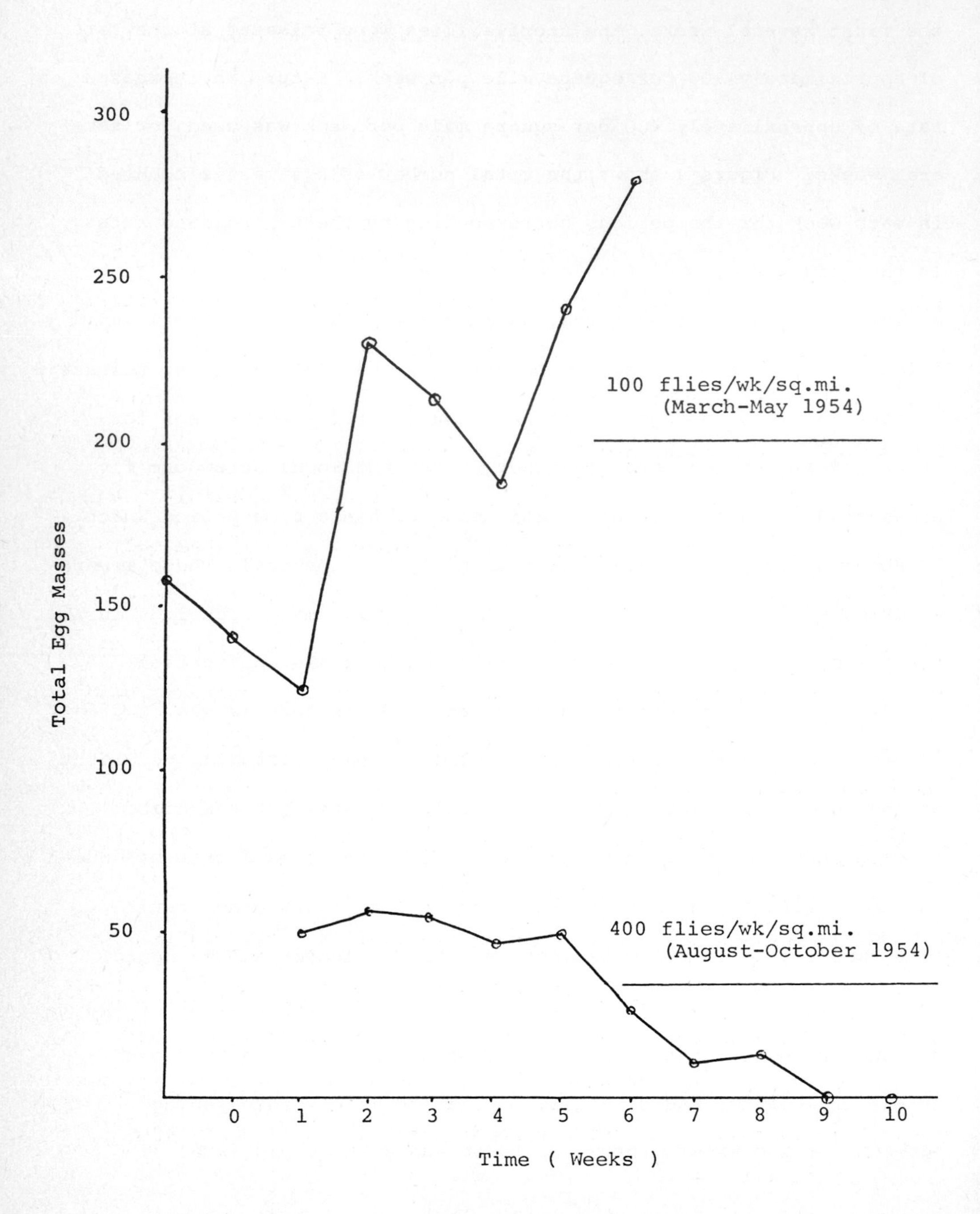

sterile males, then in what ways do other factors such as the size of the fertile population, or the environmental conditions, play in determining success or failure with this method? Such questions led us to consider mathematical models of populations having sterile members, and to formulate extensions to existing models, in order to capture some of the essential features of this phenomenon.

The Sterile Male Technique

Knipling [10] conceived the idea of releasing sexually sterile males into a natural population in order to control or reduce their numbers. The presence of sterile males partially nullifies the reproductive capacity of the female population. Whenever a fertile female mates, there is some chance that the mating will be with a sterile, rather than a fertile male, and therefore the female has a lesser chance of producing viable offspring. There will be some lowering of reproductive capacity whenever sterile males are present, but the degree of reduction depends on many factors, such as the relative numbers of steriles and fertiles, mating behavior, including the frequencies of mating and the degree of sexual polygamy, the population density, the relative sexual competitiveness of sterile and fertile males, and so on.

The reduction in reproductive capacity may or may not be sufficient to cause a lowering of population levels. If there is a decline in population, no matter how small, the effect of the sterile males becomes compounded, even if the number of steriles is maintained at approximately the same level, because the ratio of sterile to fertile males will increase. Consequently, a fertile female will

will be less likely to find a fertile mate, and thereby produce viable offspring. Other factors being constant, the presence of sterile males brings about an acceleration of any population decline. The sterile males become more and more effective as the population comes closer to extinction. This compounding or accelerating effect, which we shall call Knipling's effect, is the central idea involved in the effectiveness of the sterile male technique. Ample demonstration of the effect, in a simple setting, is provided in Table 1, copied from one of Knipling's papers [13, p. 10] along with the subsequent explanation.

TABLE 1

Trend of an Insect Population Subject to Sterile Insect Releases

Generation	Number of insects natural population	Number of sterile insects	Ratio sterile to fertile	Number of progeny
Parent	1,000,000	9,000,000	9:1	500,000
F_1	500,000	9,000,000	18:1	131,580
F_2	131,580	9,000,000	68:1	9,535
F_3	9,535	9,000,000	942:1	50
F_4	50	9,000,000	180,000:1	0

"The significant feature of this method of insect control is that the sterile insect releases become progressively more effective as the natural population declines. With a 9:1 ratio, the sterile insects provide 90% control.... However, in the F_1 generation when the same number of sterile insects is released, the ratio of sterile to fertile insects increases to 18:1. Thus, the rate of population decline accelerates. Such acceleration continues until the theoretical ratio of 180,000 sterile to 1 fertile insect is obtained and no chance of further reproduction would be expected." [The italics are ours.]

Later applications of the method

The basic insight of Knipling has been justified by success in several other cases. A direct application of the research which culminated in the critical Curacao experiments was the eradication of the screwworm fly from the Southeastern United States [1, 1962], and later the successful control of this pest in the Southwestern United States and Northern Mexico by creating a barrier zone in which sterile flies are maintained (Eddy and De Vaney [6]). This barrier severely curtailed the expansion of the flies out of their over-wintering areas. A mild and moist winter, ideal weather for screwworm reproduction, led to a breach of the barrier zone in 1972 (Tharp [22]). The very success of the sterile barrier in previous years is demonstrated by the decline of traditional methods of surveillance and treatment which has helped to exacerbate the current outbreak. Apparently, changes in factors governing the population dynamics can suddenly and dramatically transform a situation of pest control to one which is out of control.

Various fruit flies also have been studied. The melon fly was eliminated from the island of Rota, Mariana Islands using a combination of the sterile male technique and initial chemical reduction of the population (Steiner et. al.[20]). The oriental fruit fly was eliminated from the Mariana Islands using sterile males in an integrated control program (Steiner et. al.[20]). In the past few years, there has been considerable other research, with some small-scale tests.

The National Research Council [17, Chap. 15] has stated some necessary conditions for successful applications of the method:

(1) The species "can be reared economically in greater numbers than occur in nature in the infested area at that time,"

(2) the "released insects do not in themselves constitute a nuisance or source of injury,"

(3) there is adequate mixing of sterile and fertile insects in the target area before mating occurs, and

(4) it is possible to "sterilize the insects without too serious effects on their vigor, longevity, behavior, or mating competitiveness."

In addition they state that the method is "impractical against most established populations of prolific species or against species that are of little economic significance." It is obvious, from this list, that a great deal of information about a species must be collected and evaluated before the method can be used. This fact, together with the cost of mass-producing and distributing sterile insects, places a premium on achieving success in any attempt. The chance of failure must be reduced as much as possible before releases begin. The key factor determining success or failure seems to be perceived in the literature as the "overflooding ratio", the ratio of numbers of sterile males to numbers of fertile males. Insufficiency of this ratio has been pointed to as the major cause of failure in at least one case (Steiner _et. al._ [19]).

In any program the first problem is to achieve the ability to sterilize and release competitive flies. Even after this staggering research job is done, the major question remains: How many sterile males must be released in order that, with high assurance, the program will succeed? Our approach to this question lies in formulating several mathematical models that describe the dynamics of populations containing sterile members.

In what follows, we first discuss the deterministic model used by Knipling and others, and propose an alternative. We point out the assumptions used, and the type of results possible. Then, consideration of low population-level processes leads us to conclude that random effects play a crucial role in the phenomenon of extinction. We generalize the Knipling model to include stochastic elements and overlapping generations, using the so-called "birth and death" stochastic process. Our results with this model show the qualitative threshold effect observed in the Curacao experiment, and predict the number of sterile males required to achieve this threshold of success.

Deterministic Models

The most commonly used model for the sterile male technique is the one Knipling used to demonstrate the usefulness of sterile males [9]. Table 1 in this paper is an example of the output from this model. It is assumed that the sex ratio is 1 to 1, that sterile and fertile individuals are equally competitive, and that a constant number, S, of sterile males is present in each generation. If N_0 fertile males are in the parent generation and the N_0 fertile females choose mates from the entire male population, then the model postulates that only the fraction $N_0/(N_0 + S)$ of these matings will be with fertile males, and hence will produce offspring. Letting r be the rate of increase per generation, we may calculate the size of the offspring generation by:

$$N_1 = rN_0 \frac{N_0}{N_0 + S} \qquad (1)$$

Formula (1) implies that the population will decline if $rN_0/(N_0 + S) < 1$, which is equivalent to having an overflooding ratio (the ratio of sterile to fertile males) greater than $r - 1$. Thus an overflooding ration of 9:1, the value recommended by Knipling, should cover rates of increase per generation up to $r = 10$. Uncertainties in determination of N_0 and r, as well as those due to the simplicity of the model, would make it desirable to use an overflooding ratio somewhat greater than the value $r - 1$ suggested by formula (1).

The operation of Knipling's effect may be seen clearly in this model. As the population becomes smaller, the ratio $N_0/(N_0 + S)$ becomes smaller and smaller, thus accelerating the population decline. The simple model therefore incorporates the central phenomenon in the successful application of the technique. This model seems to be the one used by most researchers. It is easy to use and produces rough but intuitively appealing results in tabular form. We may characterize it as a discrete time, nonoverlapping generations, discrete population deterministic model. Berryman [4] has extended the basic model to treat cases in which the females mate more than once, the sterile males are not completely competitive, and in which multiple matings are required for a complete oviposition cycle. He refines the rate of increase, r, into component factors such as egg production, survivorship, and varying sex rations. Lawson [15] provides a general notational setting for models of this type. While some probabilistic notions are involved in the interpretation and motivation of the ideas, his model is essentially deterministic, and is of the same general nature as Knipling's and Berryman's.

The Knipling model gives results in a tabular form. In order to get a general formula for the time to extinction, so that we are not dependent on such tabular calculations, we propose a differential equation model. This model also provides a link with traditional population growth models, and allows comparison with certain features of Knipling's model. As with any differential equation model, it treats the population as a continuum and is most useful as an approximation when the population is large.

The classical theory of population growth (Costello and Taylor [5]) is usually expressed with a differential equation of the form $dN/dt = kN$, where N is the population size at time t and k is the so-called intrinsic rate of natural increase. The rate k may be represented by the difference of the population birth and death rates. If a constant number S of sterile males is maintained in the environment, we postulate--an analogy with (1)--that the birth rate is reduced by the fraction of sterile males in the total male population. This leads to the following differential equation:

$$\frac{dN}{dt} = N \left[b \frac{N}{N + S} - d \right] , \tag{2}$$

where b and d are the natural birth and death rates. Assuming that these parameters are independent of both time and population level, the solution to (2), in terms of the time to reach a specified size N, starting with N_0 individuals, is:

$$t(N) = \begin{cases} \frac{1}{d}\left[\frac{r}{r-1}\log\frac{\frac{S}{(r-1)N_0}-\frac{N}{N_0}}{\frac{S}{(r-1)N_0}-1}+\log\frac{N_0}{N}\right] & \text{if } r \neq 1, \\ \frac{1}{d}\left[\frac{N_0-N}{S}+\log\frac{N_0}{N}\right] & \text{if } r = 1, \end{cases} \tag{3}$$

where $r = b/d$ and $S \neq N_0(r-1)$. The population will remain constant if $S = N_0(r-1)$. It will decline if $r < 1$ or if $r > 1$ and $S > (r-1)N$, and will increase if $r > 1$ and $S < (r-1)N$. The time required to reduce the population size to a fraction $p(0 < p \leqq 1)$ of its original level is

$$t(p) = \frac{1}{d}\left[\frac{r}{r-1}\log\frac{\frac{S}{(r-1)N_0}-p}{\frac{S}{(r-1)N_0}-1}+\log\frac{1}{p}\right] \tag{4}$$

provided that $S/N_0 > r - 1 > 0$. When the overflooding ratio S/N_0 is extremely large, we may approximate the above value using

$$t(p) \to \frac{1}{d}\log\frac{1}{p} \text{ as } S \to \infty .$$

The critical role of the ratio $S/(r-1)N_0$ in this model is parallel to that in the Knipling model, and also to its role in the stochastic model presented below.

Figure 2 compares formula (3) with Knipling's model. For a growth rate of $r = 5$, we used the scaled results of Table 1 for Knipling's model. The time scale is in generations for Knipling's model, and in average lifetimes for the differential equation model.

The latter is in units of d^{-1}, the inverse death rate representing the average life span. If the initial overflooding ratio is 9:1, both models predict that the population will remain constant if r = 10. Plots for rates of 5 and 9 are shown, and the two different models are qualitatively alike, especially at higher population levels. We expect disagreement at low levels because we are approximating the population with a continuum in one model and discretely in the other. The models are also different in other respects, which we shall discuss below.

Discussion of deterministic models

The models presented above are essentially different. The Knipling model and its variants treats the population as a collection of individuals which exists in a discrete time setting. A cohort of parents live, reproduce and die, with no overlap of lifetimes with their offspring. On the other hand, a differential equation model such as the one we have presented treats the population as though it were a continuum evolving continuously in time. In this case the generations overlap. The implementation of the two types of model also differs. One may obtain an explicit formula for quantities such as the time to reach some population level using the differential equation model, but in general, this is not true with the Knipling model. The latter requires a simulation of the population development for each combination of model parameters. Both classes of model have desirable features as well as undesirable ones. Discrete population units would seem to be a necessary ingredient to a successful model, and yet differential equation models are often

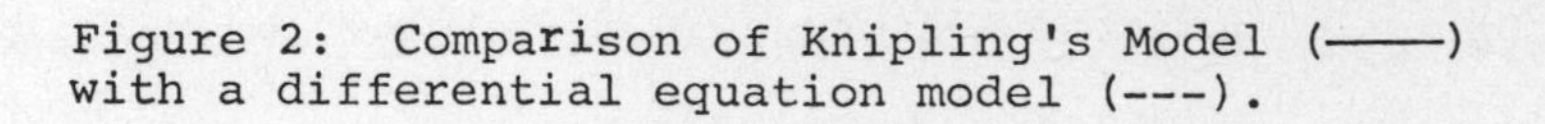

Figure 2: Comparison of Knipling's Model (——) with a differential equation model (---).

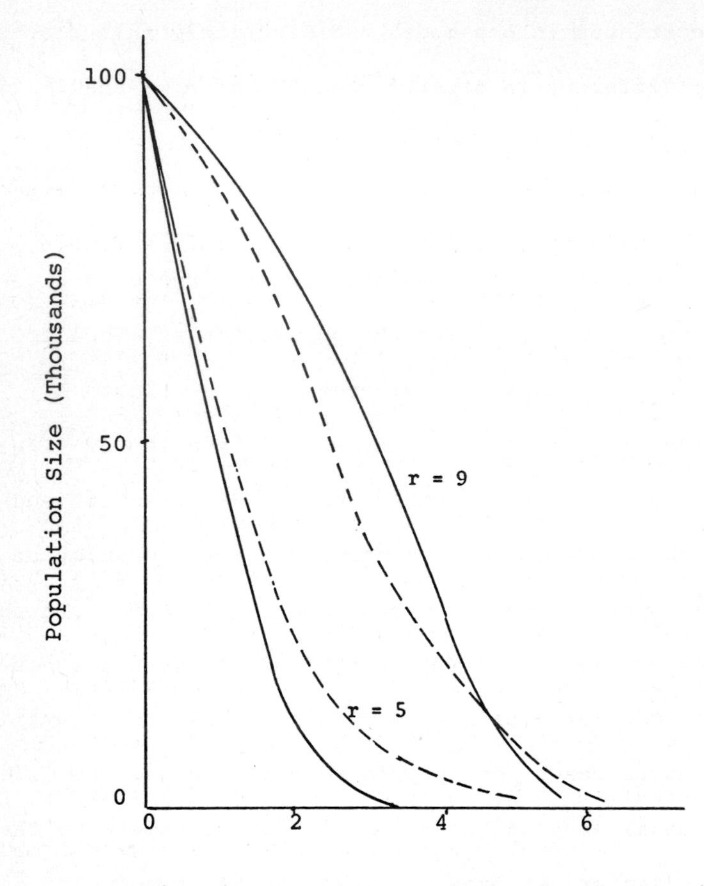

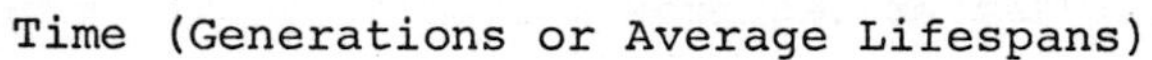

used. At high population levels, the latter may be useful approximations to discrete population dynamics. In dealing with insect pest control, however, often we are interested in bringing about extinction of the population, or at least reducing it to insignificant levels. This is especially true concerning the sterile male technique, where control is achieved through (local) extinction. The continuum model then loses plausibility and may give misleading or incorrect results. If the model predicts that the population size will be 6/7 at some time, are we to regard the population as extinct or not? Further, at low levels the method of producing integral answers, whether rounding or truncating, has greater effect on the future population trend than it would at high population levels. We must conclude that models for populations controlled to low levels should be based on discrete population units.

The choice of model time scale is somewhat less clear. We are accustomed to think of time as continuous. If, however, we are considering a species which propagates itself in generation steps, possibly yearly, with each cohort succeeding the previous one in a fairly rigid way, and without overlap, it would be rather natural to use a discrete time scale. If such conditions do not occur, then the non-overlapping generations model would be inappropriate, except possibly for rough calculations, and a continuous time model should be used.

Whether a model should provide a summary formula or a simulation-generated table or graph depends upon one's objectives. It would be nice to have both, since one can gain insight from each.

The infeasibility of obtaining a formula from Knipling's model is, however, a disadvantage.

We have discussed some advantages and disadvantages of the different models without making a clear choice between them. This was deliberate, for we believe that different circumstances or purposes often require different modeling approaches (e.g. wave versus photon models for light.) We do not believe this to hold with the shared basic element of these models, determinism. When studying low-level population processes, and especially when the phenomenon of extinction is considered, we believe that probabilistic models are required.

First, it is extremely difficult to construct a meaningful deterministic model which avoids converting non-integer population sizes to integer ones. At population levels near extinction, this leads to inaccurate and misleading results, as we have mentioned above. Obviously, the differential equation model assumes that the population size N may be any non-negative real number. Even in Knipling's model we find non-integer population sizes, because of the fraction $N_0/(N_0 + S)$ in formula (1). Calculation with that formula produces fractions which must be rounded or truncated to get the resulting population size in the offspring generation. The way in which this is done becomes critical for small population sizes, although it may be quite acceptable for large sizes.

Second, in many cases, some concept of environmental capacity plays an important role in population dynamics, and one might use models such as the Verhulst-Pearl logistic model. In such models, this environmental capacity strongly affects the rate of growth or decline of the population. Yet, whether or not extinction occurs is determined solely by the initial conditions in a deterministic model. One would like to clearly distinguish the effects of the initial population size and the carrying capacity of the environment upon the question of success or failure in an insect control program, that is, whether or not extinction will occur. But in a deterministic model, the population size will either increase to the capacity and remain there, or will steadily decline to extinction. With such models one is unable to investigate more complex effects of the capacity upon extinction. Was the second program in the Curacao experiment successful because of the higher release rate, or because the drier weather had reduced the capacity of the environment to support screwworm flies? Exactly how did the mild and moist winter lead to the 1972 breach in the barrier in the Southwestern United States? Because of increased supportive capacity or increased reproduction rates?

Third, it is our contention that the extinction of a population is essentially a random phenomenon. This randomness is reflected in the motivation of the deterministic models, and in particular in the use of the ratio $N_0/(N_0 + S)$ in formulas (1) and (2). This ratio represents the probability that a female who mates will mate with a fertile rather than sterile male. Lawson [15] makes this point well. If there are N_0 and S fertile and sterile males, respectively, an arbitrary female who mates will do so with one out of $N_0 + S$ males. Assuming each of these males is equally likely to be chosen, the chance of the female choosing any particular male is 1 out of $N_0 + S$.

Thus, the probability that a female mates with any fertile male is $N_0/(N_0 + S)$, since there are N_0 chances, each with probability $1/(N_0 + S)$. When the number of fertile males is small, the probability that a mating will be a fertile one (fertile female - fertile male) becomes very small. This factor determines what we have called Knipling's effect. But even if only a single fertile pair survives, there is a positive probability of a fertile mating, and a chance that the population may re-colonize the habitat. Low level population processes are dominated by random fluctuations which determine extinction or re-colonization, even in the presence of significant long-term trends. On her wedding flight, whether that last fertile female meets a fertile or sterile male is a chance event.

Fourth, area effects play no overt role in deterministic models. Discontinuities of population density, clustering of the population into sub-populations, dispersal of food supply, immigration and similar factors may seriously affect the likelihood of extinction. Such area effects bring an element of redundancy into the system which make extinction less likely. These factors are difficult to include in any model, but they seem to pose more difficulties in a deterministic setting than in a stochastic one.

Let us remark that the study of population processes does not always require the use of a stochastic model, for the relative variability of such processes, at normal population levels, can often be ignored. Reddingius has pointed out, in a discussion of "density-dependence" concepts [18, p. 59], that "the larger the population, the smaller the conditional variance of the mean number of offspring per individual." He continues, "This is, of course, nothing but the well-known property of means of sequences of independent random variables in an ecological disguise." On the other hand, any pest eradication model must directly confront the central phenomenon of extinction, and then probabilistic notions enter into consideration.

What type of stochastic model should be used? We have already discussed some desirable features. First, the model should treat the population explicitly and in a natural way as a collection of individuals. Second, a natural continuous time framework should be employed, and overlapping generations allowed, unless this is biologically improper. Third, there should be explicit provision for some concept of an environmental carrying capacity. Finally, the model should provide us with a formula to describe the inter-relations of the various factors in the model. The reader will not be surprised to find, in the next section, a stochastic model with these features.

Stochastic model

Our purpose here is to present and discuss a stochastic model with the broad characteristics noted in the foregoing section. We shall use one which is a natural extension of Knipling's model, in that the population growth rate is directly governed by the fraction of fertile males in the population. We have seen that this factor can account for Knipling's effect, the acceleration of a declining trend to bring about extinction. In particular, we choose to use one of the simplest of models in the above category, the birth-and-death process (Karlin [9], pp. 189-206).

Our assumed population is a collection of male and female individuals, with the sexes equal in number. Let X(t) be the number of fertile males (or females) alive at time t. For each value of t, this quantity is a random variable which may assume only non-negative integer value. We shall further assume that the population exists in some finite environment such that there is never any net flow of population members across its boundary, and such that no more than K fertile individuals of either sex may live in it at any time. In practice, there are a great number of ways in which population growth may be limited. We are excluding, for example, interspecific competition, and indeed any interaction with other species in the interests of simplicity. The environmental capacity, K, will be the sole means of expressing some natural limitation on the growth of our hypothetical population. The individuals will be assumed to be distributed uniformly in the environment, and we will not specifically consider any area or distribution effects. Reproduction and deaths occur for one individual of either sex at a time, with the sex ratio remaining 1:1. Again for simplicity, parents mate and give birth to adult offspring, and continue to live, and possibly reproduce, afterward.

Sterilized individuals are introduced into the system in such a way that there is a constant number, S, of sterile males present at all times. We make the simple assumption that a fertile female is just as likely to mate with a sterile male as with a fertile one. Only the latter matings result if offspring.

Earlier we saw a deterministic continuous time model in which, up to capacity, K, the <u>rate</u> of increase in the population was jointly proportional to the current population size n, the individual birth rate b, and a fertility ratio in the presence of steriles of n/(n+S). For the stochastic model we now introduce, we assume the <u>probability</u> of a unit increase in the population is proportional to the same factors. That is, we assume

$$\Pr\{X(t+\Delta t) = n+1 \mid X(t) = n\} = \lambda_n \Delta t + o(\Delta t)$$

where

$$\lambda_n = \begin{cases} n\, b\, n/(n+S) & \text{if } n < K \\ 0 & \text{if } n \geq K \end{cases}$$

and where $o(\Delta t)$ represents negligible remainder terms as Δt becomes arbitrarily small. In words, the probability of a unit increase in the population, given a current population size of $X(t) = n$, is jointly proportional to the population size n, the individual birth rate b, and the fertility ratio $n/(n+S)$, provided the population level is below the capacity K. We make the <u>probability</u> of a unit decrease in the population proportional to the population size n and individual death rate d, again in direct analogy with the deterministic <u>rate</u> of decrease, nd, in the earlier Equation (2). That is, we assume

$$\Pr\{X(t+\Delta t) = n-1 \mid X(t) = n\} = \mu_n \Delta t + o(\Delta t)$$

where μ_n, the population death rate, is given by

$$\mu_n = dn, \quad n = 0, 1, 2 \ldots \quad .$$

We assume that other population changes have negligible probability during arbitrarily small time intervals. These assumptions lead to the classical "birth and death" family of Markov processes. These processes have a long and fruitful history as population models.

Our application of these processes to the sterile male technique is based on specifying the dependence of the birth and death parameters on the number of sterile males present. The basic form of the parameters, without considering sterile males, is the linear growth model. This is the model used by MacArthur and Wilson [16, Chap. 4] who were interested in the case of species colonizing an island. When sterile males are introduced, and maintained at a constant level S, we shall modify the birth parameters according to the insight gained from Knipling's model. If a mating is to occur at some time when there are m fertile males present, we suppose that the probability that the female mates with a fertile male, and hence results in a birth, is $m/(m + S)$. This is multiplied by the no-steriles birth parameter mb to give λ_m for our present model. The birth and death parameters are then

$$\lambda_m = \begin{cases} mb \dfrac{m}{m + S} & \text{if } m = 0, 1, \ldots, K - 1 \\ 0 & \text{if } m \geqq K \end{cases} \tag{5}$$

$$\mu_m = \begin{cases} md & \text{if } m = 0, 1, \ldots, K \\ 0 & \text{if } m > K. \end{cases}$$

The numbers b and d are constant individual birth and death rates. We let their ratio be $r = b/d$, and let $r_m + \lambda_m/\mu_m$ for $m = 1, 2, \ldots K$, so that $r_m = rm/(M + S)$. The latter quantity is similar to the growth rate used by Knipling. The ratio r represents the average number of offspring produced by an individual female during her lifetime, that is, the net reproduction rate.

With a finite environmental capacity, K, the population is sure to eventually become extinct (Karlin [9], formula 7.9, p. 205).

Nevertheless, the event of extinction may be so remote in time that, for practical purposes, it never occurs. We are therefore interested in the (random) time of extinction, defined by:

$$T = \min\{t: t \geq 0 \text{ and } X(t) = 0\}.$$

As a criterion for evaluating sterile male release programs we choose the mean value of this extinction time, for various initial population sizes:

$$W_m = E[T|X(0) = m], \quad m = 0, 1, 2..., K.$$

In order to obtain a formula for W_m, we take advantage of a completely equivalent formulation of the birth-and-death process (Karlin [9], p. 191). Having just entered state m, the process remains there for a random length of time, called a sojourn time, until it makes the next transition. The successive sojourn times are independent random variables, each with exponential distribution, and the sojourn time in state m has mean $(\lambda_m + \mu_m)^{-1}$. At the end of a sojourn in state m, and independently of the sojourn time, the process jumps to one of the two neighboring states, with probabilities $\lambda_m/(\lambda_m + \mu_m)$ for jumping to m + 1 (birth) and $\mu_m/(\lambda_m + \mu_m)$ for jumping to m - 1 (death).

If the initial population size is X(0) = m > 0, the process sojourns in state m for a random length of time with mean $1/(\lambda_m + \mu_\mu)$, and then either a birth or a death occurs, according to the above probabilities. However, once the process jumps to the new state, it is probabilistically identical, under the Markov assumption, to one which was initially in that state. Thus, the expected time to extinction from the new state is either W_{m+1} or W_{m-1}, and we can write the expected time to extinction, starting from state m, as the sum of the mean time until the first jump and the expected time to extinction from the new state after that jump, weighted by the respective jump probabilities. This results in the basic recursion formula:

$$W_m = \frac{1}{\lambda_m + \mu_m} + \frac{\lambda_m}{\lambda_m + \mu_m} W_{m+1} + \frac{\mu_m}{\lambda_m + \mu_m} W_{m-1} \qquad (6)$$

for m = 1, 2, ..., K, with the boundary conditions $W_0 = 0$ and $\lambda_K = 0$. The solution to this system of second order linear difference equations is given in Karlin [9, pp. 204-205] for the case $K = \infty$. For finite K, we get:

$$W_m = \sum_{k=0}^{m=1} \rho_k^{-1} \sum_{j=k+1}^{K} \rho_{j-1}\mu_j \qquad m = 1, 2, \ldots, K, \qquad (7)$$

where

$$\rho_0 = 1,$$

$$\rho_k = \Pi_{i=1}^{k} r_1 = \Pi_{i=1}^{k} r \frac{i}{i + S} = r^k \frac{k!\, S!}{(k+S)!} \quad k=1,2,\ldots,K-1$$

$$\rho_K = 0 \qquad (8)$$

Since $1/\mu$ is the average individual lifetime, $G_m = dW_m$ represents the expected time to extinction, from state m, measured in average lifespans, a dimensionless quantity. We call G_m the <u>mean lifespans to extinction.</u> From (7), we get:

$$G_m = \sum_{k=0}^{m-1} \frac{1}{\rho_k} \sum_{j=k+1}^{K} \frac{1}{j} \rho_{j-1} \qquad m = 1,2,\ldots,K$$

$$= \sum_{k=0}^{m-1} \sum_{j=k+1}^{K} \frac{1}{j} r^{j-k-1} \frac{(j-1)!(S+k)!}{k!(S+j-1)!} \qquad (9)$$

Calculations using the stochastic model

A digital computer was used to compute values of G_m, the mean lifespans to extinction, for various choices of the parameters m, K, r, and S. The presence of the factorials in formula (9) suggests that G_m may be quite large under some circumstances, and this is indeed the case. This makes direct numerical computation of the products ρ_k and their summation infeasible. For one thing, the numbers often exceed the representation capacity of the computer for realistic choices of parameters. Also, the products ρ_m are ratios of exceedingly large numbers (the factorials) even when their sum is moderate. One is faced with the problem of extracting a number of reasonable magnitude from ratios which are effectively ∞/∞. In cases such that G_m itself was not too large, the following recursive procedure, based on (6), circumvented these numerical difficulties. We define

$$p_k = \sum_{j=k+1}^{K} \frac{1}{j} \frac{\rho_{j-1}}{\rho_k}, \quad k=0,1,\ldots, K-1,$$

so that $G_m = \sum_{k=0}^{m-1} p_k$. Also, letting $s_m = \sum_{k=m}^{K-1} p_k$, we then have

$$p_{K-1} = s_{K-1} = \frac{1}{K},$$

and

$$p_k = \frac{1}{k+1} + \frac{\rho_{k+1}}{\rho_k} p_{k+1} = \frac{1}{k+1} + r \frac{k+1}{S+k+1} p_{k+1}$$

$$s_k = p_k + s_{k+1}$$

for k = K-2, K-3, ..., 1, 0. Therefore $G_K = s_0$ and $G_m = s_0 - s_m$ for m = 1, 2, ..., K-1.

In Figure 3, the mean lifespans to extinction is plotted against the number of steriles, S, and initial population size, m, for fixed capacity, K, and net reproduction rate r. For each m, there is a smallest S, depending on m, for which G_m has a reasonably small valuse. The curve rises sharply for smaller numbers of steriles, but is not asymptotic, since even for $S = 0$, G_m is finite, although extremely large. As we would expect, larger initial population sizes require that higher levels of sterile males be maintained in the environment. Of at least equal significance is the prediction that, once the minimum level of S is achieved, further increases in S have little additional effect in reducing G_m.

We have extracted some specific numbers from this figure and presented them in Table 2 for further analysis.

TABLE 2

Mean lifespans to extinction as a function of the number of sterile males S and initial population level m. "$\sim\infty$" means "extremely large."

		Number of Sterile Males S			
		300	600	1000	Above 2000
Initial	100	$\sim\infty$	1000	7	5
Population	50	$\sim\infty$	10	5	4
Size	2	$\sim\infty$	1.5	1.5	1.5

The observations to be made are:

(1) Maintaining a thousand or more sterile males will cause extinction in under seven generations, on the average, no matter what be the initial population level. If seven generations to extinction is

Figure 3: Expected lifespans to extinction as it depends on the number of sterile males and the initial population size (m) for environmental capacity of 100 and net reproduction rate of 10.

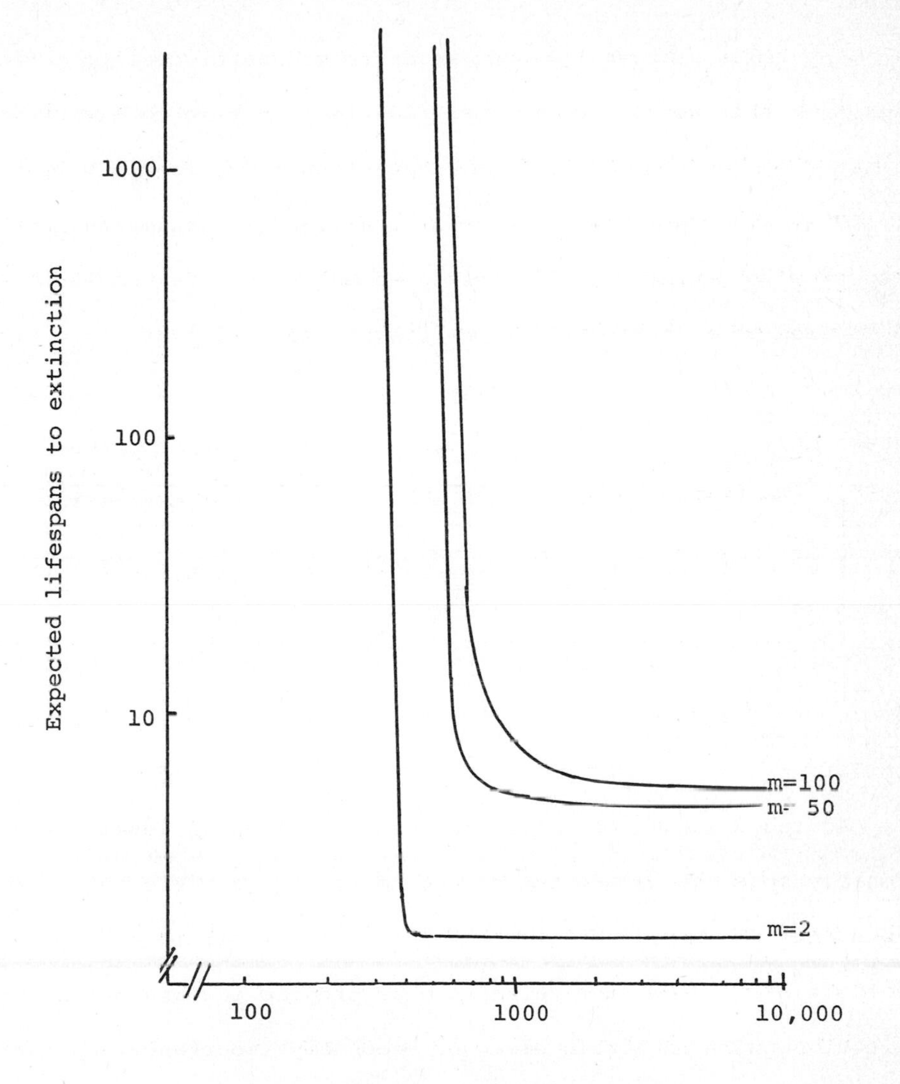

sufficiently quick to be considered feasible, then any release above S = 1000 will lead to a successful eradication program. On the other hand, increasing S above 2000 yields only marginal improvements, e.g. a reduction from seven generations, on the average, to five, at an initial population level of 100. This seems hardly significant, given the level of precision under which such a model could be applied.

Any release above S = 1000 leads to moderate times to extinction. In this range, the further reduction in time to extinction achieved by lowering the initial population level from m = 100 to m = 50 is again marginal. To achieve a significant improvement in mean time to extinction, the initial population level would have to be lowered to approximately 2% of its maximal value of 100. Pretreatment of the target population with an insecticide would be warranted only if it could lead to such a significant reduction.

(2) If the number of steriles S is less than 300, extinction will not occur in a reasonable time, no matter what the initial population level.

(3) There is a relatively narrow range of sterile male population levels, typified by the value S = 600, in which the initial fertile population level strongly affects the extinction time.

Adding a margin of safety to a planned sterile level S would probably move the values out of the narrow range where this strong dependence on initial fertile levels occurs. Therefore, from a practical point of view, the dependence of program success or failure on initial population levels seems of secondary importance, unless extremely low levels, on the order of 2% of the maximal population values, can be achieved.

Figure 4 is a plot of G_m against S and r for fixed m = K. We see that the net reproduction rate r significantly affects the minimum number of steriles which must be maintained, but has no effect on the value of G_m for large enough S. In fact, for each m we may determine the value of this horizontal asymptote by considering the limit as $S \to \infty$ in our equations. The birth parameters in equations (5) all become 0. Our birth-and-death process has then become a pure death process. The expected time to extinction from state m is the sum of the average sojourn times in states m, m-1, ..., and 1, which is $\sum_{k=1}^{m} \mu_k^{-1}$, and therefore the mean lifespans to extinction from state m becomes

$$G_m = \sum_{k=1}^{m} \frac{1}{k} . \qquad (10)$$

Formula (10) may be calculated exactly for small m, or for large m we may use the approximation:

$$\sum_{k=1}^{m} \frac{1}{k} \approx \gamma + \log m, \qquad (11)$$

where γ = .5772157... is known as Euler's constant.

Figure 5 is a plot of G_m against m and S for fixed K and r, with m along the abscissa, rather than S as in Figure 3. The diminishing returns from increasing S beyond a critical value is again apparent. Of course, for large S, the lowest of these curves is nearly logarithmic, as predicted from formula (11). For less-than-critical values of S, there is an initial population size above which the expected time to extinction becomes very large.

Figure 4: Expected lifespans until extinction, as it depends on the number of sterile males and the net reproduction rate (r) for initial population and environmental capacity of 100.

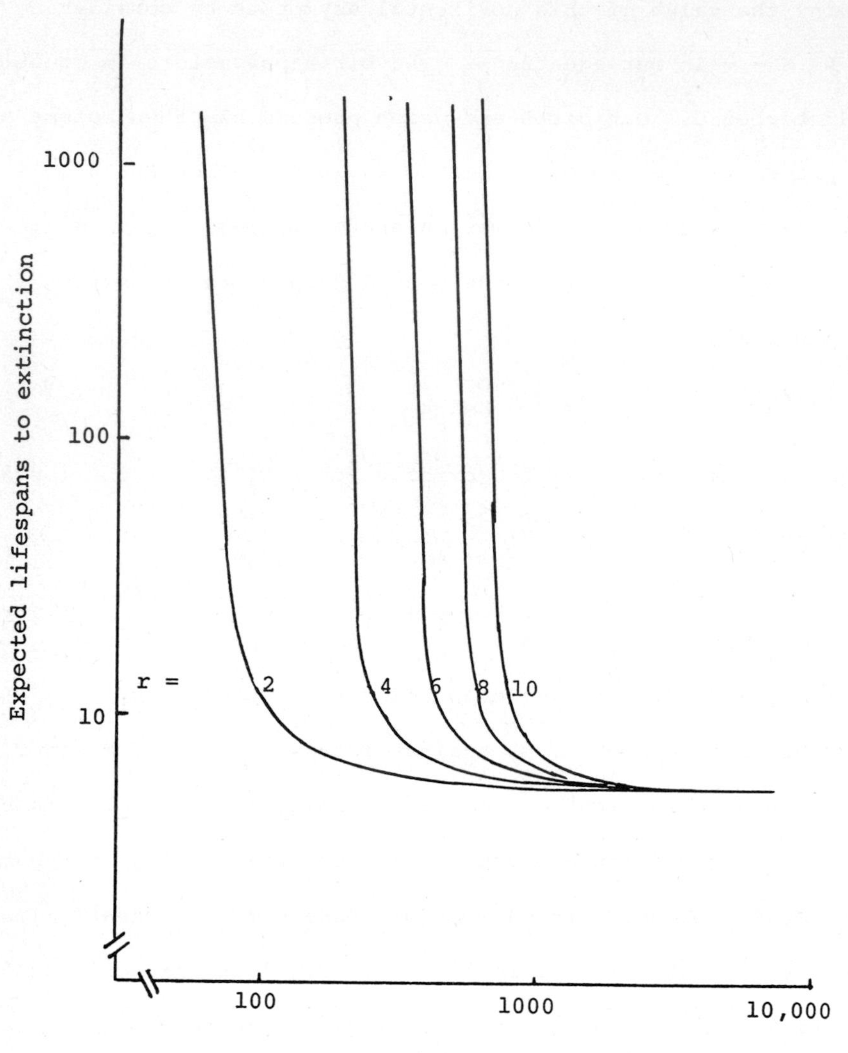

Figure 5: Expected lifespans until extinction, as it depends on the initial population size and the number of sterile males (S), for environmental capacity of 100 and net reproduction rate of 10.

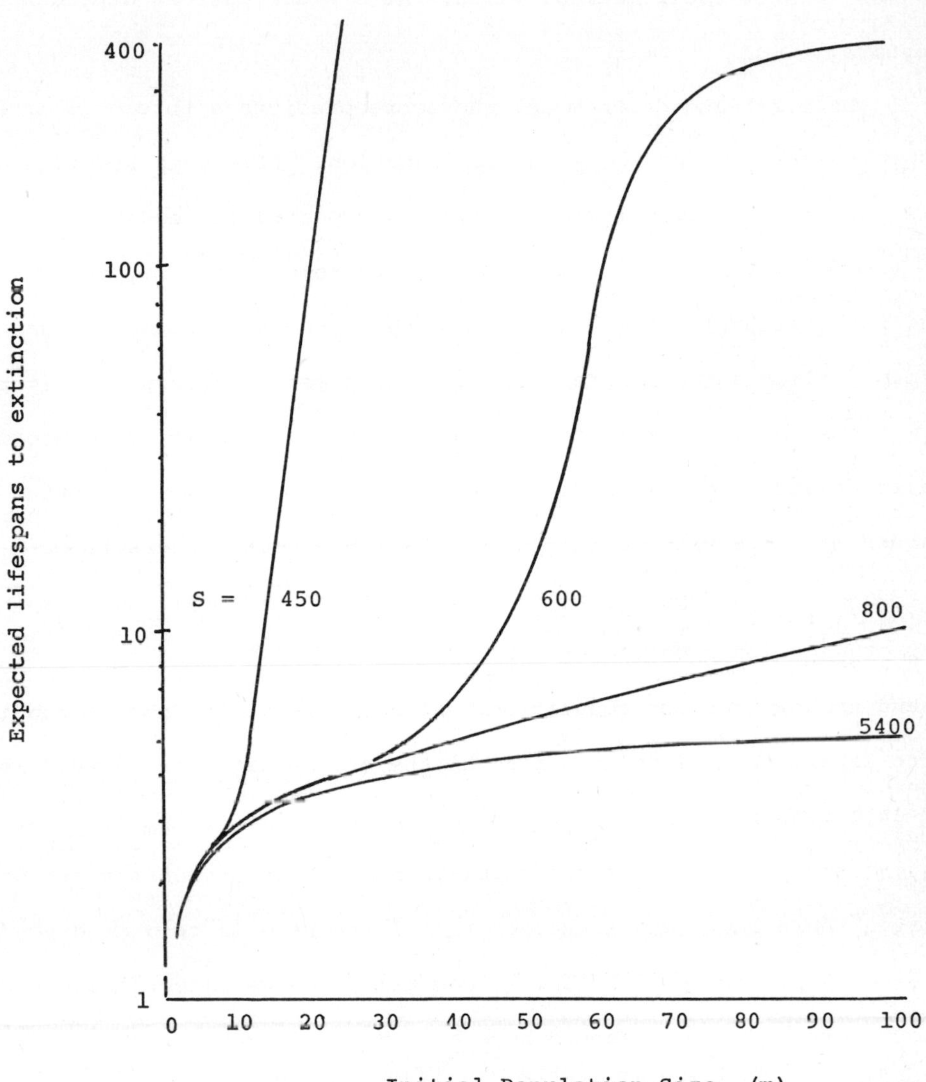

This figure clearly portrays the band of values for S in which the dependence on initial population levels is strong. If the sterile level is above S = 800, all initial population levels lead to extinction in under 10 generations. Below S = 450, the lifespans to extinction is over 10, except for population levels on the order of 15% or less of their maximal value. At S = 600, strong dependence appears.

The birth-and-death model therefore predicts a threshold in the expected generations to extinction function. If enough steriles are present, relatively quick extinction is expected (success), whereas if too few are present extinction is very remote in time, on the average (failure). Maintaining more than enough steriles brings rapidly diminishing returns in achieving greater success, and if too few are maintained, failure is expected as it would be if no sterile males at all were released. How many sterile males must be maintained in the environment in order that we expect success to be achieved? This depends on the other model parameters. Success in our eradication program certainly requires that the probabilistic trend is down for the population. At any size m, the population is more likely to decline than rise if the probability of a down jump is larger than the probability of an up jump: i.e. if $\lambda_m < \mu_m$, or equivalently $r_m < 1$. If the population by chance reaches a set of states where the trend is up, or $r_m > 1$, there will be a tendency for the population to persist for a long time, and we expect failure to occur. Therefore, we should have a down trend even for those states (population sizes) where the jump probabilities are most unfavorable to success. We don't need a large down trend for the most unfavorable case, because we depend on Knipling's effect to ultimately drive the population to extinction. The strategy of the pest management planner must therefore be to choose S, the number of sterile males in the environment, such that $r_m < 1$ for all population sizes m, including that size which maximizes r_m. The latter choice represents nature's best strategy for initial population size. By solving the inequality $r_m < 1$ for S, we see the $S > m(r-1)$ for all sizes m. Since the population size m must be less than the capacity, the planner should choose

$$S > (r - 1)\ (K - 1). \tag{12}$$

If a more complicated set of parameters, such as for a logistic process representing density dependent reproduction rates, were used, then different values would result.

With the above analysis in mind, we re-scaled the abscissa in Figure 3 so as to plot G_m against the critical ratio $S/K(r-1)$ rather than S (we used K instead of K-1 for arithmetic convenience) for different values of the net reproduction rate r. Figure 6 shows that this normalization has not only accounted for the effect of the net reproduction rate r on success or failure, but that the critical values of S which give reasonably short expected lifespans to extinction are those determined by (12), that is $S > (r - 1)\ (K - 1)$.

We see that increasing the dimensionless ratio $S/K(r-1)$ much beyond one does little to lower the mean time to extinction, while completely infeasible lifespans result if this ratio is allowed to drop much below one. In this sense the threshold is rather sharp.

Figure 6: Expected number of lifespans until extinction, as it depends on the number of sterile males (S) maintained, normalized to S/K(r-1), for initial population size and environmental capacity of 100 and net reproduction rates (r) of 2 and 10.

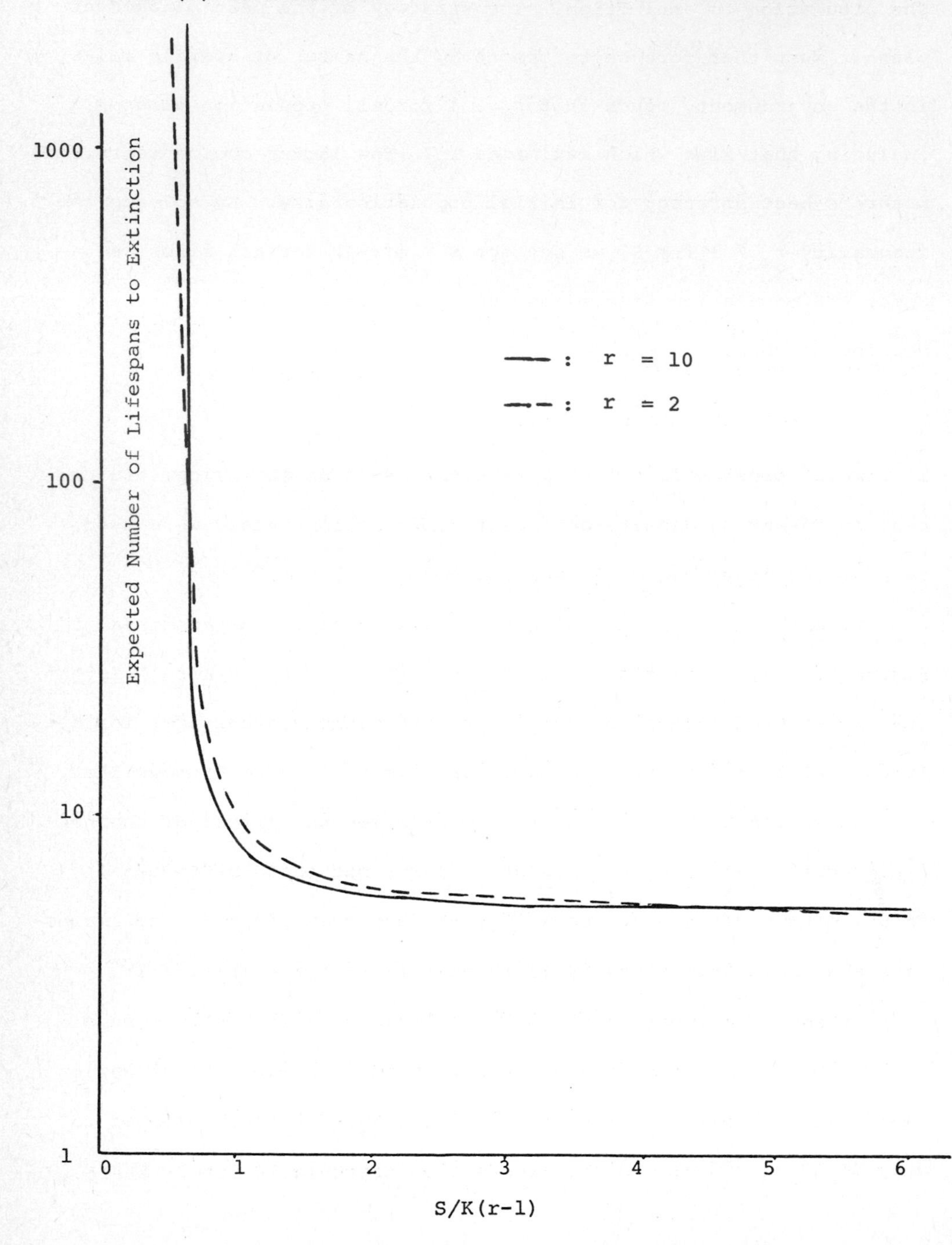

Since this chart assumes an initial population at its capacity level, the overflooding ratio is S/K. An overflooding ratio of 5 leads to S/K(r-1) = 5 when r = 2, and a mean lifespans to extinction of about 6. The same overflooding ratio leads to S/K(r-1) = .55 when r = 10, and a mean lifespans to extinction which is essentially infinite. The overflooding ratio cannot be used as a criterion unless it is weighed against the reproductive capabilities of the organism.

Conclusions from the model

Our objective was to extend existing deterministic models of the sterile male technique in such a way as to include:

1) stochastic effects involved in extinction,
2) Knipling's effect,
3) environmental carrying capacity,
4) continuous time scale and overlapping generations, and
5) discrete population units.

We did this by developing a stochastic model based on the birth-and-death process. Computations with this model show that the model predicts a threshold effect, as suggested by the Curacao experiments. Insertion of enough sterile males leads to expected success in eradicating the population, while not enough leads to failure. More than enough sterile males does not significantly reduce the expected time to extinction below that already obtained. The model suggests that, to achieve success, the number of sterile males maintained should exceed K(r-1) where r is the net reproduction rate. The critical ratio S/K(r-1) was seen to determine the threshold between success and failure, and to account for a major part of the effect of the net reproduction rate. From a practical viewpoint, some measure of environmental capacity K seems more important in determining program success or failure than the initial population size.

Comparison of Curacao data with model predictions

Our stochastic model predicts a threshold effect, which relates success or failure in eradication to population parameters and the number of sterile males present. The sharp contrast in success or failure in the two major Curacao experiments (Baumhover *et. al.* [2]), which suggested this effect to us in the first place, was qualitatively similar to that prediction. We must now ask whether or not our model would have correctly predicted the actual results in Curacao.

A prediction of success or failure requires estimates of the model parameters, net reproduction rate, initial population size, and environmental carrying capacity. It seems reasonable to assume that the initial population levels in Curacao were at or near the upper limit that the environment could support, leaving us only the net reproduction rate and initial population levels to estimate. We were unable to locate such estimates in the literature, so we set about estimating them from the Curacao data (Baumhover *et. al.* [2], pp. 222-225). These data consist of weekly values of three numbers: rate of release of sterile male flies, number of egg masses counted, and fraction of egg masses counted which are sterile.

The principle underlying our estimation procedure is quite simple. It bears many similarities with the tag-recapture method of estimating wildlife population sizes. If S sterile and N fertile males are in the population, we would expect a fraction $p = S/(S+N)$ of eggs to be sterile. If p and S are known, or estimated directly by the data, we may solve for N to get the estimate $N = S(1-p)/p$.

Suppose by this means we were able to estimate the population levels N_0, N_1 in two successive generations; then we could estimate the growth rate by solving in Knipling's equation (1):

$$N_1 = N_0 r[N_0/(N_0 + S)]$$

to give

$$r = \frac{N_1(N_0 + S)}{N_0^2} \quad .$$

These two steps illustrate the simple principle we used in forming our estimates. The application of this principle, however, was considerably complicated by the weekly data and overlapping generations. Figure 7 shows a typical life cycle for this insect, and, in particular, the significance of the generational overlap. To account for this, we developed a detailed deterministic model with overlapping generations which incorporates this typical life cycle and maintains the age structure of the population. After introducing the data concerning sterile male release rates and observed fractions of sterile egg masses, we then solved for the population levels and reproduction rates in a mean absolute deviation sense.

Our results were as follows. For the first experiment, with sterile flies released at the rate of 100 per square mile per week, we estimate that the initial population density was approximately 500 and the net reproduction rate was approximately 1.655. For the second experiment, when the higher release rate of 400 per square mile

Figure 7: Typical life cycle for the screw-worm fly showing the overlap between generations.

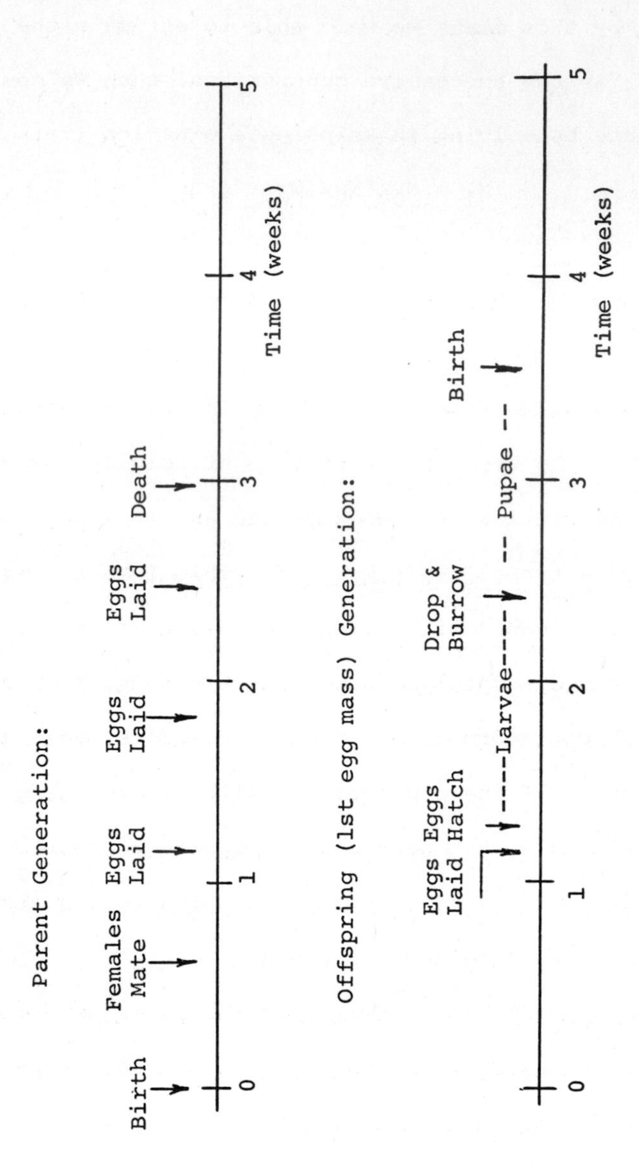

per week resulted in eradication, we estimate the initial population density to have been about 150, with a net reproduction rate of about 1.456. The initial population estimates are consistent with the rough estimates of the Curacao workers. They attribute the decline in fly population between the two experiments to a spell of dry weather (Baumhover _et. al._ [2], p. 464). Apparently, the declining trend had been partially reversed when the eradication experiment was conducted. Thus, both the smaller initial population and smaller net reproduction rate, combined with a higher sterile male release rate, contributed to the successful eradication, for $S/K(r-1)$ was approximately $800/150\ (.456) > 11$. On the other hand, the conditions of the first experiment were conducive to failure, for $S/K(r-1)$ was approximately $200/500\ (.655) < 0.6$. This is assuming that sterile males are active for about two weeks on the average so that S is roughly twice the weekly release rate. Even if the same conditions had prevailed for the eradication experiment as in the first one, this suggests that there still would have been a good chance for success because of the higher rate of release of sterile males, with a critical ratio of about $800/500\ (.655) > 1$.

What we have done is estimate certain population parameters from the Curacao experiment, use these in our threshold criterion to predict success or failure of the control program, and then compare these predictions with what seems to have happened. The event that extinction occurs during the experiment cannot be observed; we must infer it from the egg masses counted. It is possible that the population persists but lays no eggs in observed animals for several weeks, although this is unlikely. There might seem to be a circularity in estimating model parameters from the same data that is used to evaluate the model response, but this is the very nature of scientific validation of hypotheses. Since the estimation procedure is independent of the threshold suggested by the stochastic model, we can conclude whether or not the estimates and observed responses are consistent or inconsistent with the model. Our preliminary results indicate consistency of the model with the data.

Summary

We have examined existing deterministic models for the sterile male technique. Because of the nature of population phenomena at low levels, we argued for the necessity of explicitly probabilistic models. We then developed a stochastic model based on the birth-and-death process which was a direct generalization of the basic Knipling model. Our model includes most of the important general features of the sterile male process, including the Knipling effect, a term which describes the effect of the sterile males in accelerating declining population trends so as to bring about extinction. Our model predicts, qualitatively and quantitatively, the threshold effect suggusted by the fundamental Curacao experiments. It predicts the minimum number of sterile males needed for success, measured by a reasonably low expected time to extinction, based on model parameters such as the net reproduction rate, initial population size and environmental carrying capacity. It suggests that it may be very uneconomical to release a number of steriles much above this minimum number. Yet this very prediction re-emphasizes the necessity of of measuring or estimating population parameters and of fully understanding the population dynamics involved. The release of sterile males in itself is a powerful and little used tool for estimating these parameters, even as a prelude to a major release effort aimed at extinction.

REFERENCES

1. Anonymous, "Status of the screw worm fly in the United States," USDA Agric. Res. Serv. ARS 22-79 (1962).

2. Baumhover, A. H., Graham, A. J., Bitter, B. A., Hopkins, D. E., New, W. D., Dudley, F. H., and Bushland, R. C. "Screw-worm control through release of sterilized flies," J. of Econ. Ent. 48 (1955), 462-466.

3. Baumhover, A. H., Husman, C. N., and Graham, A. J. "Screw Worms," Insect Colonization and Mass Production Chap. 37, C. N. Smith (ed.), Academic Press, New York (1966).

4. Berryman, A. A. "Mathematical description of the sterile male principle," Can. Ent. 99 (1967), 858-865

5. Costello, W. G. and Taylor, H. M. "Deterministic population growth models," Amer. Math. Monthly 78 (1971), 841-855.

6. Eddy, G. W. and DeVaney, J. A. "A brief statistical review of the US-Mexico screw worm eradication program," Bul. Ent. Soc. Amer. 16 (1970), 159.

7. Geier, P. W. "Demographic models of population response to sterile release procedures for pest control," in panel proceedings Insect ecology and the sterile male technique, International Atomic Energy Agency, Vienna, STI/PUB/223 (1969).

8. Hightower, B. G. "Population dynamics of the screwworm fly, Cochliomyia Hominivorax (coquerel), with respect to control by the sterile male technique," in panel proceedings Insect ecology and the sterile male technique, International Atomic Energy Agency, Vienna, STI/PUB/223 (1969).

9. Karlin, S. A First Course in Stochastic Processes, Academic Press, New York (1966).

10. Knipling, E. F. "Possibilities of insect control or eradication through the use of sexually sterile males," J. of Econ. Ent. 48 (1955), 459-462.

11. Knipling, E. F. "Sterile-male method of population control," Science 130 (1959), 902-904.

12. Knipling, E. F. "The potential role of the sterility method for insect population control with special reference to combining this method with conventional methods," USDA Agric. Res. Serv. ARS 33-98 (1964).

13. Knipling, E. F. "Some basic principles in insect population suppression," Bul. Ent. Soc. Amer. 12 (1966), 7-15.

14. LaChance, L. E., Schmidt, C. H., and Bushland, R. C. "Radiation-Induced Sterilization," Pest Control, Biological, Physical, and Selected Chemical Methods Chap. 4, Ed. by W. W. Kilgore and R. L. Doutt, Academic Press (1967).

15. Lawson, F. R. "Theory of control of insect populations by sexually sterile males," Ann. Ent. Soc. Amer. 60 (1967), 713-721.

16. MacArthur, R. H. and Wilson, E. O. The Theory of Island Biogeography, Princeton University Press, Princeton (1967).

17. National Research Council, Agricultural Board, Committee on Plant and Animal Pests, Subcommittee on Insect Pests,"Principles of Plant and Animal Pest Control," Volume 3: Insect-Pest Management and Control, National Academy of Sciences (publication 1965), Washington, D. C. (1969).

18. Reddingius, J. "Gambling for Existence, a discussion of some theoretical problems in animal population ecology," Acta Biotheoretica XX (1971) Suppl. 1.

19. Steiner, L. F., Mitchell, W. C., and Baumhover, A. H. "Progress of fruit fly control by irradiation sterilization in Hawaii and the Mariana Islands," Int. J. Appl. Radiat. Isotopes, 13 (1962), 427-434.

20. Steiner, L. F., Harris, E. J., Mitchell, W. C., Fujimoto, M. S., and Christenson, L. D. "Melon fly eradication by overflooding with sterile flies," J. Econ. Ent. 58 (1965), 519-522.

21. Steiner, L. F., Hart, W. C., Harris, E. J., Cunningham, R. T., Ohinata, K., and Kamakahi, D. C. "Eradication of the oriental fruit fly from the Mariana Islands by the methods of male annihilation and sterile insect release," J. Econ. Ent. 63 (1970), 131-135.

22. Tharp, M. "The Screwworm Turns," Wall Street Journal (August 22, 1972), 32.

23. Vronskikh, M. D. "Some calculations for the efficiency of using sterile insects (Russian)," Zh. Obshch. Biol. 32(3) (1971), 287-298.

D.

SOCIAL AND BEHAVIORAL ANALYSIS

THE LIMITS TO GROWTH IN A NEO-CLASSICAL MODEL

Martin Beckmann
Brown University and
Technische Universitat Munchen

The following is an attempt to translate the "Limits to Growth" by Meadows et. al. into a more conventional economic framework. Under assumptions that may be more palatable to economic theorists than those of "World Dynamics", the optimal intertemporal allocation of stocks of producible goods and of the given known stocks of exhaustible resources will be investigated.

1. The following is an attempt to translate the "Limits to Growth" by Meadows et. al. into a more conventional economic framework. By a neo-classical world we mean the following.

Preferences are described by a logarithmic utility function and future utility is discounted by a constant discount factor. Production of goods and of pollution (bads) is described by Cobb-Douglas production functions with exogeneous technical progress at constant rates. In addition to producible inputs, net inputs of exhaustible resources are needed in the production of goods. Consumption of leisure is equal to the stock of persons minus the person inputs into production, and in this way labor can be treated as any other consumption good. Under these assumptions which may be more palatable to economic theorists than those of "World Dynamics", the optimal intertemporal allocation of stocks of producible goods and of the given known stocks of exhaustible resources will be investigated.

2. Notation

i, k	producible goods
j	bads
h	exhaustible resources
S_1	set of producible goods
S_2	set of bads
S_3	set of exhaustible resources
$S_{1,2} = S_1 \cup S_2$	set of bads and producible goods
x_{ik}	amount of good i used as input into production of good k
y_i	stock of good i
z_j	level of bad j
s_j	stock of exhaustible resource j
s_{ik}	amount (net flow) of resource j into production of good k
c_i	consumption of good i
$x_i = \sum_{k \varepsilon S_{1,2}} x_{ik}$	total input of good i
ρ	discount factor

3. In order to develop a dynamic programming formulation of the optimal intertemporal allocation problem we must specify the utility of consumption and the production function. The utility of a consumption plan is assumed to be the sum of discounted utilities for different periods. In any period the utility is the following logarithmic function of consumption and of the level of bads:

$$u = \sum_{i \varepsilon S_1} a_i \log c_i - \sum_{j \varepsilon S_2} a_j \log z_j \qquad (1)$$

The first sum denotes positive utility derived from the consumption of goods, the second part the disutility derived from current levels of bads (pollution).

The output of good k in period t + 1 is a function of the input x_{ik} of producible goods i in period t, of the inputs s_{hk} of exhaustible resources h in period t and of the level of bads z_j in period t as well as the state of technology denoted by the time index t:

$$y_k(t + 1) = f_k(t, \dots x_{ik}, \dots s_{hk}, \dots z_j)$$

Specifically we assume a Cobb-Douglas production function and a constant rate μ of exogenous technical change

$$y_k(t + 1) = g_k e^{\mu_k t} \prod_{i \varepsilon S_1} x_{ik}^{\gamma_{ik}} \prod_{h \varepsilon S_3} s_{hk}^{\delta_{hk}} \prod_{j \varepsilon S_2} z_j^{-\varepsilon_{jk}} \qquad (2)$$

In a limited world where some resources are fixed and nonaugmentable, returns to scale are diminishing. Therefore we assume

$$\sum_{i \varepsilon S_2} \gamma_{ik} + \sum_{h \varepsilon S_2} \delta_{hk} < 1 \qquad \text{all k.} \qquad (3)$$

The level of bads is a function of the contemporaneous use of certain inputs in various production activities and of special control activities. Let x_j be the level of the activity which seeks to control z_j. The production function for z_j is assumed to be Cobb-Douglas of the following form

$$z_j = g_j x_j^{-1} \prod_{i,k} x_{ik}^{\upsilon_{ikj}} \qquad (4)$$

υ_{ikj} is the elasticity of the supply of bad j with respect to input x_{ik} in the production of k. Control activity x_j is also assumed to have a Cobb-Douglas production function

$$x_j = h_j \prod_{i \varepsilon S_1} x_{ij}^{\gamma_{ij}} \qquad (5)$$

In principle, it would be possible to let the production of x_j also involve exhaustible resources and the level of bads as well as be subject to technical change. This complication would not affect the following analysis except by changing the values of certain coefficients. Summarizing we have

$$u = \sum_{i\varepsilon S_1} a_i \log(y_i - \sum_{k\varepsilon S_{1,2}} x_{ik})$$

$$- \sum_{j\varepsilon S_2} a_j \log[g_j \prod_{i,k} x_{ik}^{\upsilon_{ikj}} h_j^{-1} \prod_{i\varepsilon S_1} - x_{ij}^{\gamma_{ij}}] \qquad (6)$$

$$y_k(t+1) = g_k e^{\mu_k t} \prod_{i\varepsilon S_1} x_{ik}^{\gamma_{ik}} \prod_{h\varepsilon S_3} s_{hk}^{\delta_{hk}}$$

$$\prod_{j\varepsilon S_2} [g_j \prod_{i,k} x_{ik}^{\upsilon_{ikj}} h_j^{-1} \prod_{i\varepsilon S_1} x_{ij}^{-\gamma_{ij}}]^{-\varepsilon_{jk}} \qquad (7)$$

$$s_i(t+1) = s_i(t) - \sum_k s_{ik} \qquad (8)$$

4. The optimum allocation is described by the following dynamic program. Let $V_n(\ldots y_i, \ldots s_h)$ denote the utility that can be achieved by an optimal allocation of initial stocks y of commodities and s of resources. Then

$$V_n(\ldots y_i, \ldots s_h) = \underset{x_{ik}, x_{ij}, s_h}{\mathrm{Max}} \{\sum_{i\varepsilon S_1} a_i \log(y_i - \sum_{k\varepsilon S_{1,2}} x_{ik})$$

$$- \sum_{j\varepsilon S_2} a_j \log z_j + \rho \mathbf{V}_{n-1}(\ldots g_k e^{\mu_k t} \prod_{i\varepsilon S_1} x_{ik}^{\gamma_{ik}} \prod_{n\varepsilon S_3} s_{nk}^{\delta_{nk}} \prod_j z_j^{-\varepsilon_{jk}},$$

$$\ldots s_h - \sum_{k\varepsilon S_1} s_{nk})\} \qquad (9)$$

where z_j is given by (4).

Since $\rho < 1$ it can be shown that $\lim_{n\to\infty} V_n(\ldots y_i, s_h)$ exists. In the following we assume that $n \to \infty$ and drop the subscript n from V_n.

The specification of logarithmic utility and Cobb-Douglas production function is interesting because it permits us to write $V(\ldots y_i \ldots z_h)$ in closed form. In fact

$$V = b_o + \beta t + \sum_{i \varepsilon S_1} b_i \log y_i + \sum_{h \varepsilon S_3} b_h \log s_h \qquad (10)$$

5. This may be verified as follows. Substitute (10) in the right hand side of (9) and maximize first with respect to x_{ik}, observing (4). The result is

$$0 = \frac{-a_i}{y_i - \sum_{k \varepsilon S_{1,2}} x_{ik}} + \frac{-\sum_j a_j \upsilon_{ikj} - \sum_r b_r \upsilon_{ikj} \varepsilon_{jr} + b_k \gamma_{ik}}{x_{ik}}$$

For brevity write this as

$$\frac{a_i}{y_i - x_i} = \frac{\lambda_{ik}}{x_{ik}}$$

where

$$x_i = \sum_{k \varepsilon S_{1,2}} x_{ik} \qquad (11)$$

$$\lambda_{ik} = b_h \gamma_{ik} - \sum_j a_j \upsilon_{ijk} - \rho \sum_{j,r} b_r \upsilon_{ikj} \varepsilon_{jr} \quad k \varepsilon S_1 \qquad (12)$$

i.e.,

$$x_{ik} = \frac{\lambda_{ik}}{a_i} (y_i - x_i) \qquad (13)$$

Next maximize the right hand side of (9) with respect to x_{ij}, $j \varepsilon S_2$:

$$0 = \frac{a_j \gamma_{ij}}{x_{ij}} + \frac{\rho \sum_k b_k \gamma_{ij} \varepsilon_{jk}}{x_{ij}} - \frac{a_i}{y_i - \sum_{k \varepsilon S_{1,2}} x_{ik}}$$

or

$$\frac{a_i}{y_i - x_i} = \frac{\lambda_{ij}}{x_{ij}}$$

where

$$\lambda_{ij} = a_j \gamma_{ij} + \sum_{k \varepsilon S_1} b_k \gamma_{ij} \varepsilon_{ik} \qquad j \; \varepsilon \; S_2 \qquad (14)$$

$$x_{ij} = \frac{\lambda_{ij}}{a_i} (y_i - x_i) \qquad j \; \varepsilon \; S_2 \qquad (15)$$

Summing (13) and (15),

$$x_i = \frac{\lambda_i}{a_i} (y_i - x_i)$$

where

$$\lambda_i = \sum_{k \varepsilon S_{1,2}} x_{ik} \qquad (16)$$

or

$$x_i = \frac{\lambda_i}{\lambda_i + a_i} y_i \qquad (17)$$

Substituting (17) in (13) and (15),

$$x_{ik} = \frac{\lambda_{ik}}{\lambda_i + a_i} y_i \qquad i \varepsilon S_2 \quad k \varepsilon S_{1,2} \qquad (18)$$

A fixed proportion $\frac{\lambda_{ik}}{\lambda_i + a_i}$ of every resource i should thus be allocated to the production of every commodity k and to the control of every bad j. (For some good k or bad j and some input i, this proportion may be zero).

The constant proportion

$$\frac{y_i - x_i}{y_i} = \frac{a_i}{\lambda_i + a_i}$$

should be consumed. It is larger, <u>ceteris paribus</u>, the larger a_i.

Substitute now these values of x_{ik} and x_{ij} in the utility and production functions on the right hand side of (9). The result is that the right hand side and hence V on the left hand side is linear in the logarithm of y_i. A comparison of coefficients on both sides shows that

$$b_i = a_i + \sum_j a_j \left[\sum_k \upsilon_{ikj} - \gamma_{ij}\right] + \rho \sum_k \gamma_{ik} b_k$$

$$- \rho \sum_{j,k,r} b_r \, \upsilon_{ikj} \varepsilon_{jr} + \rho \sum_{j,k} \varepsilon_{jk} b_k \gamma_{ij} \qquad (19)$$

6. Finally the right hand side of (9) must be maximized with respect to irrecoverable resource inputs s_{hk}. This yields

$$\frac{\rho \, b_k \, \delta_{hk}}{s_{hk}} - \frac{\rho \, b_k}{s_k - \sum_k s_{hk}} = 0 \qquad \text{or}$$

$$s_{hk} = \frac{b_k \delta_{hk}}{b_h} (s_h - \sigma_h) \qquad (20) \text{ where}$$

$$\sigma_h = \sum_{k \varepsilon S_1} s_{hk} \qquad (21)$$

Summing (20),

$$\sigma_h = \frac{\sum_k \delta_{hk} b_k}{b_h} (s_h - \sigma_h)$$

$$\sigma_h = \frac{\sum_k \delta_{hk} b_k}{b_h + \sum_k \delta_{hk} b_k} \, s_h .$$

Substituting in (20),

$$s_{hk} = \frac{\delta_{hk} b_k}{b_h + \sum_k \delta_{hk} b_k} \, s_h . \qquad (22)$$

Now substitute (22) in the right hand side of (9) and compare coefficients on both sides

$$b_h = \rho b_h + \rho \sum_k b_k \delta_{hk}$$

$$b_h = \frac{\rho}{1 - \rho} \sum_k \delta_{hk} b_k \tag{23}$$

Substituting (23) in (22) and summing we finally have

$$\sigma_h = \frac{1}{\frac{\rho}{1 - \rho} + } s_h \qquad \text{or}$$

$$\frac{\sigma_h}{s_h} = 1 - \rho \tag{24}$$

Thus the same fraction $1 - \rho$ of every exhaustible resource should be used up in each period. If r is the implied rate of interest,

$$\rho = \frac{1}{1 + r}$$

Then

$$1 - \rho = \frac{r}{1 + r} \approx r \tag{25}$$

is approximately the interest rate.

7. Under this policy the stock of resource h at time t is

$$s_h(t) = s_h(0) \rho^t. \tag{26}$$

Equation (24) serves to verify that on both sides of (9)

$$V(\ldots y_i, \ldots s_h) = \sum b_h \log s_h + \ldots$$

The terms involving t on either side are

$$\beta t = \rho [\sum_{k \varepsilon S_1} b_k \mu_k + \beta] t \qquad \text{so that}$$

$$\beta = 1 - \rho \sum_{k \varepsilon S_1} b_k \mu_k . \tag{27}$$

Consider now the development of commodity stock $y_i(t)$ over time. Using (18) and (20) in the production function for bads (4) and for goods (2) we have, in logarithmic terms:

$$\xi_j(t) = \sum_i [\sum_k \upsilon_{ik} - \gamma_{ij}] \eta_i(t) + w_j \qquad (28)$$

where

$$\xi_j = \log z_j \qquad (29)$$

$$\eta_i = \log y_i \qquad (30)$$

and w_i is a constant.

For the output of goods we have

$$\eta_k(t + 1) = w_k + \mu t + [\log \rho \cdot \sum_h \delta_{hk}] t$$

$$+ \sum_i (\gamma_{ik} - \sum_j \varepsilon_{jk}[\sum_r \upsilon_{ir} - \gamma_{ij}]) \eta_i(t). \qquad (31)$$

In vector and matrix notation we have the linear system of difference equations:

$$\eta(t + 1) = w + [\rho\mu_k - \delta| \log \rho|] t + \Theta'\eta(t) \qquad (32)$$

where $e' = (1, \ldots 1)$

$$\delta' = (\ldots, \sum_k \delta_{hk})$$

$$\Theta = (\Theta_{ik}) \qquad \Theta_{ik} = \gamma_{ik} - \sum_j \varepsilon_{jk}(\sum_r \upsilon_{ir} - \gamma_{ij}).$$

Output levels will converge to a stationary state if

$$\mu_k = |\log \rho| \sum_h \delta_{hk} \qquad \text{all } k, \qquad (33)$$

and will be increasing provided

$$\mu_k \geqq |\log \rho| \sum_h \delta_{hk} \qquad k \varepsilon S_1 \qquad (34)$$

This means the following. In order to assure that future generations will not be worse off due to resource depletion, a depletion rate $1 - \rho$ for exhaustible resources must be set at such a level that inequalities (34) applies for all producible commodities k. The permissible depletion rate is thus proportional to the rate of technical progress. Chosing a depletion rate is tantamount to setting a value on future as compared to present utility.

If current valuation of future welfare differs from this, then this implies that the optimum policy is one of living well now and facing future decline--*apres nous la deluge.* Even when confronted by such decline there is an optimum way to decline and fall. Who is to say that a stationary state is preferable?

REFERENCE

Meadows, D. H., Randers, F., Behrens III, W. W., "The Limits to Growth", a report for *The Club of Rome's* Project on the Predicament of Mankind, Universe books, New York 1972.

DYNAMIC OPTIMIZATION OF THE TRADE-OFF BETWEEN PRODUCTION AND POLLUTION

Osman Coskunoglu
The Scientific and Technical Research Council of Turkey
Operations Research Unit, Ankara

Environmental pollution is an undesirable by-product of an essential process--production activities. Every allocation of resources for the control of this undesirable by-product reduces the rate of production increase, since environmental pollution control processes are assumed to be unproductive. Thus, in order to determine the amount and timing of resources which will be allocated for the control of environmental pollution, it is necessary to resolve the conflict between production and pollution control in an optimum manner. In this study the problem is viewed as a time-continuous decision making problem. In mathematical terms the problem is that of choosing decision variables. Here decision variables are the amounts (and timing) of resources which are to be allocated to pollution control and production. The model is formulated as a control problem, the objective being to maximize the discounted present value of the profit achieved throughout the planning horizon plus the present value of the terminal values of production capacity and penalty (or subsidy) due to the pollution level reached at the end of the horizon.

1. INTRODUCTION

Decision problems in ecology are complex in nature due to the interdisciplinary character of the problems. The factors that should be considered are qualitative rather than quantitative. These characteristics necessitate an approach that focuses on an entire problem or system rather than on each of its components, with the purpose of providing a common framework.

The purpose of this paper is to present a model which will contain the qualitative factors as parameters. Parametric solution of the model will provide the necessary relations which contain these parameters. From these relations it is possible to see the consequences of assigning various values to these parameters. The effect on the general economy of investments into pollution control has been selected as a measure of effectiveness. This selection also makes it possible to relate the ongoing research on antipollution policies (Stahr [16])to the general economy and furthermore, to define economic and social costs with less subjective units than units like "willingness to pay"(Cemborowicz [5]).

Investments in environmental pollution control are called "external diseconomies". Mathematical models are "urgently" necessary "for the analysis of environmental pollution, urban congestion, landscape deterioration, and the host of other externality (adverse environmental quality) phenomena which accompany economic growth" (Kneese _et_. _al_.[9]). Dr. Robert Anderson has made an analysis (in Rose [14]) of air-pollution control by assuming annual anti-air-pollution investments of a fixed amount by industry, and measuring the effect of this on the rate of change of the GNP.

Since production (economic growth) and pollution are interdependent and dynamic processes, an attempt has been made in this paper to allocate the available resources between them in a planning horizon so as to optimize a certain objective, rather than to assign a fixed amount to either of them. This allocation of resources is the direct output of the model presented here. Indirect outputs are the shadow prices of qualitative measures.

In the formulation of the model, control theory, which has been recently accepted as a latent potential in studying dynamic economic models [1, 2, 3, 6, 7], has been utilized. Thompson and George's [6] paper encompassing a firm's operation and investments which is formulated as a control problem is taken as the basis of the model and pollution control investments and costs are introduced into the model, and solved by using Pontryagin's Maximum Principle [13].

In Section II the formulation and description of the model is presented. After the solution and discussions of the solution are given in Section III, a discussion of the model has been made in Section IV. At the end of the paper a description of the control theory is presented in the Appendix.

II. THE MODEL

The model presented below is a general one which is not limited to any specific pollution or production. Although it can be formulated for a firm, a municipality, or a country, for the sake of consistency, the term "firm" is used throughout the paper.

Before presenting the model, an explanation of the notations is given.

State variables (variables describing the system at any given time):

$x(t)$: the production capacity of the firm

$d(t)$: the net debt of the firm

$y(t)$: the level of pollutant emission per unit time

Control variables (decision variables):

$u(t)$: rate of purchase of units of new pollutant emission abatement capacity

$v(t)$: rate of purchase of units of new production capacity

Input variables:

$r(t)$: profit per unit of output

$g(t)$: output of a unit of capacity in a unit of time

$c(t)$: price of a unit of production capacity

$i(t)$: market rate of interest

$a(t)$: rate of attrition of capacity

$s(t)$: rate of change of pollutant emission from unit production capacity

$k(t)$: price of a unit of pollutant emission abatement capacity

y_s : specified level of pollutant emission per unit time, for the end of the planning horizon

z : penalty (or subsidy) per unit of pollutant emission per unit time above (or below) the specified level at the end of the planning horizon.

$N(t)$: upper bound on the rate of purchase of new production capacity

$M(t)$: upper bound on the rate of purchase of new pollutant emission abatement capacity.

T : planning horizon

$\phi(t)$: $\exp(-\theta t)$

θ : constant rate of discount

The formulation of the model is presented below:

$$\max_{u,v}: \quad J = \int_0^T [r(t)\, g(t)\, x(t) - C(t)\, v(t) - i(t) d(t)$$

$$-k(t) u(t)]\, \phi(t) dt + x(T) . c(T) \phi(T)$$

$$+ z[y_s - y(T)] \qquad \text{(I.1)}$$

subject to:

$$\dot{x}(t) = v(t) - a(t)x(t) \qquad \text{(I.2)}$$

$$\dot{d}(t) = c(t)v(t) + i(t)d(t) + k(t)u(t) - r(t)g(t)x(t) \qquad \text{(I.3)}$$

$$\dot{y}(t) = s(t)x(t) - u(t) \qquad \text{(I.4)}$$

$$0 \leqq u(t) \leqq M(t) \qquad \text{(I.5)}$$

$$0 \leqq v(t) \leqq N(t) \qquad \text{(I.6)}$$

Here, the objective function (I.1) is the discounted value of net production profits less interest charges and costs of new production and pollution control capacity over the period. This maximization is constrained by three differential equations (I.2), (I.3) and (I.4) and the two sets of inequalities (I.5) and (I.6).

Equation (I.2), describes the rate of change of production capacity. Equation (I.3) describes how the firm is financed. The firm is assumed to have one financial account from which it can borrow or into which it can deposit; thus $d(t)$ can be either positive or negative.

Initial and terminal times as well as initial capacity ($x(o) = x_o$), debt ($d(o) = d_o$) and pollution level ($y(o) = y_o$) are fixed.

The following is assumed: x_o and y_o are positive; the functions $M(t)$, $N(t)$, $r(t)$, $c(t)$, $i(t)$, $g(t)$, $a(t)$ are continuously differentiable and positive on $[O, T]$.

III. THE SOLUTION

In the following, the "function of time" notation is omitted, i.e., instead of $x(t)$, only x is used.

The Hamiltonian function is:

$$H = (r.g.x-c.s-i.d-k.u)\ \phi + P_1\ (v-ax) + P_2(c.v + i.d + k.u - r.g.x) + P_3(s.x-u) \quad \text{(III.1)}$$

To maximize H with respect to control variables u and v, terms in H containing u and v are separately presented in (III.2) and (III.3) respectively,

$$[P_2.k - \phi .k - P_3]u \quad \text{(III.2)}$$

$$[P_1 + P_2.c - \phi .c]v \quad \text{(III.3)}$$

Since the only term common to both expressions is ϕ and it has the same sign in both expressions, maximizing H is equivalent to maximizing (III.2) and (III.3) individually. Thus from the Maximum Principle:

$$[p_2^*k - \phi k - p_3^*]u^* \geq [p_2^*k - \phi k - p_3^*]u \quad \text{(III.4)}$$

$$[p_1^* + p_2^*c - \phi c]v^* \geq [p_1^* + p_2^*v - \phi c]v \quad \text{(III.5)}$$

Considering inequalities (I.5) and (III.4), the optimum value of u(t) is as follows:

$$u^* = \begin{cases} M \text{ when } [p_2{}^*k - \phi k - p_3{}^*] > 0 & \text{(III.6.a)} \\ O \text{ when } [p_2\ k - \phi k - p_3{}^*] < 0 & \text{(III.6.b)} \\ \text{unspecified when } [p_2{}^*k - \phi k - p_3{}^*] = 0 & \text{(III.6.c)} \end{cases}$$

Similarly, considering inequalities (I.6) and (III.5), the optimal value of v(t) is as follows:

$$v^* = \begin{cases} N \text{ when } [p_1{}^* + p_2{}^*c - \phi c] > 0 & \text{(III.7.a)} \\ O \text{ when } [p_1{}^* + p_2{}^*c - \phi c] < 0 & \text{(III.7.b)} \\ \text{unspecified when } [p_1{}^* + p_2{}^*c - \phi c] = 0 & \text{(III.7.c)} \end{cases}$$

where, the values of the costate variables, $p_i{}^*(i=1,\ 2,\ 3)$, at the optimum are as follows

$$p_1{}^*(t) = -\frac{\partial H}{\partial x} = -r(t)g(t)\phi(t) + p_1(t)a(t) + p_2(t)r(t)g(t) \qquad \text{(III.8)}$$

$$p_2{}^*(t) = -\frac{\partial H}{\partial d} = i(t)\phi(t) - p_2(t)i(t) \qquad \text{(III.9)}$$

$$p_3{}^*(t) = -\frac{\partial H}{\partial y} = 0 \qquad \text{(III.10)}$$

where

$$p_1{}^*(T) = \frac{\partial F}{\partial x_T} = c(T)\phi(T) \qquad \text{(III.11)}$$

$$p_2{}^*(T) = \frac{\partial F}{\partial d_T} = 0 \qquad \text{(III.12)}$$

$$p_3{}^*(T) = \frac{\partial F}{\partial y_T} = -z \qquad \text{(III.13)}$$

Solution of (III.8)

$$p_1^*(t) = e^{\int_0^t a(S)dS} \left[\int_0^t e^{-\int_0^s a(s)ds} [p_2(s) - \phi(s)] r(s) . g(s) ds + A \right] \qquad \text{(III.14)}$$

where, A is an integration constant, $\phi(s) = e^{-\theta s}$ and considering (III.9), $p_2(t)$ is as follows

$$p_2^*(t) = e^{-\int_0^t i(s)ds} \left[\int_0^t e^{\int_0^s i(s)ds} [i(s) . \phi(s)] ds + B \right] \qquad \text{(III.15)}$$

assuming the market rate of interest to be constant and equal to 2θ, i.e., $i(t) = 2\theta$, and substituting $\phi(s) = e^{-\theta s}$, (III.15) reduces to:

$$p_2^*(t) = e^{-2\theta t} [2e^{\theta t} - 2 + B] \qquad \text{(III.16)}$$

From equation (III.12),

$$B = 2(1 - e^{\theta T}) \qquad \text{(III.17)}$$

Substituting (III.16) and (III.17) into (III.14) and assuming constant rate of attrition of capacity, $a(t) = b$, final form of equation (III.14) is as follows:

$$p_1^*(t) = 2e^{bt} \left[\int_0^t [e^{-(b+\theta)s}(½ - 2e^{-\theta s}) + e^{(-bs-\theta T)}(e^{-\theta T} - 1)] r(s) g(s) ds + A \right] \qquad \text{(III.18)}$$

Constant A can be found from (III.11). The expressions for the remaining two costate variables are as follows:

$$p_2^*(t) = 2e^{-\theta t} [1 - e^{\theta(T-t)}] \qquad \text{(III.19)}$$

$$p_3^*(t) = -Z \qquad \text{(III.20)}$$

Substituting equations (III.18), (III.19) and (III.20) into (III.6) and (III.7), optimum values of control variables can be found.

To illustrate, expressions for the optimum value of u are presented below:

From (III.6), let,

$$G(t) = p_2^*(t)k(t) - \theta(t)k(t) - p_3^*(t)$$

$$= 2\, e^{-\theta t} [\tfrac{1}{2} - e^{\theta(T-t)}]\, k(t) + z \qquad \text{(III.21)}$$

then,

$$u^* = \begin{cases} M \quad \text{when } G(t) > 0 & \text{(III.22.a)} \\ 0 \quad \text{when } G(t) < 0 & \text{(III.22.b)} \\ \text{unspecified when } G(t) = 0 & \text{(III.22.c)} \end{cases}$$

The behavior of G(t), therefore u*(t) depends upon k(t). Similarly, v*(t) depends upon r(t), g(t) and c(t). Since the main purpose of this paper is to present a mathematical model using control theory, specifications of these functions will not be considered here. This can be made the subject matter of a follow-up work leading to the explicit determination of u* and v* and sensitivity analyses of the type given in the Appendix.

IV. DISCUSSION OF THE MODEL

The model presented in this paper is rather simple in structure, thus easy to solve. This convenience is due to the fact that functions and variables are assumed to change by time regularly and continuously. If these assumptions are dispensed with a discrete dynamic optimization problem with irregular transformation by time appears. A tabular solution technique for this type of problem, utilizing the dynamic programming method can be employed. But the disadvantage of a tabular formulation is the high dimensionality of the problem (see Nemhauser [12]).

Another point which needs further research is the relation function between production and pollution. This function differs according to the medium polluted (air, water or land) and according to the pollutant itself (CO, sludge etc.)

After resolving these points, especially the second one which is purely technical, the next necessary step is dispensing with the assumption that pollution control is unproductive and encountering the benefits of not disturbing the balance of nature (Koenig _et. al._[10]).

APPENDIX

THE CONTROL PROBLEM AND THE MAXIMUM PRINCIPLE

The mathematical formulation of the control problem, with known initial conditions and final time (which is the case in this paper): $i = 1, 2, \ldots, n$; $j = 1, 2, \ldots, r$

$$\max \quad J = \int_{o}^{T} I(x_i, u_j, t)\,dt + F(x_{iT}, T) \tag{A.1}$$

such that:

$$\dot{x}_i(t) = f_i(x_1, x_2, \ldots, x_n, u_1, u_2, \ldots, u_r; t) \tag{A.2}$$

$$x_i(0) = x_{io} \qquad \text{all } i \tag{A.3}$$

$$x_i(T) = x_{iT} \qquad \text{all } i \tag{A.4}$$

$$u_j(t) \in U(t) \qquad \text{all } j \tag{A.5}$$

where:

$x = (x_1(t), x_2(t), \ldots, x_n(t))$: state variables

$u = (u_1(t), u_2(t), \ldots, u_r(t))$: control variables

T: time horizon

U(t): constraints imposed on control variables

I: intermediate function

F: final function

Proposition 1 (Pontryagin's Maximum Principle) [4,7,8,13,15]:

Let u* be a choice of control variables ($0 \leqq t \leqq T$) which maximizes

$$\int_0^T I(x, u, t)\, dt + F(x_T, T) \tag{A.6}$$

subject to the conditions

(a) $\dot{x} = f(x, u, t)$

with some constraints on the choices of control variables, and initial conditions on the state variables. Then there exist costate variables, functions of time, $p_i(t)$, such that, for each t

(b) u* maximizes H(x, u, p, t), called the Hamiltonian function, if

$$H(x^*, u^*, p^*, t) \geqq H(x^*, u, p^*, t) \tag{A.8}$$

$$0 \leqq t \leqq T, \; u \in U$$

where,

$$H(x, u, p, t) = I(x, u, t) + \sum_{i=1}^{n} p_i f_i(x, u, t) \qquad (A.9)$$

and the functions p_i satisfy the differential equations

$$\text{(c)} \quad \dot{p}_i = - \frac{\partial H}{\partial x_i} \qquad (A.10)$$

Optimum control, u*, is determined from (A.8) in terms of state and costate variables. To determine these 2n unknown variables plus the 2n integration constants of equations (A.7) and (A.10), there exist 2n differential equations, (A.7) and (A.10), and n initial conditions given in equation set (A.3).

Proposition 2 (see Pontryagin [13]) (Transversality Conditions):

The solution to Proposition 1 also satisfies the condition

$$p_i(T) = \frac{\partial F}{\partial x_{iT}} \qquad (A.11)$$

The conditions of Proposition 2 supply the n additional condition needed to determine completely the solutions of the differential equations in Proposition 1 (a) and (c).

The conditions given in Propositions 1 and 2 are in general necessary but not sufficient for a global optimum solution. They are, however, necessary and sufficient if the Hamiltonian is linear in the control variables (Rose [14]) or if the maximized Hamiltonian is a concave function of the state variables (Stahr [16]). In the problem considered in this paper, the Hamiltonian is linear in control variables, hence, these conditions are used as necessary and sufficient for a global optimum solution.

The interpretation of the costate variables is similar to the interpretation of Lagrange multipliers of static economizing problems, i.e., the extremal costate is the sensitivity of the maximum value of the objective function to changes in the corresponding state variable.

Sensitivity analyses (see Intriligator [7]):

(i) The sensitivity of the optimal value of the objective function (J^*) to a change in the initial time (t_o) is given by:

$$\frac{\partial J^*}{\partial t_o} = -[I(x^*, u^*, t)]_{t_o} \qquad \text{(A.12)}$$

that is, by the negative of the initial value of the intermediate function.

(ii) The sensitivity of J^* to changes in T is given by:

$$\frac{\partial J^*}{\partial T} = [I(x^*,u^*,t)]_T + \frac{\partial F}{\partial x_T} \cdot \frac{\partial x_T^*}{\partial T} + \frac{\partial F}{\partial T}(x_T^*,T) \qquad \text{(A.13)}$$

that is, by the terminal value of the I plus the increase in the F.

(iii) The sensitivities of J^* to changes in the x_o are given by:

$$\frac{\partial J^*}{\partial x_o} = p^*(0) \qquad \text{(A.14)}$$

That is, by the initial value of the corresponding optimal costate variable.

REFERENCES

1. Adams, G. Gerard, Burmeister, Edwin "Economic Models", IEEE Transactions on Systems, Man, and Cybernetics, V. SMC-3, N.1 19-27 (January 1973).

2. Arrow, Kenneth J. "Applications of Control Theory to Economic Growth", Lectures in Applied Mathematics 12 (Mathematics of the Decision Sciences, Part 2), Providence, Rhode Island: American Mathematical Society (1968).

3. Arrow, Kenneth J., Beckmann, Martin J. and Karlin, Samuel "The Optimal Expansion of the Capacity of the Firm", Studies in the Mathematical Theory of Inventory and Production, Kenneth J. Arrow, Samuel Karlin and Herbert Scarf (eds.) Stanford, California: Stanford University Press (1958).

4. Arrow, Kenneth J., Kurz, Mordecai "Public Investment, The Rate of Return, and Optimal Fiscal Policy," Baltimore, Maryland: The Johns Hopkins Press (1970).

5. Cemborowicz, Ralf G. "Models of Water Supply Systems", Paper presented at the NATO Advanced Study Institute on: Systems Analysis for Environmental Pollution Control, Baiersbronn, Germany (December 9-16, 1972).

6. George, Melvin D., Thompson, Russel G. "Optimal Operations and Investments of the Firm", Management Science 15, 1, 49-56 (September 1968).

7. Intriligator, Michael D. "Mathematical Optimization and Economic Theory", Englewood Cliffs, New Jersey: Prentice Hall, Inc. (1971).

8. Kirk, Donald E. "Optimal Control Theory", Englewood Cliffs, New Jersey: Prentice Hall, Inc. (1970).

9. Kneese, Allen V., Ayres, Robert U., and d'Arge, Ralph C. "Economics and the Environment - A Materials Balance Approach" Washington, D.C.: Resources for the Future, Inc. (1970). (Distributed by The Johns Hopkins Press, Baltimore, Maryland).

10. Koenig, Herman E., Cooper, William E. and Falvey, James M. "Engineering for Ecological, Sociological, and Economic Compatibility", IEEE Transactions on Systems, Man and Cybernetics, V. SMC-2, N. 3, 319-331 (July 1972).

11. Mangasarian, D. L. "Sufficient Conditions for the Optimal Control of Nonlinear Systems", SIAM Journal on Control 4 139-152 (1966).

12. Nemhauser, George L. "Introduction to Dynamic Programming", New York: John Wiley and Sons, Inc. (1967).

13. Pontryagin, L. S., Boltyanskii, V. G., Gamkrelidze, R. V. and Mishchenko, E. F. (K. N. Trirogoff, translator and L. N. Newstand, editor) "The Mathematical Theory of Optimal Processes", New York: Interscience (1962).

14. Rose, Sanford "The Economics of Environment", Fortune 81, 2, p. 186 (February 1970).

15. Rozonoer, L. I. "L. S. Pontryagin Maximum Principle in the Theory of Optimum Systems--I, II, III", Avtomatika i Telemekhanika 20, 1320-1334, 1441-1458, 1561-1518; translated in: Automation and Remote Control 20, 1288-1302, 1405-1421, 1517-1532 (1959).

16. Stahr, Elvin J. "Antipollution Policies, Their Nature and Their Impact on Corporate Profits", Economics of Pollution (The Charles C. Moskowitz Lectures, Number XI), New York: New York University Press (1971).

A "POLLUTION" GAME: A THEORETICAL AND EXPERIMENTAL STUDY

Denis Bayart
Bertrand Collomb
Ecole Polytechnique

Jean-Pierre Ponssard
Ecole Polytechnique and CESMAP

The present paper investigates a 5-person-multistage non-cooperative game developed in the psychological literature. This game is presented here as an economic system with externalities.

The first part of the paper is devoted to a theoretical study to analyze the basic structure of the model and its connections with other economic representations of externalities such as pollution. In particular we characterize the Nash equilibria and the core of the game. Moreover, we show that these two solutions concepts generate the same points for the infinite stage game. Some other aspects, such as the introduction of a discount factor, are also investigated.

In the second part of the paper, we analyze, in view of these theoretical insights, the results of an experiment which was led among a population of scholars in France. We study group cohesiveness and individual behavior in four distinct situations: two types of externalities combined with two different initial situations. Significant results are compared with those obtained previously.

It is thought that this paper may be of some interest for psychologists and economists who have been developing models of pollution in their own fields.

FOREWORD

This paper is devoted to a theoretical and experimental study of a "pollution" game which has been introduced in the psychological literature [Dana, Rubenstein]. Informally stated, one might say that the Game simulates a dynamic economic environment with limited resources. These resources are commonly owned by the community and each agent continuously uses some of them for his private comsumption. Once used they are degraded and the agent has the choice between regenerating them or not. Thus each agent may contribute either to the preservation or even the amelioration of the environment (at a certain cost) or to its degradation.

This situation has been abstracted by the following rules: The game is a succession of stages. Each stage begins with a deal of seven cards to each player from a fifty-card deck of blue or red cards. If a player receives less than three red cards he wins 4 units, otherwise he loses 1 unit. Then he has the choice between three moves as regards the composition of the central deck: change a blue card into a red card at no cost, make no change at a cost of 2 units, change a red card into a blue card at a cost of 4 units. The actual composition of the central deck for the next stage is the global result of each player's move.

This Game has been extensively experimented in different countries and some specific hypotheses concerning pollution seem to be verified [Watzke, Doktor, Rubenstein, Dana].

In these conditions it appeared interesting to investigate the model from a theoretical point of view, comparing its features with "economies with externalities" as modeled in the economic literature.

Moreover, in view of the theoretical insights gained by the analysis, it was decided to run another experiment among a population of scholars in France. These results are reported in the second part of the study.

FIRST PART

§ 0 Introduction

This first part of the paper gives a theoretical treatment of the "pollution" Game. The concepts used in the analysis are drawn from economics and from game theory. Several hypotheses are studied, such as finite or infinite horizon as well as different discount rates. For convenience, the interpretations of the main results are gathered at the end of each paragraph.

§ I Formal Description of the Game

The Game may be described in extensive form; that is: the moves, the information sets and the final payoffs. However, instead of describing each move, it will be more convenient to see the Game as a succession of stages.

I - 1 The Stages

Let N be the number of stages. Each stage n ($0 < n \leq N$) consists of four steps.

Step 1: Chance deals seven cards to each player from a deck of fifty cards which contains $r_n (0 \leq r_n \leq 50)$ red cards and $50 - r_n$ blue cards.

Let r_n^i be the number of red cards dealt to player i, $i \in I$ where I is the set of Players. If $r_n^i \geq 3$, player i wins - 1 unit otherwise he wins + 4 units.

Step 2: Each player i, $i \in I$, selects a decision variable x_n^i

which can only take three values. Each value is associated with some immediate cost to player i.
Namely, if $x_n^i = -1, 0, +1$, player i has to pay 4, 2, 0 units respectively.

Step 3: Change the number of red cards in the deck, so that

$$r_{n-1} = r_n + \sum_{i \in I} x_n^i$$

(Except at the boundary of (0, 50) where we have:

$$r_n + \sum_{i \in I} x_n^i < 0 \rightarrow r_{n-1} = 0,$$

$$r_n + \sum_{i \in I} x_n^i > 50 \rightarrow r_{n-1} = 50).$$

Step 4: If $n-1 = 0$, the Game is over, otherwise play stage $n-1$.

I - 2 The Information Sets

We shall assume that each player knows N and, at the beginning of each stage, r_n. However player i does not know $\bar{x}_n^j$, $j \neq i$, the decision made by the other players.

I - 3 The Final Payoff

The final payoff of the Game to each player may be seen as his cumulative payoff over the N stages.

§ II Analysis of the Game as a Finite Horizon Non Cooperative Game

II - 1 Formulation

If the number of stages N is finite, we may take advantage of the stage structure of the Game to use a Dynamic Programming approach. We shall be specifically interested in "end play" considerations, since we shall develop a direct approach for the infinite horizon case.

In order to derive the Nash equilibria, we shall make use of a recurrence equation. Let $x_n = (x_n^i)_{i \in I}$ be the vector of decisions at stage n.

Then, if $F_n^i(r_n, x_n)$ denotes player i's cumulative payoff at the n-th stage depending on the number of red cards r_n and on the players' decisions x_n:

$$(i \in I)\ (0 < n \leqq N)\ F_n^i(r_n, x_n) =$$

(Step 1) $\quad 4 \text{ Prob } (r_n^i < 3 | r_n) - 1 \text{ Prob}(r_n^i \geqq 3 | r_n) +$

(Step 2) $\quad -2(1 - x_n^i) +$

(Stage n - 1) $\quad F_{n-1}^i(r_n + \sum_{i \in I} x_n^i, x_{n-1})$*

(in which: $(i \in I)\ F_o^i \equiv o$)

As such, the Game appears as a deterministic Markov game. It may be solved using a computer program which combines a dynamic component with the resolution of $|I|$-person matrix game.

II - 2 The Nash Equilibria

In spite of the overall simplicity of the Game, it is already a difficult computational task to characterize its Nash Equilibria. We shall present here some specific results obtained in the case of a two-person game. It is hoped that these results will throw some light on the general problem as well as give an idea of its complexities.

---

*This last term should be corrected at the boundary of (0, 50). see cf: I-1.

We briefly recall that a Nash equilibrium is, in this context, a vector

$$x = (x^i_{n,r})_{i \in I}; \quad n = 1, \ldots . N, \ r = 0, \ldots . 50$$

in which $x^i_{n,r}$ is a probability distribution over (-1, 0, +1) which specifies player i's move at the n-th stage if $r_n = r$, such that each player has no incentive to change alone his own decisions.

We shall represent a Nash equilibrium as a set of paths in a (r,n) diagram (starting at some (r,n), there are 5 possible next positions of which only some are part of an equilibrium.) Randomized decisions will be taken into account with dotted lines. Moreover we shall label the points depending on which player has the higher cumulative payoff at this point. In case of equality, there will be no label.

There actually are a very large number of equilibria. In figure 1, we have represented the most advantageous for Player A ($I = \{A,B\}$). For instance, at the point (7, 11) A plays +1 and B -1, so that A is better off by 4 units; at the point (8,11) A plays 0 and B -1 so that A is better off by 2 units, however at the point (9,11), both players play -1. It is actually a simple matter to construct the players' moves from any point of the diagram except those few ones which include randomization.

If the number of stages before the end of the game is less or equal to 10 then both players play +1. However if the number of stages is larger than 10 then depending on r, the players may

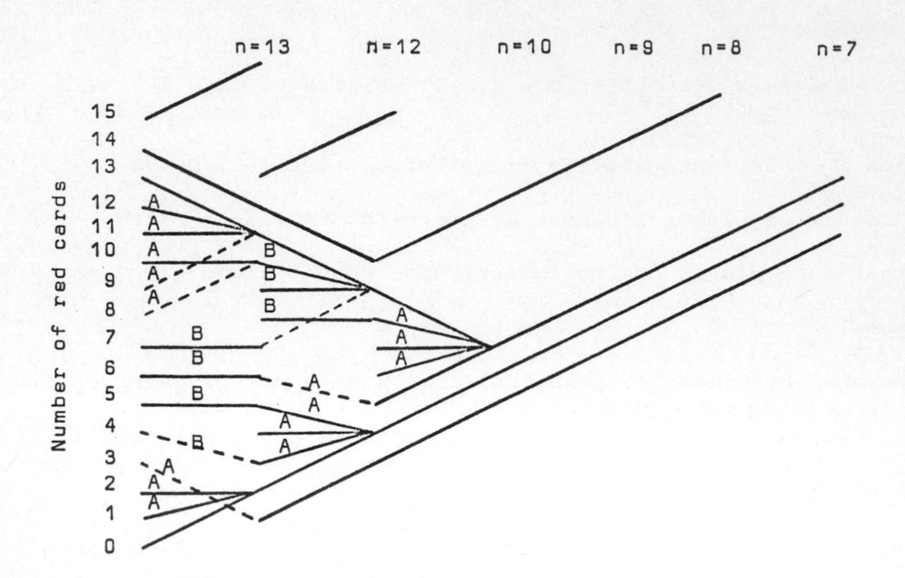

Figure 1 : Equilibrium Paths of the Two Player Game

Interpretation:

play +1 or -1. Two paths (14, 13) to (7, 10) and (0, 13) to (6, 10) have the property that both players get the same equilibrium payoff. It is interesting to note that between these two paths the equilibrium paths are not symmetric, one player having a higher payoff, and unstable, the same player is unable to keep a higher payoff for two successive stages. Moreover, it seems that these equilibria are moving towards the two symmetric paths.

A behavioral interpretation of these results may be as follows: Starting at some (r,n) it may make sense to "pollute" (+1) since the pair of moves (+1,-1) is an equilibrium and there are two "good reasons" to do so, first it is a way to get a higher payoff than the other player, second (-1,+1) is also an equilibrium. If the two players are "hard players", then the number of red cards is likely to go up until some point at which it does not make sense any more to pollute: even if the other player is depolluting it is also its own interest to depollute. From then on, the game is played symmetrically. In an experiment, both players may not "realize" at the same stage that now is time for depollution so that there is the risk to pass through the "break-even path" ((14, 13) to (7, 10)) and then it does not make sense any more to depollute: it is too late!

§ III Analysis of the Game as an Infinite Horizon Game

While the preceding section was devoted to "end play" considerations, we shall now be interested in the infinite horizon game. We shall use a direct approach as opposed to an indirect one which would consist in letting N, the number of stages, go to infinity. This will lead to a reformulation of the Game, keeping only the salient features in a simplified model. Then, the model will be studies using both cooperative and non-cooperative concepts.

III - 1 Selection of a Criterion: the Significance of Discounting

The previous criterion, the maximization of the expected payoff, does not remain operational in an infinite horizon model. It will be replaced by the maximization of the average payoff per stage. Another criterion which is conveniently used is the maximization of the discounted payoff.

This paragraph investigates the significance of discounting for the one-player game. Let $\rho (0 \leq \rho \leq 1)$ be the discount factor and $p(r)$ the probability of winning*. Then the optimal policy is determined by comparing the discounted values associated with each decision.

We obtain for all r in (1, 50):

- if $p(r) - p(r+1) < 2(1-\rho) / 5\rho$

  or if $r = 0$, play $x = +1$,

- if $p(r-1) - p(r) > 2(1-\rho) / 5\rho$

  play $x = -1$

- otherwise play $x = 0$

It follows that, after a finite number of stages, the number of red cards will remain stationary (r_∞). It may be noticed that we have:

$\rho = 1 \rightarrow r_\infty = 2$

$\rho = \rho^* \rightarrow r_\infty = r^*$

and $\rho < \rho^* \rightarrow r_\infty = 50$

Thus the function $r_\infty(\rho)$ is not continuous at ρ^* (cf Fig. 2 & 3).

$$*p(r) = \left[\frac{(50-r)!}{7!(43-r)!} + \frac{r(50-r)!}{6!(44-r)!} + \frac{r(r-2)}{2} \frac{(50-r)!}{5!(45-r)!} \right] \Big/ \frac{50!}{7!43!}$$

Interpretation:

This is easily interpreted in terms of pollution. As might be expected, the more the future is discounted, the more the present is polluted; however if the discount factor gets too high (higher than $\rho*$ in the model) then the present situation suddenly gets completely deteriorated!

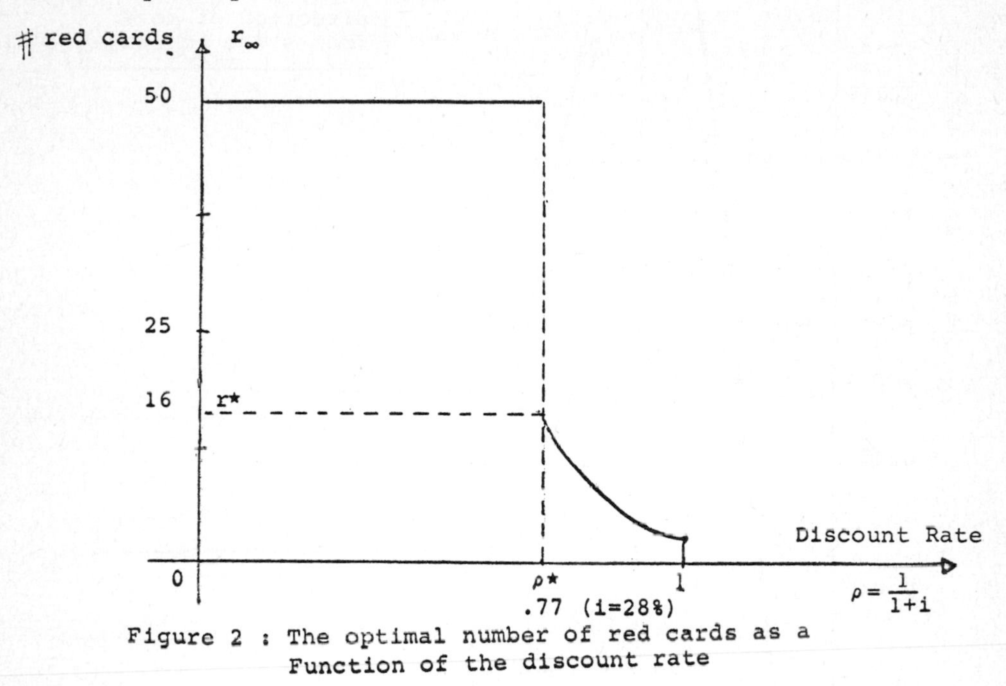

Figure 2 : The optimal number of red cards as a Function of the discount rate

III - 2 Formulation of a Simplified Model:

In order to facilitate the study of the infinite horizon game, only the two extreme states ($r = 0$ and $r = 50$) will be kept in the model. A possible justification to do so is that in a stationary solution the other states of the Markov game are certainly going to be transient.

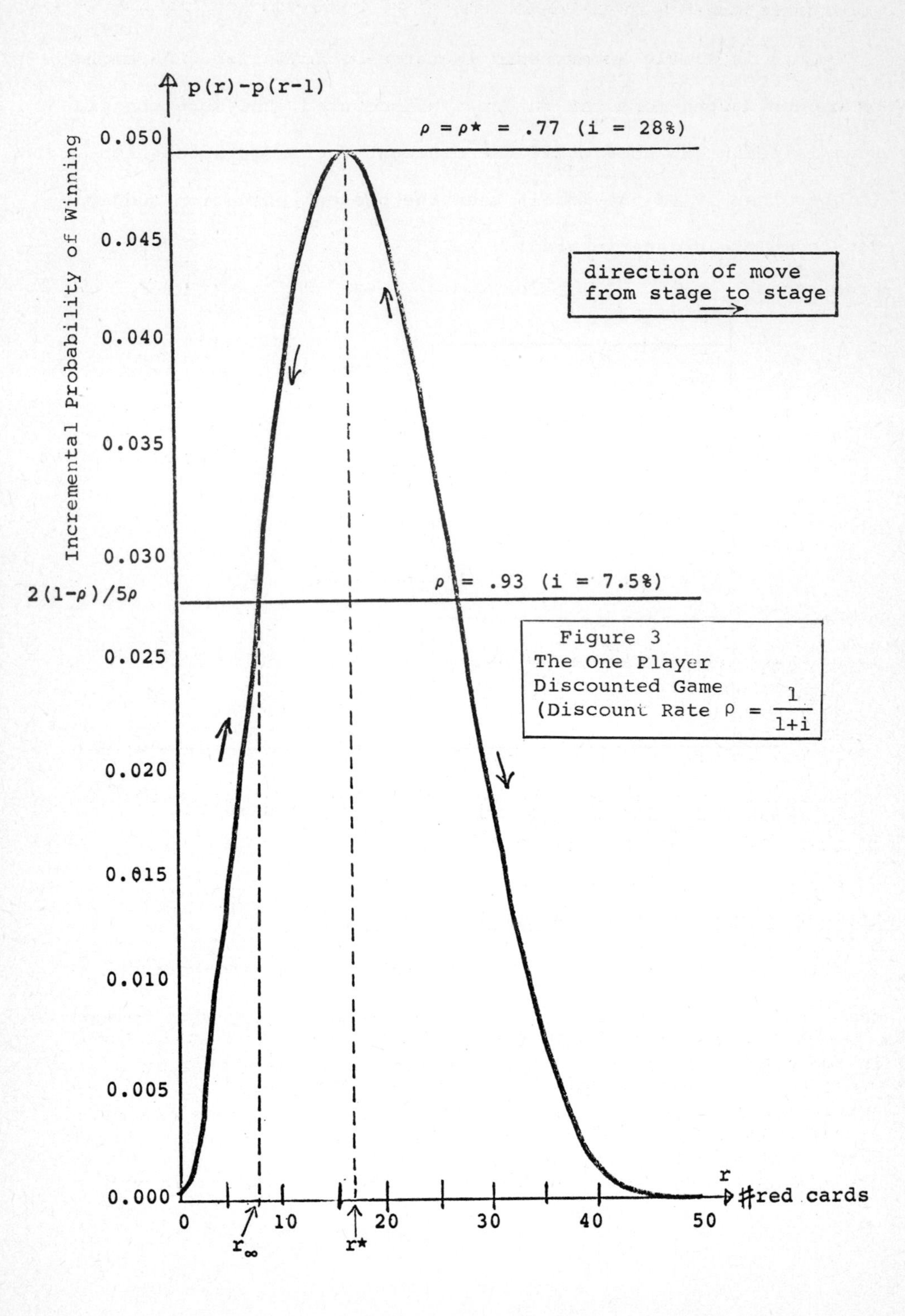

Figure 3
The One Player
Discounted Game
(Discount Rate $\rho = \frac{1}{1+i}$

This simplified model is described in Figure 4 (For specificity, it will be assumed that the players start in the winning state but this will turn out to be of little significance).

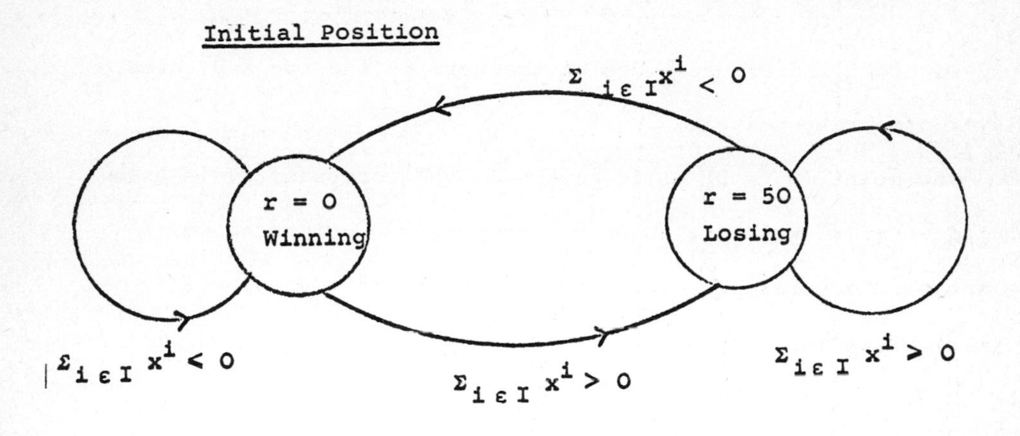

Figure 4 - The Simplified Model

III - 3 The "Core" of the Game

We recall that the criterion is the maximization of the average payoff per stage. In these conditions, the characteristic function v is found to be (assuming there are more than two players $|I| > 2$):

$$(S \subset I) \quad |S| < |I|/2 \quad v(S) = -|S|,$$

$$|S| \geqq |I|/2 \quad v(S) = 2(2|S| - |I|).$$

It follows that there is no incentive to join a coalition except if the size of this coalition is exactly $|I|/2 - 1$ if $|I|$ is even and $(|I| - 1)/2$ if $|I|$ is odd.

Let w belong to $R^{|I|}$. The core of this convex game, that is the set of average payoffs per stage which are feasible and undominated via any coalition, is such that:

$$\text{core} = \{w \mid \Sigma_{i \varepsilon I} w^i = 2|I|,\ (i_o \varepsilon I)\ \Sigma_{i \varepsilon I - i_o} w^i \geqq 2|I| - 4$$

Figure 5 and Figure 6 depict the core of the two and three player game respectively.

The point $(w^i = 2)_{i \varepsilon I}$ is always in the core, and it is associated with $(x^i = 0)_{i \varepsilon I}$. However, unsymmetric points as well belong to the core: for instance the point $(w^1 = 4, w^2 = 2, w^3 = 0)$ which is associated with $(x^1 = +1, x^2 = 0, x^3 = -1)$ in the 3-player Game.

Interpretation:

It is interesting to note that Shapley and Shubik (1969) presented an example of an economic system with externalities called "the lake" which has a very similar characteristic function. Indeed, their remarks on the unsymmetry and bigness of the core also apply here. These formal connections may give some support to the interpretation of the Game in a pollution context.

III - 4 The Nash Equilibria:

The preceding analysis, while motivated by a theoretical interest, is not directly related to the experiment since the possibility of enforceable agreements is not explicitly recognized in the rules. Nevertheless, we shall argue that the points in the core are also Nash equilibria so that they actually can be obtained by self-enforceable agreements. That this is so is

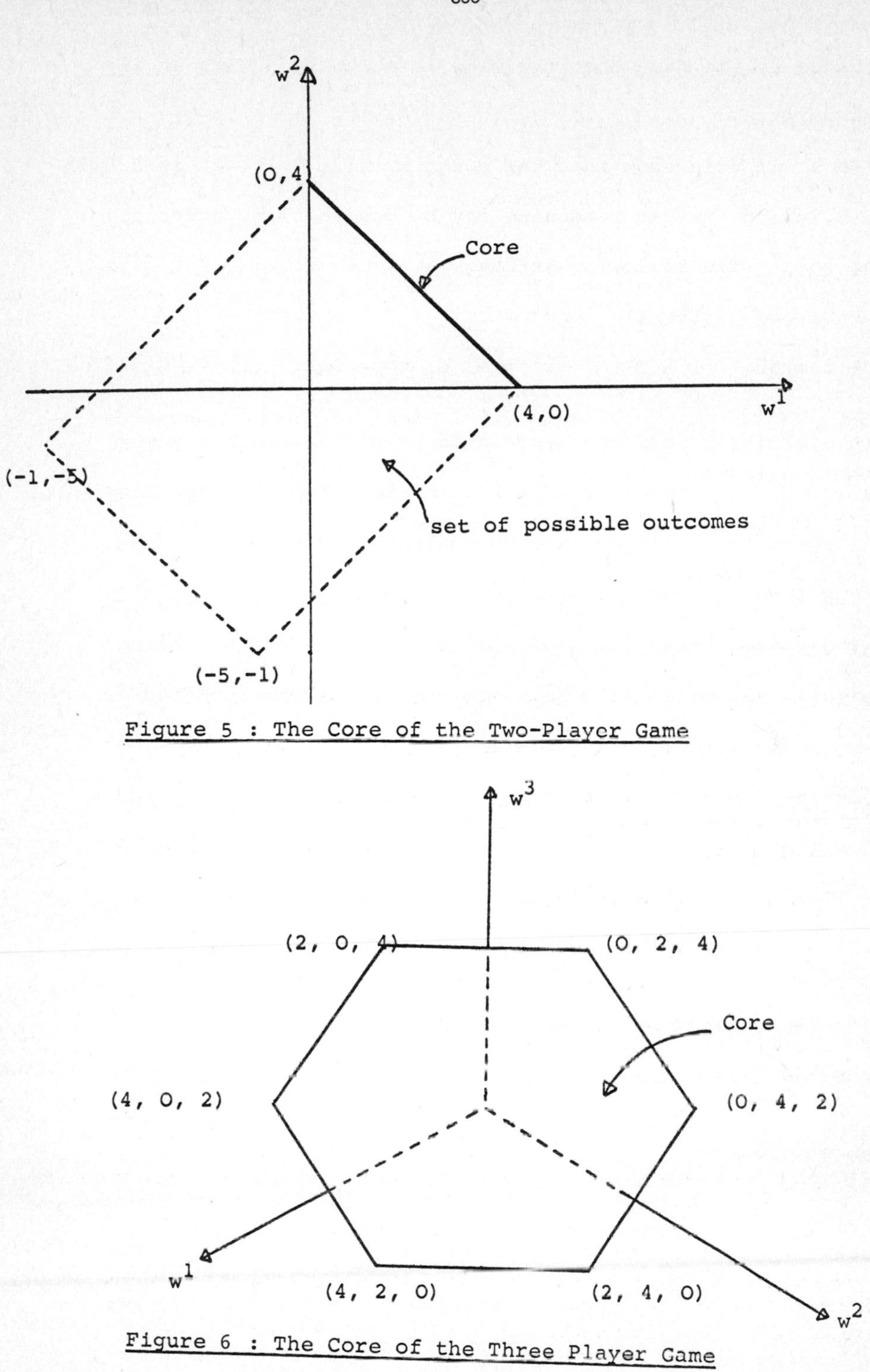

Figure 5 : The Core of the Two-Player Game

Figure 6 : The Core of the Three Player Game

simply due to the fact that the Game is now played with an in-finite horizon.

We shall only show that the point $w = (2, 2 \dots 2)$ is a Nash equilibrium; a similar reasoning may be used for any other point in the core. The following strategy:

- $r = o$ $(i \varepsilon I)$ $x^i = o$,
- $r = 50$ $(i \varepsilon I)$ $x^i = +1$ or $x^i = 0$.

are in equilibrium and the associated payoff is precisely $w = (2, 2 \dots 2)$. Notice that if no player deviates then the Game will stay in the winning state so that the second part of the strategy ($r = 50$, $x^i = +1$ or $x^i = o$) is only a threat. Once a player deviated, then the Game will either stay in the losing state ($x^i = +1$) or it will be up to the player who deviated to play so as to return to the winning state ($x^i = 0$).

Besides the points in the core there is another Nash equilibrium that is:

- $r = o$ $(i\ I)$ $x^i = +1$,
- $r = 50$ $(i\ I)$ $x^i = 0$.

The associated payoff $w = (-1, -1 \dots -1)$ is not "group rational".

Interpretation:

This last paragraph suggests that the lack of communication between the players should not prevent the realization of some point in the core. The symmetric outcome $w = (2, 2 \dots 2)$ may be interpreted as follows: each player is not polluting because he is convinced that the other players would "punish" him (and themselves) if he did so. The strength of such a threat seems very questionable in the context of the Game and certainly depends on psychological attitudes. A possible way to explore this direction may be to allow direct communication from time to time.

The equilibrium point w = (-1, -1 ... -1) is to be compared to a "blocked society" going to ruin by lack of confidence in its members.

Finally, the connections between the cooperative and the non-cooperative solution may to some extent illustrate the role of the law in society as regards pollution: though the points in the core are actually "self-enforceable" the risk of deviation may actually be too high to be supported without the recourse to some "enforceable" agreement contracted by law.

SECOND PART

§ 0 Introduction

This part is devoted to the analysis of the results obtained in the experimentation of the "pollution" game.

The precise conditions of the experiment which was led among a population of French sixteen year old scholars, are described in the first section.

Then, in the second section we present a qualitative analysis of two typical runs of the game. In the third section, we study the statistical significance of some behavioral hypotheses. At this point, the previous results obtained in (W-D-R-D) are compared to ours.

Finally, in the last section some general conclusions, in particular about the interest of the Game as a pedagogical tool, are drawn.

§ I Description of the Experiment

The general conditions of the experiment were those already used in (W-D-R-D); that is: during the Game, the rules described in the Foreword whereas at the beginning of the Game the number of red cards would be 0 or 25 and at the end of the Game either each player would receive an amount of money according to his own score (Keep condition) or all the scores would be added and all the players would receive the same amount (Pool condition). Thus we obtained four distinct kinds of groups: 0 or 25 red cards and Keep or Pool Conditions (to be called briefly 0 Pool, 25 Keep,...)

The players were not allowed to communicate and they were sitting in circle facing outside in order to prevent non verbal communication. Every four deals they would be informed of the number of red cards in the deck at that stage.

The score sheets had room for 40 deals but the Game was stopped without warning at the 25th deal. The Players started the Game with 30 units and after the 25th deal they received a monetary profit according to the following rate: 1 unit = .20 FF.

At the end of the session the following written questions were asked to the players:

(1) What was your strategy at the beginning of the Game?

(2) What did you think about the probable evolution of the number of red cards in the deck?

(3) Had you any idea (and to what extent) of the other players' behavior?

(4) Did you change your own behavior at some moment and why?

(5) Would you behave the same way if you had to play again?

The answers to this questionnaire turned out to be very useful in order to pursue a qualitative analysis of the Game.

§ 2 A Detailed Behavioral Study of Two Runs of the Game

Two typical examples (a 25-Keep and 0-Pool) serve as an illustration of the kind of reactions which can be raised in such a situation and of the ideas which may underline the players' attitudes. The first graphic shows the number of red cards r as a function of the number of stages k. This curve $r(k)$ draws the evolution of the situation resulting from the players' behavior; it also shows with which probability the players may score +4 with a good hand (the probability of winning is .88 for 10 red cards in the deck, .42 for 20 red cards and so on).

Figure 2 represents for each stage of the Game the number of red cards which were added by each player since the beginning of the Game; thus one curve is drawn for each player. At each stage the (algebraic) sum of this variable over all players gives the actual number of red cards in the deck, thus the first graphic.

Description of the Results for Group 1 (25 Keep)

As to the attitude of the whole group, Figure 1 shows that the evolution of the deck of cards was fairly well controlled, though the high proportion of red cards did not allow the players to make substantial benefits: each player won in the average from 6 to 8 times, that is nearly 35 units, as compared to an average individual expense of 47 units; their expenses were badly

Figure 1 :
Evolution of the number of red cards in the deck.

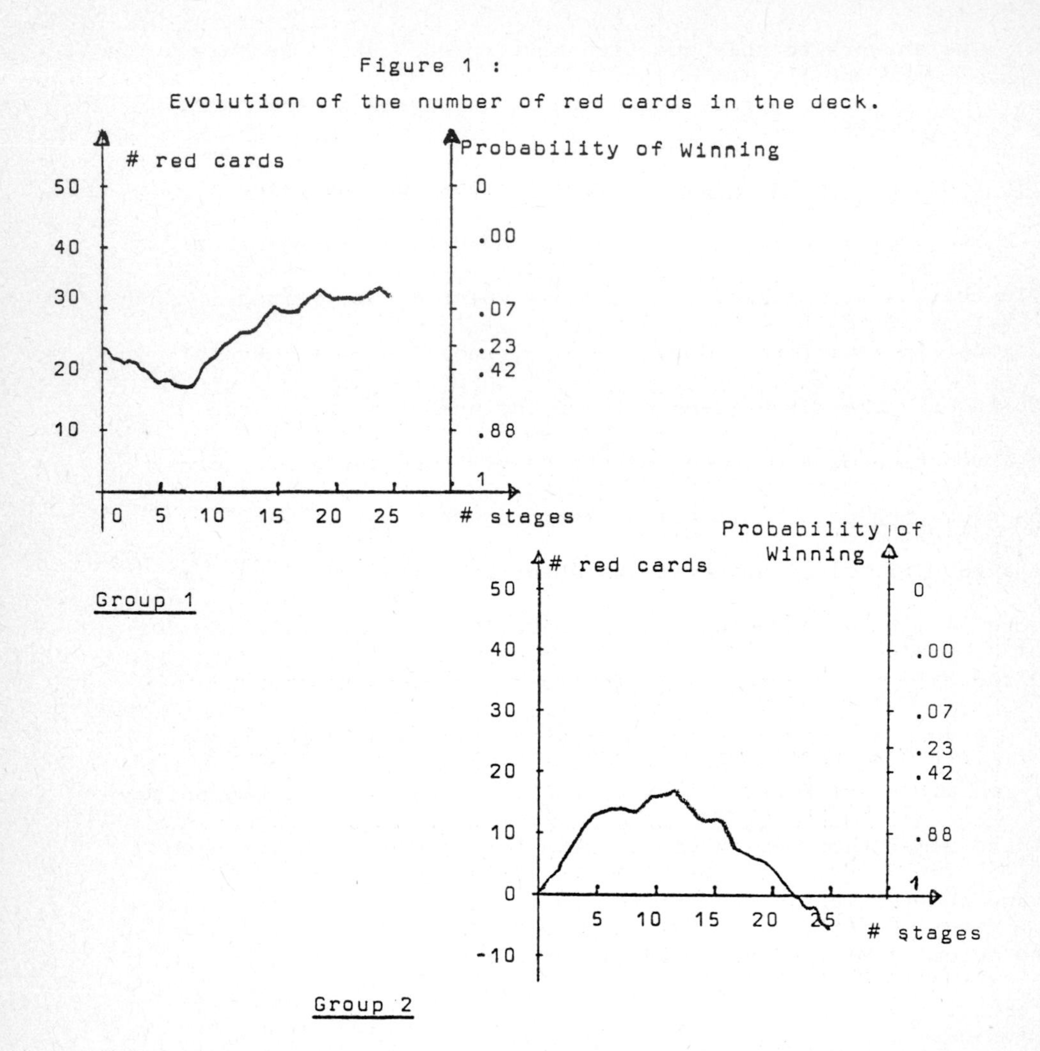

spread over time for with these 47 units correctly employed they could have won at least 13 times: 20 units spent over the first 5 deals would have brought the level of red cards down to 0, and this level might have been held for 13 deals, spending 2 units, but winning 4 by the same time at each deal.

As to the individual behavior (Figure 2), there appears a sharp distinction between two attitudes: to "pollute" (i.e. play +1) vs. to "depollute" (i.e. play -1 or 0); most of the players however changed their strategy at least once in the Game.

Individual differences between players are well depicted by the share they held in the total expenses; here they are in percentage:

Player	Units spent	Share of total expenses in percentage
1	22	9.3
2	70	29.6
3	78	33.1
4	30	12.7
5	36	15.3
Total	236	100.0
Mean	47.2	20.0

Group 1 (25 Keep)

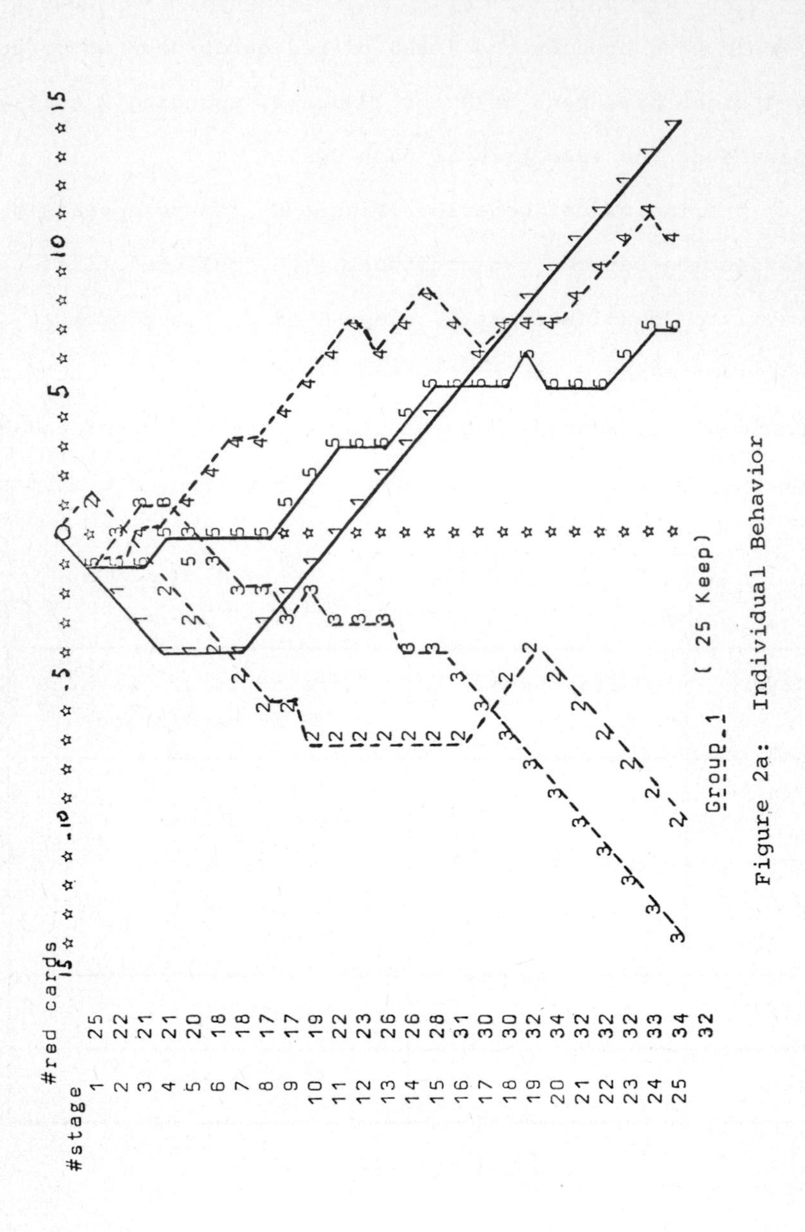

Figure 2a: Individual Behavior

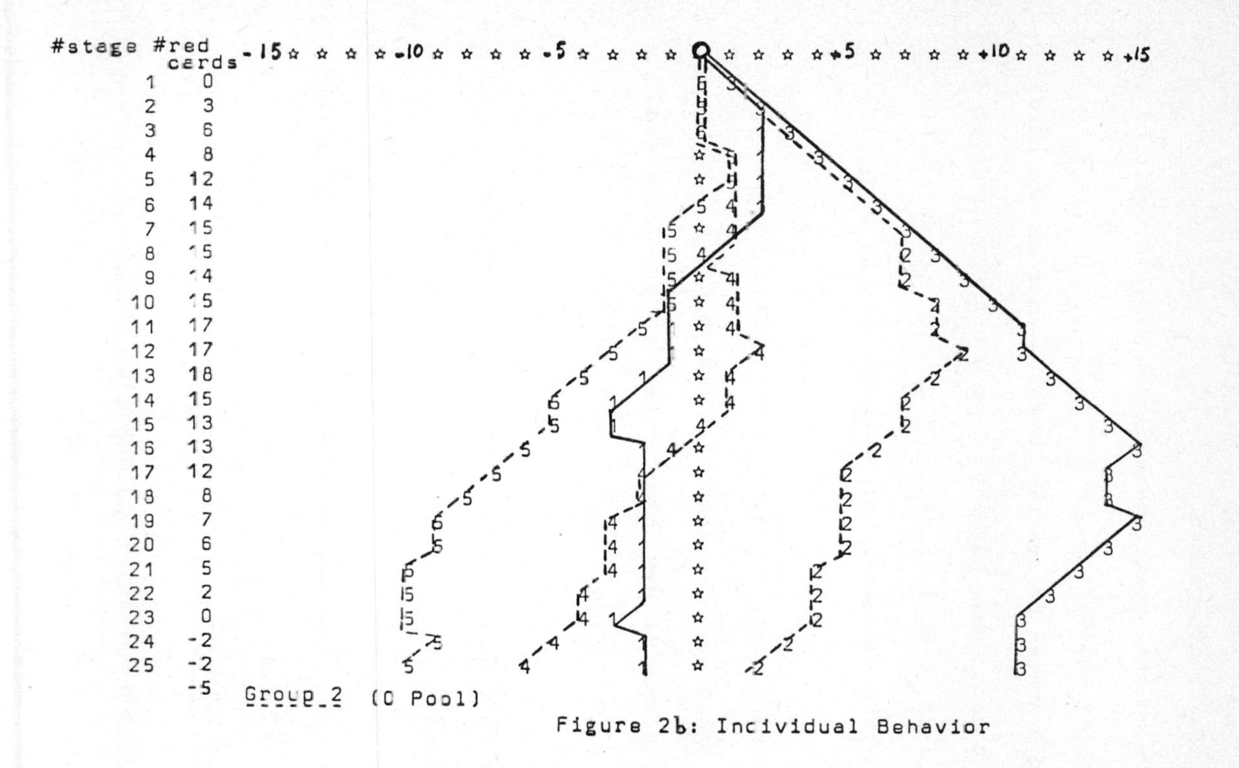

Figure 2b: Incividual Behavior

Figure 2 shows that players 2 and 3 mostly depolluted whereas players 1, 4, 5 mostly polluted; these results are consistent with the figures above.

According to the questionnaires, the players describe their own strategies as follows:

- Player 3 states he reacted to the number of red cards he received at each deal, playing -1 or 0 when he received less than 5 blue cards, 0 or 1 otherwise. He did not change his behavior all along the Game, which induced him to play -1 very often.

- Players 1, 2 and 5 understood that the interest of the group was to bring down the number of red cards, and they say they began to play that way in order to win more afterwards. All of them however declare they changed their attitude to polluting at some moment because they had individually the feeling they were the only ones to depollute and they did not want to be taken advantage of by the others.

- Player 4 wanted to "reach the highest possible score, maintening at the same time more blue cards than red ones in the deck", which led him to play -1 once in a while from the 13th deal up to the 20th.

Description of the Results for Group 2 (0 Pool):

The number of red cards went up to a maximum of 18 (at the 13th deal), then came down to a final -5 (the players did not notice there was no red card left).

The average of the expenses was 52 units, for an average 19 winning hands.

Here are the players' shares of the total expenses:

Player	Units spent	Share of total expenses in percentage
1	54	20.8
2	46	17.7
3	28	10.8
4	62	23.8
5	70	26.9
Total	260	100.0
Mean	52.	20.0

Group 2 (0 Pool)

According to these figures, the main pollutors were players 3 and 2, the depollutors 4 and 5, whereas player 1 was nearly neutral.

- Players 4 and 5 understood that the group's interest was to keep the number of red cards down from the beginning of the Game. As they were however forecasting a fast increase in the number of red cards, player 5 chose systematically to depollute, whereas player 4 reacted to the number of red cards in hand, playing -1 with 6 or 7 blue cards, 0 with 5 blue cards in order to minimize losses. It is worth noting that these two players thought of the others playing the same way, which is in contradiction with their opinion about the evolution of the deck.

- Player 1 intended to play +1 until he received 2 red cards, and not to change anything from then on. He actually played -1 when he realized that the proportion of red cards was growing.

- Players 2 and 3 wanted to win the maximum by playing +1; they admit however they had not a clear understanding of the Game at the beginning. Player 2 notes that he understood by the 8th deal the implication of his attitude upon the chances of receiving a winning hand, and that he changed then to play 0 or -1. Player 3's only comments are that he played -1 about the 17th deal to bring back some blue cards in the deck.

Discussion of the examples:

The statements of all 105 players involved in the experiment enabled us to establish the following classification of the players according to their level of understanding of the situation at the beginning of the Game; the reader may check this classification with the examples above.

(1) Players who do not understand that an increase in the number of red cards will lower with certitude their probability of winning; some players of this kind rest upon chance to receive a winning hand, in spite of probabilistic laws.

(2) Players who understand the basic connection between the number of red cards in the deck and the probability of winning, but who do not realize that their choice of -1 or +1 has an

effect upon the deck, e.g. we noted the following comment in a group 0 Pool: "We had to win the most we could at the beginning of the run, as long as the deck was blue, for it would very quickly turn red."

(3) Players who understand that red cards generate losses and that they are able to control the evolution of the deck, i.e. that they must not allow the number of red cards to grow beyond a certain level. However these players disagree about the estimation of that level; we found three types of attitudes:

- intuitive estimation of the level, which might stand between 20 and 35 red cards.
- implicit determination of the level through reacting to the number of red cards in hand.
- reasoned estimation, which might be nearly true (0 to 2 red cards) or false (2/7 red cards against 5/7 blue cards).

Thus we see that players' attitudes may vary considerably. A general consequence of this state of fact is that, most likely, we should not observe the same behavior over a whole group for more than a few deals, unless some tacit agreement takes place between the players.

Let us now examine some consequences of this diversity over players' behavior in both Keep and Pool conditions.

In the Keep condition, conflict arises between individual and group interest for the reason that if a member of the group keeps playing -1 and not the others, only losses are generated for this player. As a consequence, players will tend to be prudent and if their attempt to depollute does not yield quick perceptible results, they will think the others take advantage of them and very few will carry on. Now considering the wide range of all possible reactions of the players at some moment of the Game, it will seldom happen that the number of red cards goes down by more than 2 units from one deal to another; Figure 3 (part 1) shows that this quantity has little influence upon the probability of winning. Reactions similar to those of players 1, 2, 5 in group 1, which were disappointed though the actual number of red cards in the deck was slowly going down, may occur. However some players like player 3 in group 1 do not seem to feel being exploited, which led them to be neglectful of their individual interest.

In the Pool condition, on the contrary, there should not be such a conflict of individual vs. group interest; some players however keep polluting; we may infer that, either they have not understood the Game situation i.e. they belong to the first 2 classes, or they would tolerate a high level of red cards. If such players are in the minority, the others are theoretically able to balance their polluting action by playing -1: no obstacle is raised by individual interest, for a player who is depolluting is thereby contributing to the total benefits which will be shared even if his score is low as compared to the others'. We observe however that no player in Pool condition spent more than 70 units to maintain the resource (the maximum being 100 units, cf. Table 1), even when the bad state of the deck was demanding such an action.

Thus it seems that an obstacle stands in the way of a throughout depolluting behavior: as stated in the questionnaires, some depolluting players reacted angrily towards others in the group who were polluting: "Next time, I'll let the others do the job (i.e. adding blue cards) and I'll keep the score." This brings the hypothesis that the play is seen as a game where, despite the final monetary profit, satisfaction and feeling of success are tightly connected with the individual score. Frustration of the depolluting players might then arise from the fact that, aware of their superiority in understanding the Game situation, they do not reach a high score, which would put it in concrete form, nor receive actual acknowledgment for their sacrifice. Hence the main obstacle to a cooperative behavior in Pool condition seems to be that some players do not clearly understand the consequences of their choices and think in terms of minimizing short-term losses, which in the Game is not equivalent to maximizing long-term profit. This obstacle to cooperation remains true of course in Keep condition, but added to which comes the lack of confidence between the players.

§ 3 Some Statistical Results on Individual and Group Behavior

The preceding section appears as a case study which reveals some salient psychological attitudes of the players. These insights, interesting as they may be, remain qualitative and their predictive value is subject to personal judgment. Therefore, we shall now test some specific behavioral hypotheses so as to have some general conclusions on the experiment.

We shall systematically investigate whether the imposed external conditions (0 or 25 red cards and Keep or Pool) do have some significant impact on the players' behavior.

The experiment was led with 21 groups of 5 players distributed as follows:

	0 Red	25 Red
Pool	5	5
Keep	5	6

The experimental data consists of the following variables:

- Total number of units spent in the group (choosing -1 costs 4; 0 costs 2; +1 costs 0),
- Final number of red cards, as resulting from the last (25th) deal,
- Units spent by the player who spent the most in the group expressed in units and in percentage of the total expense,
- Units spent by the player who spent the less, also in units and percentage,
- Difference between the two previous measures, equally in units and percentage. This variable is to be called "dispersion".

Numerical values for these variables are gathered in Table 1.

External conditions	Group	Final # red cards	Total units spent	Highest expense		Lowest expense		Dispersion	
				Units	%	Units	%	Units	%
0 Pool	Po1	9	232	50	21.5	42	18.1	8	3.4
	Po2	10	230	60	26.1	24	10.4	36	15.7
	Po3	34	182	58	31.9	0	0.0	58	31.9
	Po4	36	178	68	38.2	12	6.7	56	31.5
	Po5	-5	260	70	26.9	26	10.8	44	16.2
Average		16.8	216.4	61.2	28.9	21.2	9.2	40.0	19.7
25 Pool	P11	17	266	60	22.6	48	18.0	12	4.6
	P12	12	276	70	25.4	42	15.2	28	10.2
	P13	30	240	64	26.7	24	10.0	40	16.7
	P14	4	292	64	21.9	50	17.1	14	4.8
	P15	0	300	62	20.7	58	19.3	4	1.4
Average		12.6	274.8	64.0	23.5	44.4	15.9	19.6	7.6
0 Keep	Ko1	22	206	58	28.2	26	12.6	32	15.6
	Ko2	22	206	64	31.1	22	10.7	42	20.4
	Ko3	4	242	62	25.6	36	14.9	26	10.7
	Ko4	58	134	40	29.9	16	11.9	24	18.0
	Ko5	19	212	76	35.8	26	12.3	50	23.5
25 Keep	K11	31	238	60	25.2	40	16.8	20	8.4
	K12	28	244	74	30.3	38	15.6	36	14.7
	K13	6	288	82	28.5	46	16.0	36	12.5
	K14	32	236	78	33.1	22	9.3	56	23.8
	K15	24	252	72	28.6	2	0.8	70	27.8
	K16	8	284	80	28.2	44	15.5	36	12.7
Average		21.5	257.0	74.3	29.0	32.0	12.3	42.3	16.7

TABLE 1
The experimental data

Tested condition	Initial Number of Red Cards		Reward Mode	
Population 1 Population 2	0 Pool 25 Pool	0 Keep 25 Keep	0 Pool 0 Keep	25 Pool 25 Keep
Tested variable :				
- Total units spent	1<2 p<.008	1<2 p<.009	no difference	1>2 p<.123
- Final number of red cards	no difference	no difference	no difference	1<2 p<.123
Highest expense in units	no difference	1<2 P<.041	no difference	1<2 p<.033
Lowest expense in units	1<2 p<.028	1<2 p<.142	no difference	1>2 p<.063
Dispersion in units	1>2 p<.075	no difference	no difference	1<2 p<.063
Highest expense in %	1>2 p<.048	no difference	no difference	1<2 p<.009
Lowest expense in %	1<2 p<.111	no difference	1<2 p<.111	1>2 p<.009
Dispersion in %	1>2 p<.111	no difference	no difference	1<2 p<.041

TABLE 2 : Results for Mann-Whitney Test

Note : (1) "1<2 $p<\alpha$" means that the values of the tested variable over population 1 are lesser than those over population 2 at the level of significance α.

(2) "no difference" means that the level of significance is greater than .20.

We shall successively apply the Mann-Whitney test to these variables for each following pair of external conditions: 0 Pool, 25 Pool; 0 Keep, 25 Keep; 0 Pool, 0 Keep; 25 Pool, 25 Keep. A positive conclusion to the test has the following form: with probability of error $p < \alpha$, the values of the tested variable over the first population are lesser (resp. greater) than over the second population. A population is the set of all groups playing in the same external conditions, from which the present observation constitutes a sample.

The results for the test are given in Table 2.

Discussion of the Statistical Results

The first two variables are representative of group behavior in the whole: the total units spent is an evaluation of the efforts of the group to maintain the resource, whereas the final number of red cards represents the results these efforts have yielded after 25 deals. (WDRD) have also been using these variables in the same sense; we shall compare our results to theirs.

Comparing all-blue with half-blue initial conditions, we observe that, in both Pool and Keep condition, the groups starting with 25 red cards spent significantly more than those starting with 0 red card. On the other hand, there is no such difference on the final number of red cards. We may put this the following way: groups starting with a half-degraded resource end up after 25 deals in situations which might be generated by groups starting with a pure resource; it follows that they spent much more to maintain the resource. (WDRD) come to the same conclusion, observing moreover that groups starting with 25 red cards tend to end up in a better situation than groups starting with an all-blue deck.

Comparing Pool and Keep condition, the test yields no difference for the group starting with 0 red card, and only a tendency ($p < .123$) for 25 Pool to spend more than 25 Keep groups. **(WDRD)** obtained more decisive results showing that Pool groups spend more than Keep groups, both with 0 and 25 red cards.

The difference between all-blue and half-blue initial condition being clearly established, we will now focus on the comparison between Pool and Keep conditions, in order to specify the behavioral differences which do not appear if we only use the former variables. We studied then the extreme attitudes in each group (i.e. highest and lowest expense), which will be expressed in units and in percentage of the total expense of the group.

Starting from the idea that on account of their short-term interest, players should individually spend less in a competitive situation than in a cooperative one, we shall submit to test the two following hypotheses:

(1) the strongest pollutors in Keep situation still pollute more than the strongest pollutors in Pool situation.

(2) the strongest depollutors in Keep situation do not depollute as much as the strongest depollutors in Pool situation.

The variables tested are (1) the lowest expense in units (2) the highest expense in units. Table 2 shows that the first hypothesis is confirmed for the groups beginning with 25 red cards, but not for the groups beginning with 0 red cards. The second hypothesis is not confirmed for any type of group; on the contrary, the opposite hypothesis is justified at a significant level ($p < .033$) for the groups starting with 25 red cards. Thus we may not conclude that our hypotheses are valid: the extreme attitudes in the groups do not agree with what we might expect, taking into account the criterion of individual interest. We already noticed this fact, in the analysis of the examples.

Note also that some players in competitive situation (25 Keep) spent more than all the players in cooperative situation (25 Pool): good will may withstand a monetary loss. On the other hand the only player that has not spent one unit is to be found in a 0 Pool group; in this case we see that the eventuality of some earnings does not always generate good understanding.

Behavior in 25-Pool groups seems to be very cooperative as opposed to 0 Pool: no strong pollutor appears and group dispersion is very low, whereas the total units spent are much higher than in 0 Pool condition. The main difference lies in the fact that at the beginning of the Game nearly all the players feel an urge to improve the state of the resource; thus cooperative behavior is achieved in most cases as soon as the first few deals. The effect of this action becomes quickly perceptible to the players, who become aware that their collective interest is best served by going on the same way.

Further investigation in the experimental situation should be necessary to clarify the psychological attitudes of the players in various types of external conditions, especially to determine whether the level of understanding of the Game situation varies with the type of initial conditions.

§ III Conclusion

We may summarize the results of this study in the following comments.

The initial number of red cards appears as a secondary factor as regards the determination of the situation at the end of the Game. Indeed, considering the wide variation on the final number of red cards among the groups playing in the same external conditions, it seems that other factors are prevalent such as a spread feeling of confidence or a certain level of understanding or some unexplored internal conditions.

The influence of the reward mode is somewhat more tangible for the group beginning with 25 red cards. However the results are far less significant than those obtained with the U. S. data. Here one may see the consequence of some cultural factor, which confirms the tendency already observed in (W-D-R-D) for the Swedish data.

Finally one may add that the players were generally quite enthusiastic about the experiment, feeling that the Game was demonstrative and worth playing. The same Game was actually used in a training session for senior managers. The goals were to introduce economic reasoning and to show the insufficiency of such considerations to produce a group rational behavior. Though not yet tested, significant differences with the sixteen-year-old population might be observed. First the final situations were more polluted, second the players would hardly deviate or twice from a stationnary strategy fixed at the beginning of the Game.

ACKNOWLEDGEMENTS

The authors are highly indebted to Mrs. Keiser, Director of Lycee Francois Villon, Paris, for her assistance in making the experimentation possible.

REFERENCES

Dana, J. M., Rubenstein F. D. "The Psychology of Pollution and other Externalities," Unpublished paper, Stanford, Calif.: Stanford University, Graduate School of Business (1970).

Shapley, L. S., Shubik, M. "On the Core of an Economic System with Externalities," The American Economic Review 59, 4 678-684 (September 1969).

Watzke, G. E., Doktor, R. H., Rubenstein, F. D., Dana, J. M. "Individual vs. Group Interest: An Experimental Study in Three Cultures:, Preprint Series International Institute of Management, I/72-23 (August 1972).